Construction as Depicted in Western Art

Construction as Depicted in Western Art

From Antiquity to the Photograph

Michael Tutton

Routledge
Taylor & Francis Group
LONDON AND NEW YORK

First published in 2021 by Amsterdam University Press Ltd.

Published 2025 by Routledge
4 Park Square, Milton Park, Abingdon, Oxon OX14 4RN
605 Third Avenue, New York, NY 10158

Routledge is an imprint of the Taylor & Francis Group, an informa business

ISBN: 9789462982550 (hbk)
ISBN: 9781041177456 (pbk)
ISBN: 9781003693093 (ebk)
NUR 684

Cover illustration: 1587, German, unknown master, The Tower of Babel, oil or tempera on panel. Z 2249, Germanisches Nationalmuseum, Nuremberg.

Cover design: Coördesign, Leiden

DOI: 10.5117/9789462982550

For Catherine, Emma and Samuel

Acknowledgements

I wish to acknowledge the following for help and support in producing this book: Catherine Tutton, Emma Tutton, Nicole Blondin, James W P Campbell, David Davidson, Monica Knight, Eleonora Scianna, Catherine Yvard, and my editors at AUP.

Table of Contents

List of Figures

Introduction

The aim in writing this book is to provide a comprehensive handbook for the professional, academic and general reader interested in art, architecture, construction and the process of building as it has been depicted in Western or European Art. The origins lie in my own deep-seated interest in art and my professional involvement in working on the repair and maintenance of historic buildings. The genesis goes back to the exhibition *The Art of Invention* at the Science Museum, London, which ran from October 1999 to April 2000.[1] Particularly two images relating to Filippo Brunelleschi's dome of Florence cathedral, Santa Maria del Fiore. These are Biagio di Antonio's (attributed), *c.*1470, *Tobias and the Archangels*,[2] where in the distant background is seen the dome with scaffolding around the lantern, and Gherardo Mechini's 1601 drawing showing scaffolding and hoisting machines used in that year to repair lightning damage to the lantern. (Fig. I)

Scope

The scope of the book explores how the process of building, with all its various trades and operations, has been depicted. This exploration into 'The Art of Building' addresses a gap in research by focusing on the history of building construction from antiquity to the advent of widespread photography at the end of the nineteenth and the beginning of the twentieth centuries. It investigates the representation of the building construction process, trades and materials as depicted in paintings, illuminated manuscripts, watercolours, prints, drawings and sculpture, although the latter is limited to the Roman period only. Building operations and activities are analysed and explained including appropriate technical terms in the glossary, with descriptions of the works and, here and there, insights into the artists' sources of inspiration. Not many paintings are entirely or principally devoted to construction sites, but many give tantalizing glimpses of such activity within the composition. How do these images carry meaning, how much is imagined and how much can we interpret as being taken from reality? Why do artists depict buildings under construction and what can this tell us about their perception and rendering of this process? What can it tell us about how the process has changed over time? Images of building construction in art are not common but there are more than readily meet the eye when looked for. Those selected here, and it is only a selection, will hopefully give a comprehensive picture of the genre available.

There are limitations in researching and writing the book: It is not an intrinsic history of building, does not cover architectural styles, or architectural and technical drawings; nor does it attempt to cover architectural treatises, building accounts or contracts, or the various Guilds and their rules. Many images and models in the *The Art of Invention* exhibition were of construction machinery, however these works are deemed technical drawings and models rather than artistic creations. The nearest images to purely technical representations are the progress drawings and watercolours produced by Sir John Soane's pupils and assistants and George Scharf's sketches,[3] nevertheless these works lean more heavilly towards, and should be considered as, works of art rather than technical representations.

Many images of construction are biblical scenes from Medieval and later manuscripts, where paintings or illuminations depicting the construction of the Tower of Babel, the rebuilding of the Temple of Soloman, the construction or reconstruction of cities such as Rome or Troy are common. Such subjects were generally accepted as historical fact, however it must have been accepted as common knowledge that these constructions took place a millenia earlier and their physical form would have been unknown so how else would these artists have been able to illustrate such a scene without recourse to their own experience? This is aptly summarised by Francis B Andrews:

> [F]or there can be no question that the artist who used the costume and other items of his common daily experience, gave also the current practice in building operation as he saw it being done.[4]

Current Literature and Sources

Together with *The Art of Invention* exhibition, and the accompanying catalogue with its plentiful references and extensive bibliography,[5] the other source of inspiration was Günther Binding's 2004 English edition of *Medieval Building Techniques*, a copiously illustrated book with some 650 line drawings copied from Medieval manuscripts together with 26 monochrome photographic plates.[6] This was a result of decades of painstaking research by Binding and his team from the University of Cologne where he was Professor of Art History. Other notable texts that I have made use of are: Andrea Louise Matthies's unpublished 1984 Ph.D., thesis *Perceptions of Technological Change: Medieval Artists View*['s] [*of*] *Building Construction*[7] and Dr Frieda Van Tyghem's 1966 *Op en Om de Middeleeywse Bouwwerf*,[8] written in Flemish, where the 286 plates speak eloquently. Other texts that I should mention here are L F Salzman's 1952 *Building in England Down to 1540 A Documentary History*[9] and among John Harvey's many books his *Medieval Craftsmen* of 1975.[10] Salzman's work is essential for anyone wanting to delve deeper into building contracts and accounts in the Medieval and early Renaissance periods and Harvey for craftsmen, including the building crafts, in general. In addition many other sources are recorded in the notes and bibliography.

Images for the works reproduced were gathered together from prior knowledge, visiting galleries and museums and online databases. Of particular note in the latter category are the British Library resources[11] and the French website 'Enluminures', devoted to images of illuminations in manuscripts held in French municipal libraries, where every image has been examined.[12]

Definitions

A note on terms is essential, what craftsmen called their tools and processes varied enormously, they differ from country to country, from region to region, from workman to workman and from period to period. Technical terms were generally in Latin in documents up until the seventeenth century, but these were typically Latinised versions of local terms. Therefore for simplicity, and those unfamiliar with Latin or the many scripts to be found in early documents and manuscripts,[13] the terms used in this book are modern English usage.[14] The reader is referred to the full glossary which gives definitions of technical and unusual terms.

The terms architect, clerk of works and foreman are used loosely. These terms before, and well into, the Renaissance are controversial as they did not really exist as seperate and distinctive jobs or professions. With the master carpenter or master mason, they morph into what appears to be a 'person in charge' in many images. This is not to mention the owner or person paying for the work, however we can be sure that one or other responsible for these duties was present to advise, take responsibility, or oversee the work.

The terms primitive and naïve are used in the text, but these are not used in their art historical categorisations but because in comparison to contemporary or similar images of better or higher quality they are simply just that in the literal sense.

Layout of the Book

Images are arranged broadly chronologically: Roman practice is seen before Medieval, and in turn, Medieval before Renaissance and later. Although where it is convenient for comparision this chronology is not strictly adhered to.

The first chapter is on the work of the carpenter and is divided into two parts. Part One deals with temporary access scaffolding, formwork (also temporary) and the permanent timberwork incorporated into the building. The second part looks at cranes and other lifting devices, largely made of wood or timber and also temporary; although some lifting machines were left inside larger buildings such as cathedrals and survive today. Much of this temporary carpenters work was used almost exclusively by masons; it was built by the carpenter to aid them. Scaffolding is the backbone of the construction site and little is achieved without it. Some will argue that scaffolding is little understood because of its temporary nature; it starts to disappear towards the end of a project and is soon forgotten. However the scaffold structure is removed to the scaffolder's yard for use elsewhere. It is self-resurrecting equipment, is moved from one site to another and so on. Individual elements may wear out or get damaged and be discarded but the basic process is rotational. The hardware is too valuable to abandon at the completion of a building project, whether it be poles or other products from the management of woodland, or today's steel and timber products. Many large cathedral and castle building projects had their own permanent scaffold yards. Formwork is temporary wooden or timber structure erected to support masonry whilst construction proceeds and the mortar cures.[15] Masons depended on it when building arches, door or window heads and vaults. It would often need to be in place for several days, or even weeks. On dismantling it would be set aside for alternative uses. These components when finally finished with, particularlly in remote areas where transport was expensive, may well have been sold on as fire wood or for other rural purposes such as fencing. Finally cranes and lifting devices were invariably made by the carpenters, either at the point of use or in the yard and then moved to where they were needed.

The second chapter deals with masonry and this includes both stone and brick. Masons and their ancillary workers are the most widely depicted workers in this particular genre of art works. The crafts of the stonemason and brickmason go back to early antiquity. The basic raw materials are however quite different. Stone is a natural geological product, which is part of the structure of the earth's crust. It is quarried at the surface or mined, depending on how and where it outcrops or is found. Virtually any stone that is easilly accessible above ground, has been used as a building material in the past.[16] Brick however, is a manufactured product, albeit also from a geological material, clay. It is formed into blocks, or bricks, allowed to dry and then fired in a kiln.[17] In antiquity, before kiln technology was perfected, bricks would have been fired in direct sunlight. The masons other vital material is mortar; in the time-line of this book all mortar would invariably have been produced by firing limestone in a kiln,[18] although in antiquity and the pre Roman world, clay-earth mortars would also have been used.[19]

Following these two excursions into carpentry and masonry the third chapter looks at the role of the blacksmith, without whom neither the carpenter or mason could work, for they depended on the blacksmith for their tools and indeed transport. There are images of blacksmiths in art but very few relating directly to the construction site.

This book comes at a time when the representation of architecture in painting is becoming more widely and intensively researched. Recent scholarship over the last twenty five years or so has led to important exhibitions on the depiction of buildings in art. Recently at the National Gallery in London was 'Building the Picture – Architecture in Italian Renaissance Painting'. This was preceded by four other recent exhibitions exploring different aspects of architectural representation in art. Those in Madrid in 2010; in Princeton also in 2010; in Lisbon in 2012 and Urbino in 2012.[20] These in turn followed the earlier exhibition *The Renaissance from Brunelleschi to Michelangelo: The Representation of Architecture*, which toured Venice, Washingon, Paris and Berlin in 1994-1995. In the forward of the accompanying book, Feliciano Benvenuti wrote, '[t]he exhibition [...] is therefore, as always, not merely a gift to its visitors, but a prompt to scholars and historians to carry their personal enquiries a stage further, and thereby raise awareness of our cultural and scientific legacy.'[21] It is in the furtherance of this legacy that the current book has been researched, for it is a natural and logical progression to include not only the buildings but also their very mode of construction.

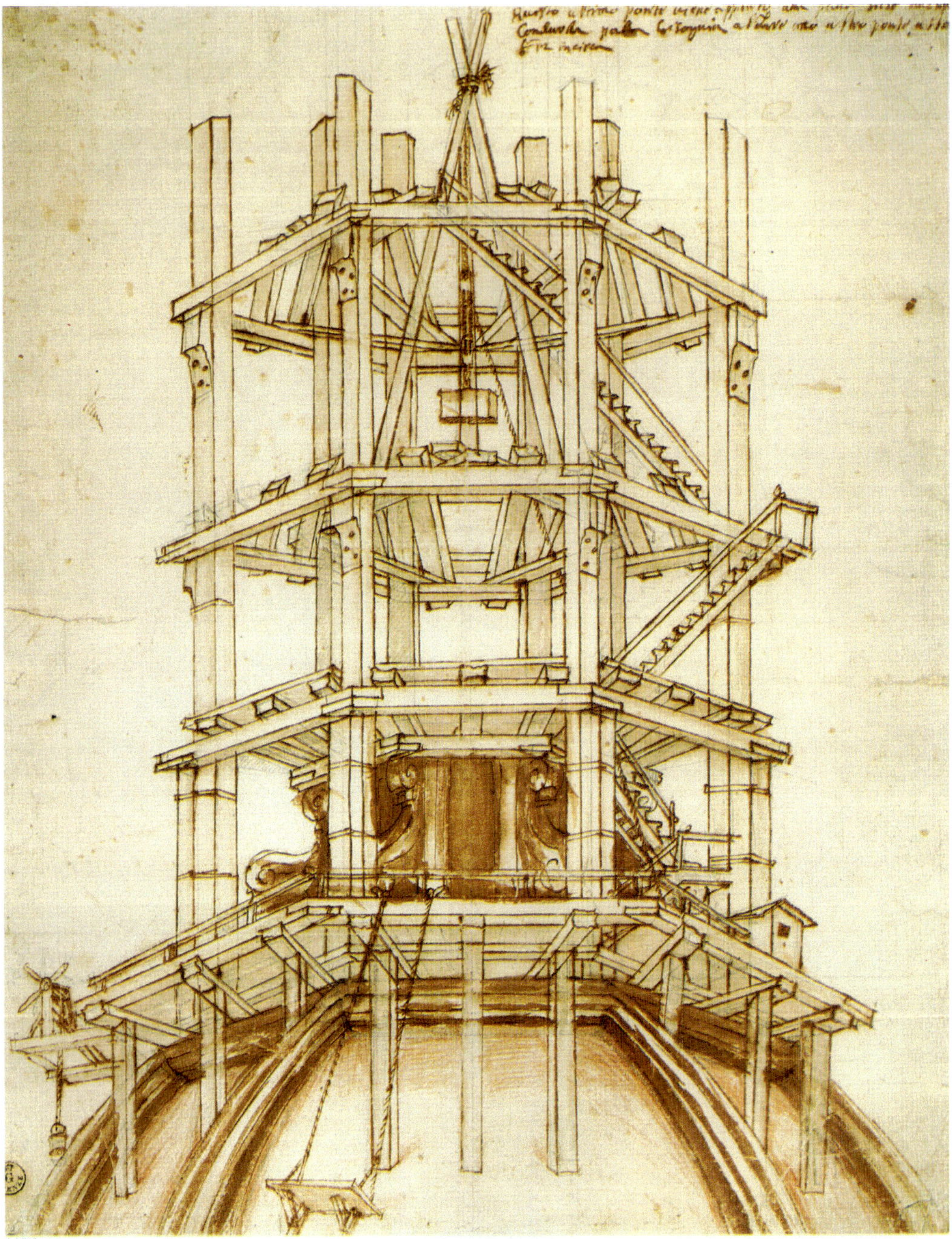

Figure I: Gherardo Mechini (?), 1601, The dome of Florence cathedral: *Scaffolding and hoisting devices used in 1601 to repair lightning damage to the lantern*, drawing, pen, brown ink and wash, 248A, Gabinetto Disegni e Stampe, Galleria degli Uffizi, Florence. (Courtesy of the Uffizi Galleries)

1. The Carpenter

PART 1

In the Medieval construction world the carpenter or wright had the highest standing along with the master mason, with whom they often worked together and enjoyed equal status. It is certainly true that the craft of the wright, including the scaffolder, formwork erector, and carpenter, arose hand in hand with that of military engineering. Such craftsmen were held in high regard and achieved considerable status and sometimes great wealth.[22] The carpenter was responsible for all aspects of access scaffolding for the masons as well as roof structure, framing and formwork.[23]

Access: Scaffolding, Ladders and Ramps, the Raw Materials

Up to the nineteenth century, and well beyond into the twentieth, scaffold was a wood product. Produced in the main from managed woodlands, this can be established from at least Roman times onwards.[24] The products or components came from the underwood of woodlands, that is, areas surrounding 'timber' trees that were regularly coppiced, pollarded, or left after felling to produce suckers. The craft of this production process is known as woodmanship. The terms timber and wood represent different products: 'Wood is rods, poles, and logs, used for fencing, wattlework, and many other specialized purposes [...] Timber is the stuff of beams and planks.'[25] Rods and poles were produced in vast numbers. In the *Grant of Burgred, King of Mercia, 866*, an Anglo-Saxon Charter, the perambulation states, 'pasture for 70 pigs in that wooded common [...] and 5 wagons of good rods (*virgis*) and every year one oak for building [...] and wood [...] for the fire as necessary'[26]. (Fig. 2) The 'oak for building' would have been a 'timber tree', invariably oak, not coppiced or pollarded, but left to stand to produce large-scale timbers for roof, floor structures and timber framing.[27]

The products of woodmanship included scaffolding components comprising: poles for standards, the principle upright members; ledgers, the horizontal members parallel to the building and connecting the standards; putlogs supported on the ledgers and spanning back to the building where they were either built into, or inserted into, the face of the building. The term for the resulting hole, which was often left open in-situ when the scaffold was struck or dismantled, was also putlog. Poles could be anything from approximately 2 to 10 inches (5-25cm) in diameter, depending on species and the length of the harvesting cycle. The working platform, or 'lift' was usually made from hazel

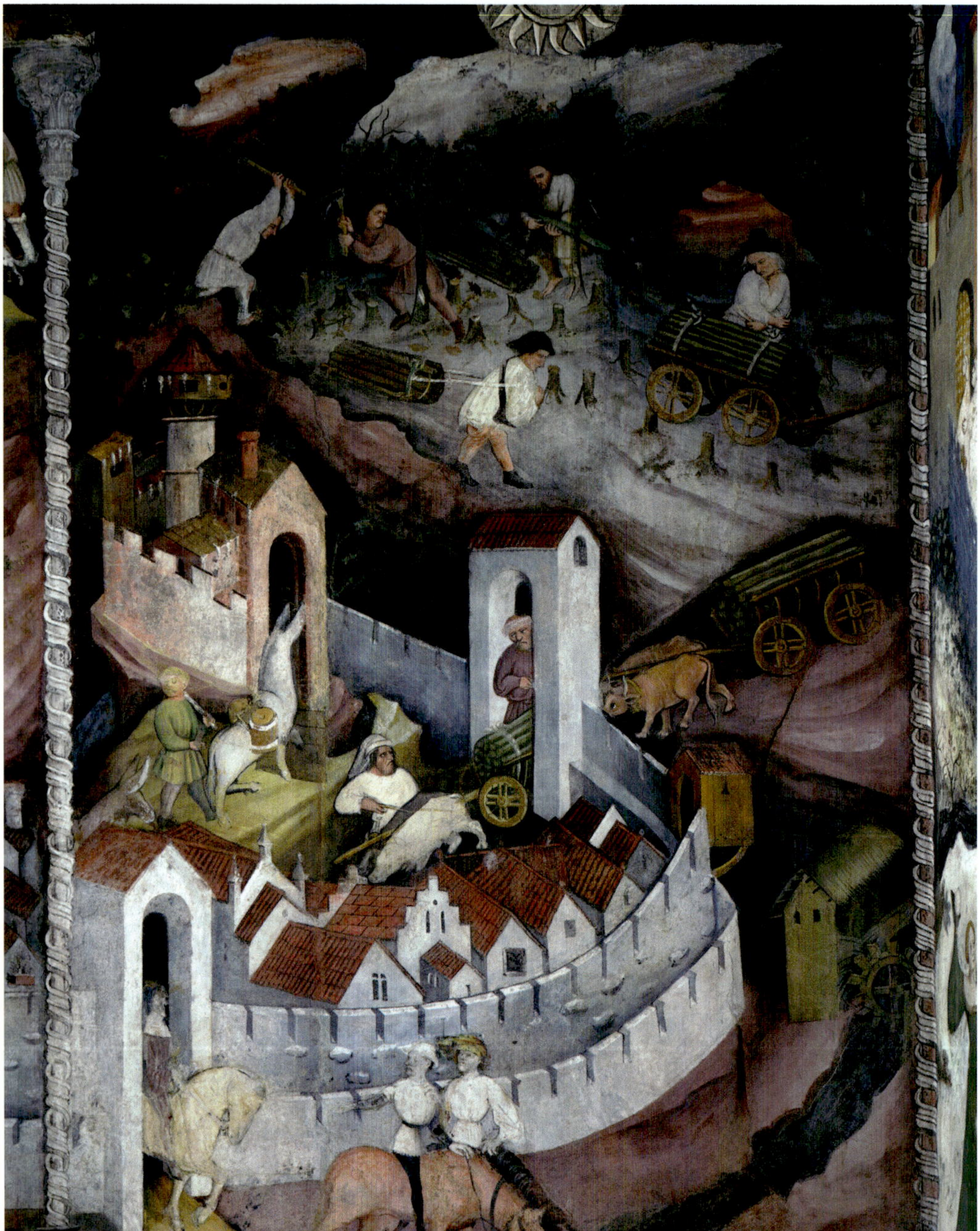

Figure 2: 1390-1400, Gathering poles from coppicing, Master Wenceslas, Bohemian, *Month of December, Cycle of Months*, fresco, Torre Aquila, Buonconsiglio Castle, Trento, Italy. De Agostini Picture Library, A De Gregorio / Bridgeman Images.

rods closely woven into wattle hurdles, also known as 'clays'.[28] Sometimes planks were utilised, but these were expensive and usually kept for use in the actual structure of the building itself. The building accounts of Conway Castle and Town Walls in 1286 record the purchase of 2000 poles (*lignis*) and 1180 'clayes' for scaffolding.[29] Species of tree were various, although alder was common; ash, lime, poplar and sometimes oak were also used.[30] These poles were bound, or lashed together with rope, twine or cords, collectively known as withies. Withies were often made from another product of the woodlands, known as bast or *bastropps*.[31] Bast or bass is the tough and fibrous inner bark of trees, particularly Lime, which can be woven into twine.[32] The withies were not normally adequate on their own because they would tend to loosen, and wedges were driven in to tighten them.

The Evidence for Early Scaffolding, Ladders and Ramps to the End of the Sixteenth Century

Within the eighteenth dynasty, New Kingdom tomb of Rakhmire, Vizier to Thutmosis III, at Thebes in Upper Egypt, is an image of a scaffold erected around a sculpture, of modest size by Egyptian standards, dated to *c.*1450 BCE. (Fig. 118) This is a scaffold of wooden poles tied together with cords; the composition of the platform is not clear. For heavy building work the Ancient Egyptians favoured ramps, images of which can also be seen at the tomb. For such work as this however, finishing and carving *in situ*, scaffolds were far more servicable and easilly constructed.

Post Ancient Egyptian antiquity, one of the earliest depictions of a scaffold is the Roman wall painting of builders at work (Fig. 3) from the Hypogeum of Trebius Justus in Rome. Located on Via Giuseppe Mantellini it dates to the fourth century CE and depicts builders at work on a construction site. Within the same chamber there are, among others, depictions of the transportation of building materials, agricultural products and Trebius Justus in conversation with a worker who is possibly a mason. From these paintings it is postulated that Trebius Justus was a builder and rich landowner.[33]

The painting shows two bricklayers at work on a scaffold building a brick wall. Although the details of the scaffold are a little muddled, the standards, ledgers and some putlogs are clearly defined. Below the ladder a brace can be seen. Other workers are a hod carrier climbing the ladder with what appears to be mortar; a worker to the right approaching the ladder with a basket of bricks and a worker to the left mixing mortar with a hoe.

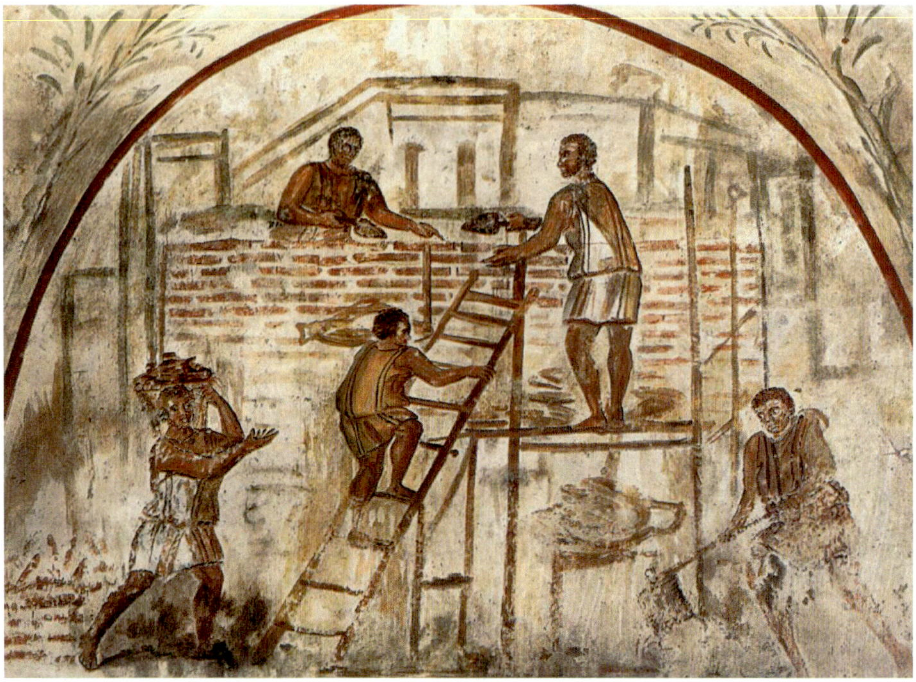

Figure 3: Early Christian Painter (active fourth century in Rome) Tomb of Trebius Justus, Rome, fourth century AD. https://www.wga.hu/frames-e.html?/html/zearly/1/2mural/5vialati/latina7.html.

Most early representations of scaffold can be read and understood, but some early illuminations resort to a fanciful conception where workers are suspended mid-air. In figure 4, an illustration from the ninth century Golden Psalter of St Gall, it is obvious from the draughtsmanship that the artist was capable of drawing the scaffold but it has been left out for the purposes of clarity and precedence. A scaffold would have partially obscured the more important elements of the composition; the 'building of the church' and the church itself are of greater importance. Three workers are shown, two of whom stand on leaf like representations of clays or scaffold platforms. One is a carpenter with a chisel in his left hand and a large axe in his right. The other two are roofers, one of whom passes tiles to the other on the next highest platform. Similarly, in the illumination of the building of the Tower of Babel, in the eleventh century Old English Hexateuch, (Fig 5) although here there is shown just one lift of actual scaffold at the bottom left. A standard with a forked top, probably a natural cleft in the wooden pole, bears a putlog which spans back into the wall of the tower. The rest of the scaffold, which is certainly present but invisible, is left to the viewer's imagination. Of the ten workers shown, four, or possibly six, are on

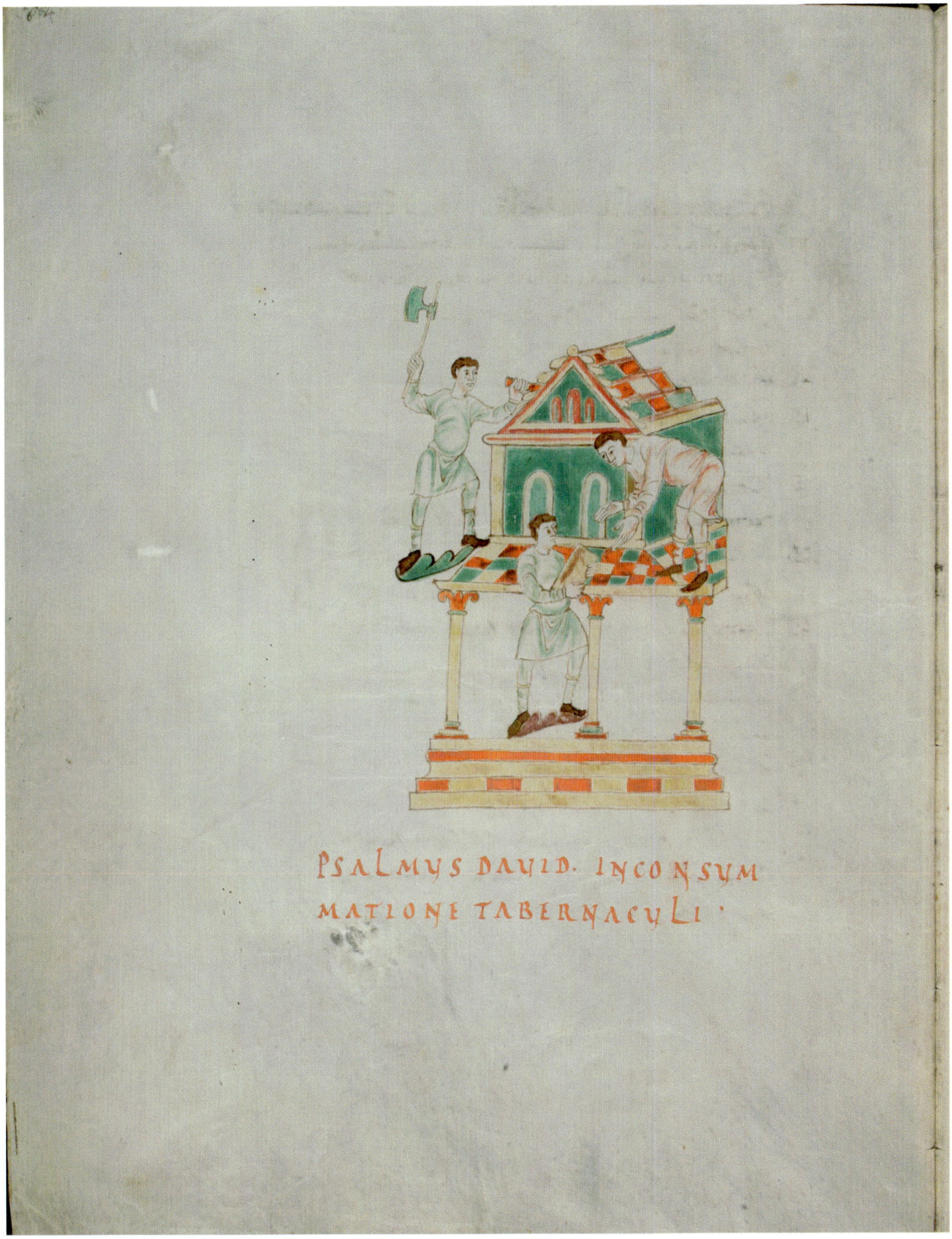

Figure 4: 890-900, French (West Franconian Empire?), *Building a Church*, The Golden Psalter (Psalterium aureum) of St. Gall, St. Gallen, Stiftsbibliothek, Cod. Sang. 22, p.64.

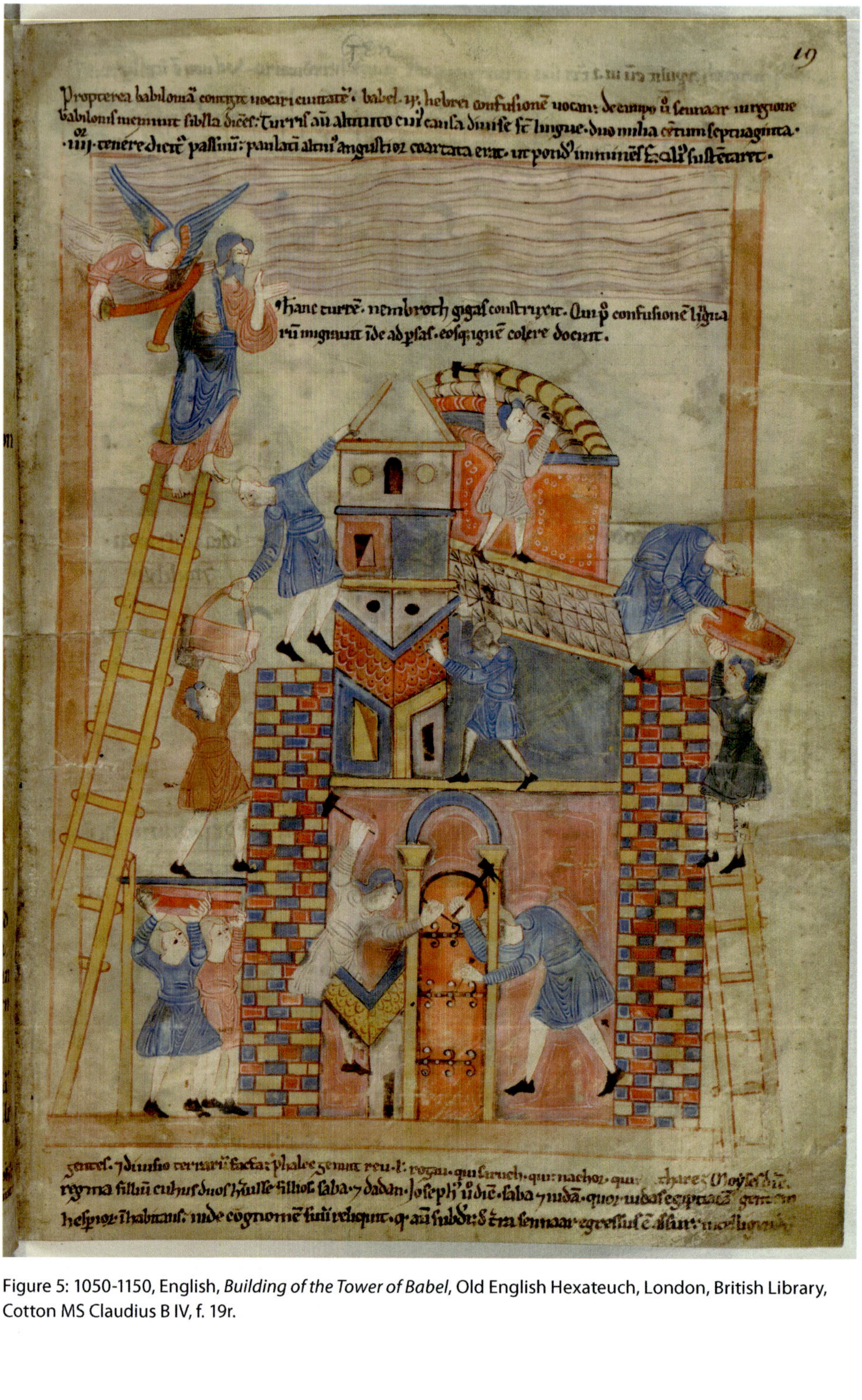

Figure 5: 1050-1150, English, *Building of the Tower of Babel*, Old English Hexateuch, London, British Library, Cotton MS Claudius B IV, f. 19r.

imaginary scaffold lifts; four of these are masons or carpenters using hammers and chisels; of the other two, one receives a hod possibly containing mortar from another on a ladder whilst the other takes a bucket from a worker standing on the actual scaffold. The remaining two workers are on the ground lifting what appears to be another hod of mortar. God is at the top of a symbolic ladder, the *scala Dei*, or *scala Paradisi*, with an attendant angel, contemplating when to bring the whole operation to a halt.

The eleventh century Frankish or Italian miniature in figure 6 depicts, albeit faintly in light green, a scaffold erected to build a tower, probably the tower of Babel. This shows actual trees used for the standards and felled for the purpose. The branches have been lopped close to the trunk, but with sufficient stump left to support the ledger, at least to the one on the right. The opposing one to the left is hidden behind the ladder. Ladders were also made from the products of coppicing. Here the rungs are mortised through the stiles or stringers to leave the ends projecting. The illumination shows a series of hod carriers, two on the ladder taking mortar to a mason, who perches precariously one-legged on the ledger and holds an up-ended hod in one hand and a trowel in the other. There must have been a platform supported on putlogs but the artist has left these out. One expects a degree of primitive depiction in these early works, but the rendering of the standards as small trees is compelling and supported by later examples.

The 'Morgan Bible' a French manuscript of the 1240s, shows, in the upper right quadrant of folio 3r, (Fig. 7) the construction of the Tower of Babel again taking place. There is no scaffold depicted, but there obviously must be one if only to support the ladder, which is not against the building but hangs in mid-air. This ladder is shown, even for the thirteenth century, at a dangerous angle, too dangerous for the pair carrying the barrow with a heavy load of stone or bricks. The assumption must be that it is a ramp with the infilling panels between the rungs left off or forgotten; there must have been some confusion in the studio between both items of equipment. It cannot be imagined, even for the period, that two workers with such a burden, would attempt to walk up a ladder at such an angle. However, the figure before them is more convincing, he climbs the ladder with a hod strapped to his back and has both hands to steady himself. So this is how to climb a ladder with at least one hand for steadying, otherwise he will fall off, but not at this angle because his weight is not properly distributed on the rungs, as is true for the pair behind him. A ramp will work at this angle but not a ladder. A ladder must be at a much steeper angle, conversely a ramp will not work at the normal pitch for a ladder which can be near vertical if safely held in position. Another anomaly is, if the worker with the hod on his back were more vertical the mortar would simply fall out of the container.

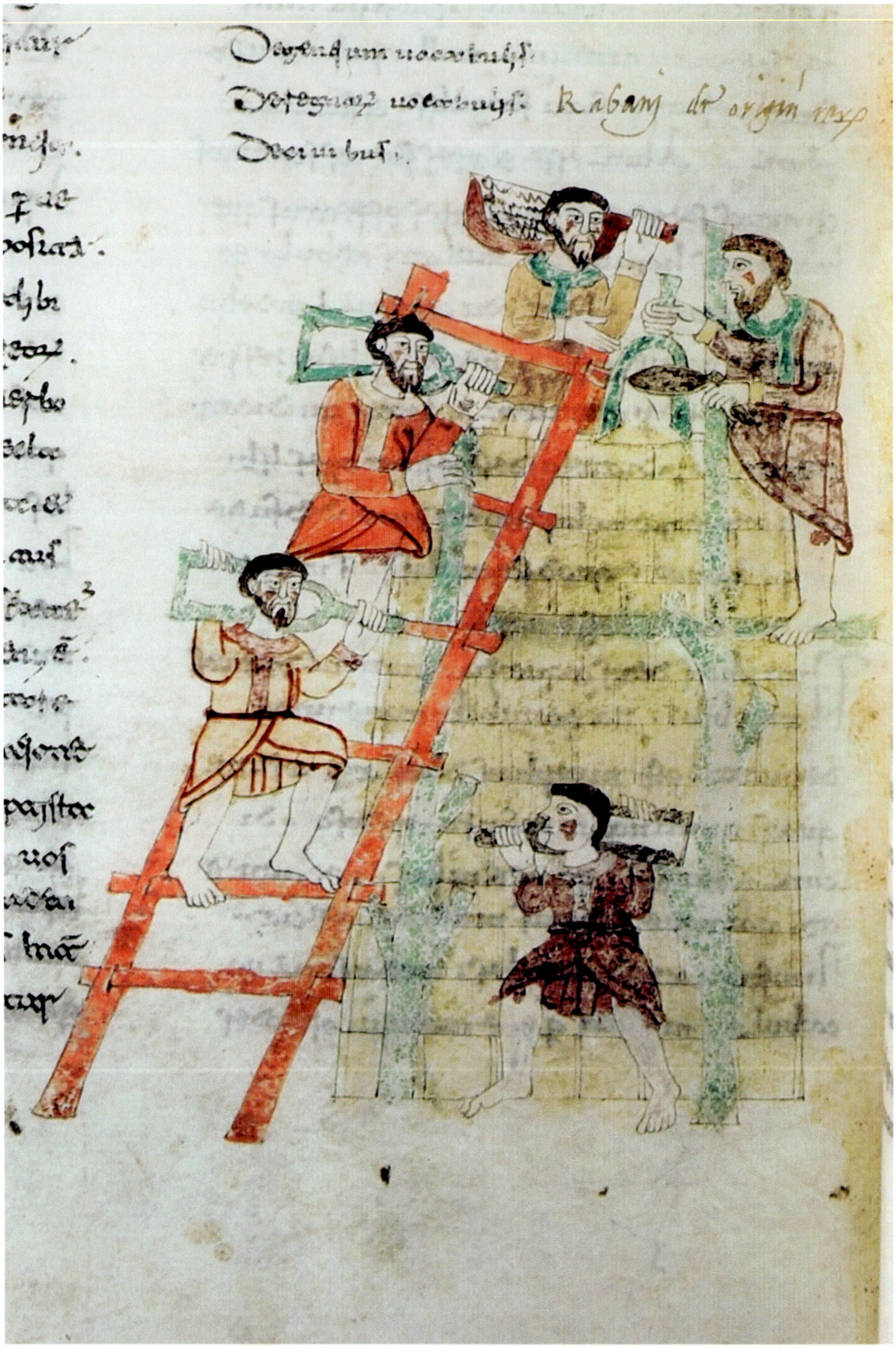

Figure 6: *Building a Tower*, eleventh century, Frankish or Italian, from Rabanus Maurus, *De Universo*, cod.132, f. 394, Romanesque edition of the Carolingian original. (Abbey of Monte Cassino, Italy)

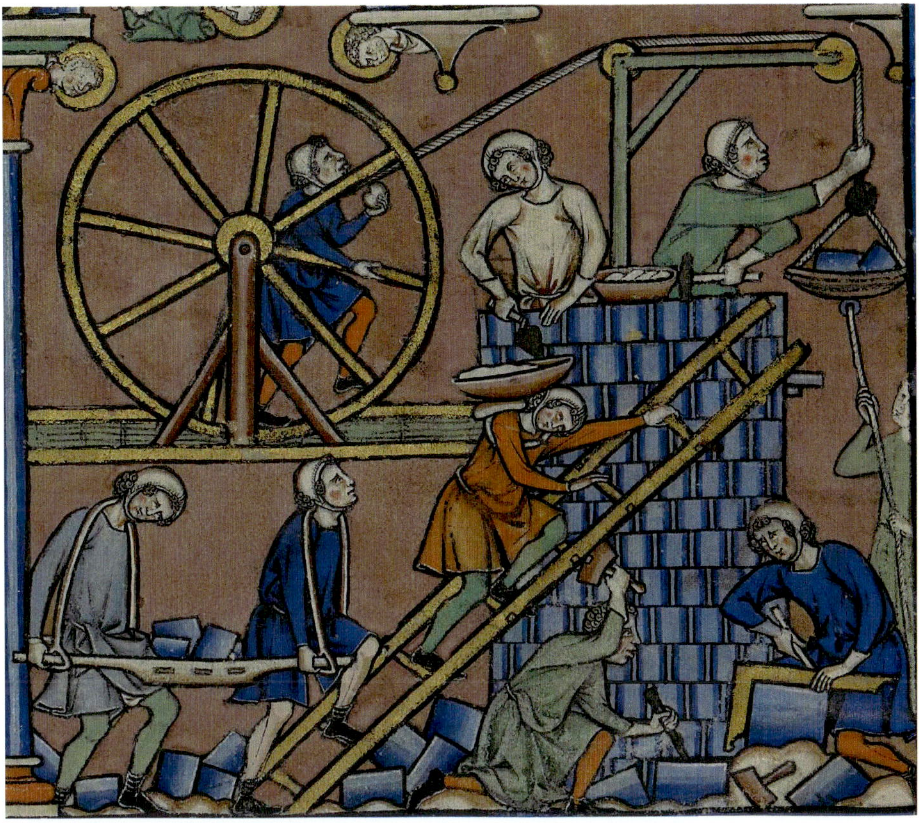

Figure 7: *c.*1244-1254, French, *Construction of the Tower of Bable*, illumination from the 'Morgan'. Bible possibly commissioned by Louis IX before embarking on his first crusade in 1248, The Morgan Library and Museum, New York. MS M638, f. 3r. Purchased by J P Morgan (1867-1943) in 1916.

Ramps were common on Medieval and Renaissance building sites. In a similar illumination of the same period, although far less refined in its execution (Fig. 8) we see two workers walking up what is certainly a ramp. Again no scaffold is visible but one must be assumed. Ramps were made with boards or wattlework strengthened by stiles or lengths of wood either side and with cross pieces or battens, similar to ladder rungs, nailed to the boards which helped to prevent slipping and gave a firm foothold.

Two fourteenth century miniatures show ramps for access along with fairly extensive scaffolding. The first, a miniature of 1385, shows a tower under construction, again a depiction of the building of the Tower of Babel. It could just as well be some other building operation of the late fourteenth century which the artist has observed and recorded. (Fig. 9) The building of the Tower of Babel, from Genesis 11:4 (*'Come let us build ourselves a city, with a tower that reaches to the heavens'*),

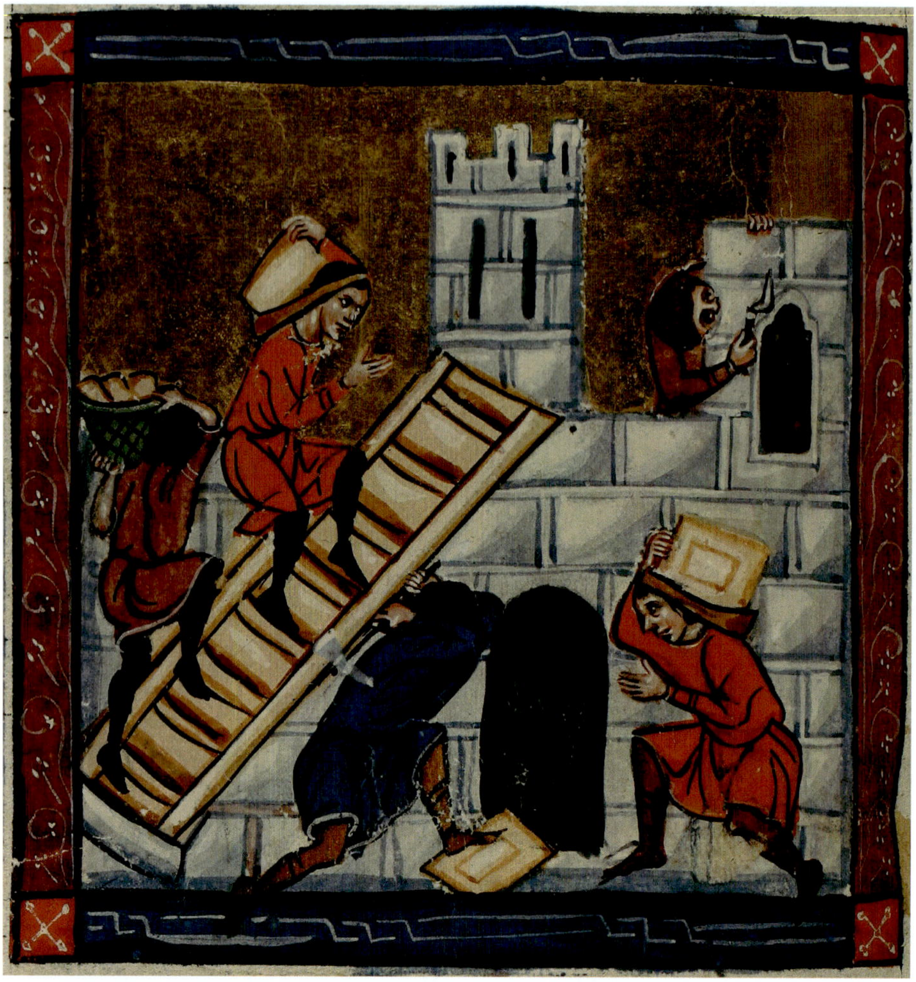

Figure 8: 1260-1270, Holy Land (Acre ?), *Construction of Rome*, from 'Histoire ancienne jusqu'à César', ms. 0562, f. 009, Bibliotheque Municipale de Dijon, France.

is a popular theme for illustration in Medieval and later manuscripts. Here the scaffold is keenly observed, although there are some inconsistences. The artist has lost his way where the return of the tower is visible on the left and cannot deal with the perspective. Nevertheless there is a lot of information. The standards, ledgers and putlogs are poles from coppicing, and, as one would expect, have greater diameter at the base. To gain height, as is universally necessary with most scaffolds, the standards are joined and the binding is clearly visible, as it is on the ledgers and putlogs. The ramps, as well as the working lifts, have handrails. The cross battens are clearly visible but the angle of both is perhaps a little steep

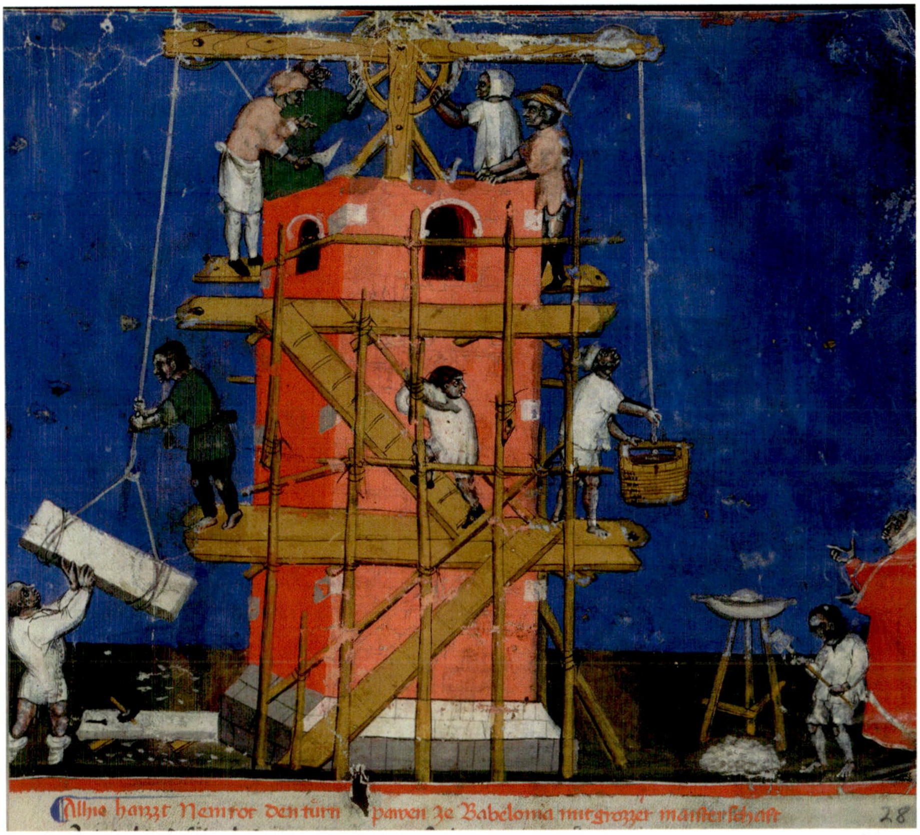

Figure 9: Unknown artist, *Building the Tower of Babel*, 1385, German, Bavaria, illustration from the *Weltchronik (Chronicle of the World)* by the Austrian epic poet Rudolf von Elms. An illuminated manuscript page: Shelf mark: 2 ° Ms. Theo. 4 f. 28r Universitatsbibliothek Kassel. Germany (Landesbibliothek und Murhardsche Bibliothek der Stadt Kassel).

for comfort. The lifts or working platforms are stout planks, these would have been expensive as they required sawing along the length, a labour intensive and expensive process.

The second image, (Fig. 10) shows a ramp at a slightly more comfortable angle, a labourer climbing it with a hod of mortar on his left shoulder. With his right hand he steadies himself by holding the handrail clearly visible in the image. This perhaps emphasises the impossibility of the situation in Fig. 5. The artist has lost his way with the complicated grid of scaffolding with putlogs at various angles and overhanging the standards. The only lift boarded, with what appears to be a plank, is that to which the hod carrier is making his way towards. The other platforms have been removed. Two men carry a barrow of mortar whilst two others mix

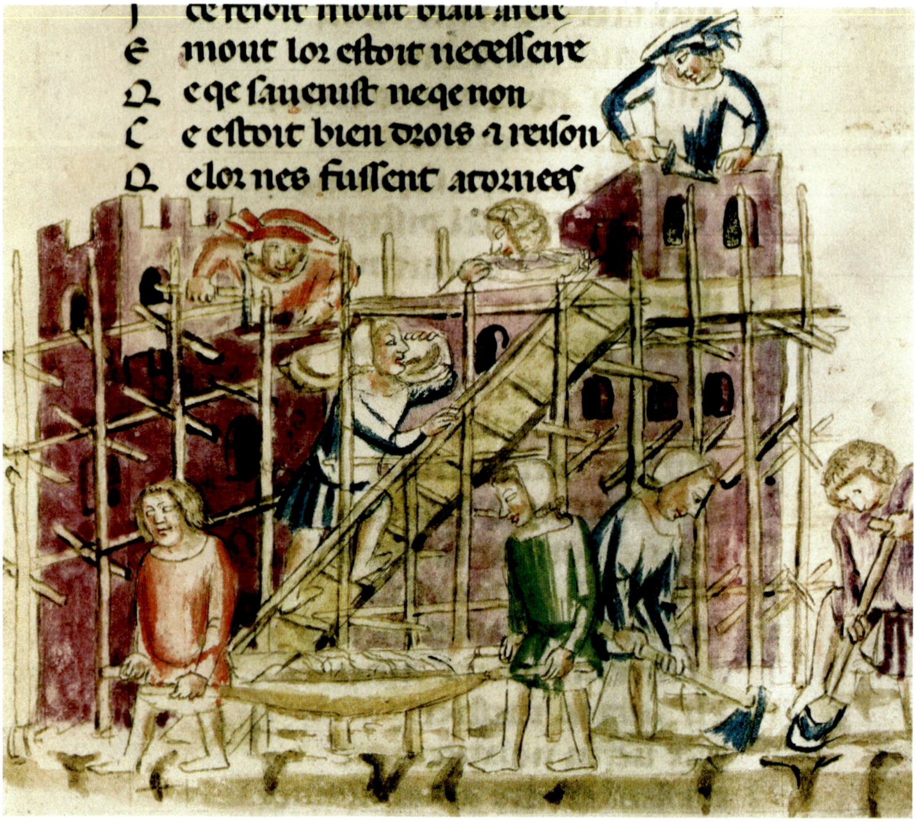

Figure 10: French, *Construction of a Fortress*, page from a manuscript in the Bibliotheque Nationale de France. Mary Evans Picture Library.

mortar. The central figure behind the wall handling another hod of mortar, stands on a scaffold within the building, two standards of which protrude above the wall.

Another hybrid ramp, because it is at an impossible angle, (Fig. 11) is to be found in an Italian fourteenth-century copy of *Histoire Ancienne Jusqu'à César.* This is a manuscript rich in scenes of construction, all in the lower margins or 'bas-de-page'. In this particular image, the *Building of Lavinium*, there is no scaffold, just the ramp tuned through 90°, so that we see it is definitely a ramp and not a ladder.[34] The figure using it balances treacherously with both hands supporting the burden on his shoulders. Similarly in the same manuscript, in the wonderfully capricious *Building of Carthage*, a labourer carries a piece of stone up a ramp with over-sized cross pieces to a seemingly detached part of the building suspended mid-air. (Fig. 12) In the same series a definite ladder is in use in *The Rebuilding*

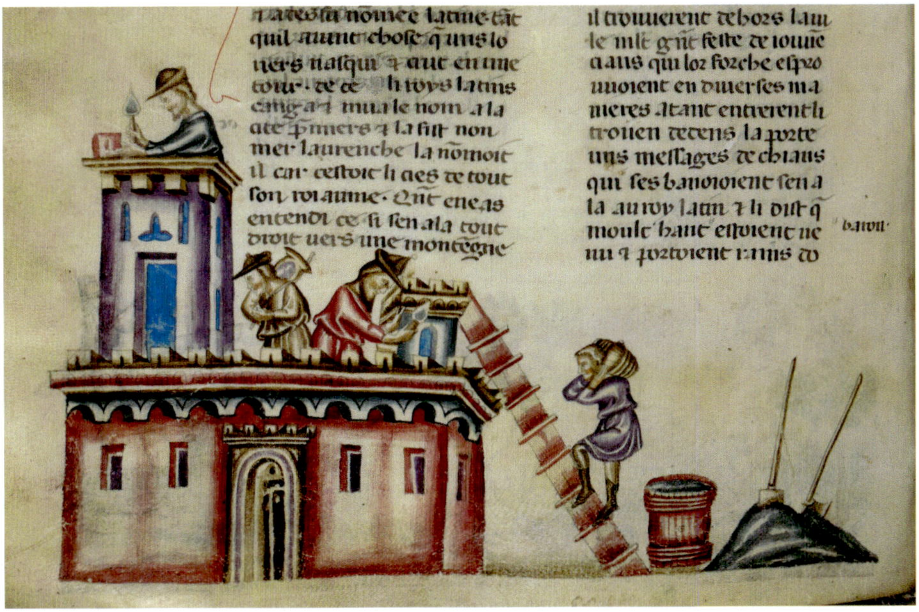

Figure 11: Italian, Naples, 1325-1350, *Building of Lavinium*, from *Histoire Ancienne Jusqu'à César*, © British Library Board, London, Royal MS 20 DI, f. 202v.

of Troy (Fig. 13) and in the image of *Ulysses Building a Castle* (Fig. 14) we wonder if the platform the two workers are standing on is a scaffold supported only on putlogs and brackets, or a balcony, which is part of the building. In all five of these images although construction is clearly depicted, all of the buildings are seemingly complete.

Among the most important works of Ambrogio Lorenzetti are the frescoes dating to 1338-9, in the Palazzo Pubblico, Siena, in the Hall of which are his famous scenes of Good and Bad Government. In his *Effects of Good Government in the City*, the middle and far ground are occupied by a cluster of buildings rendered with striking attention to detail. At the very top, just to the right of centre is a busy building scene with masons and labourers constructing the outer wall of a house. (Figs. 15a & b) The scaffold on which four of the five figures depicted are working is supported entirely on putlogs. These are squared off poles neatly inserted into putlog holes, and arranged exactly above the holes from the former, lower lift. Onto these are set transverse poles, on which are placed wattlework hurdles or clays, a common form of scaffold platform. None of the other buildings in the fresco show putlog holes, but buildings in Siena are peppered with them, particularly the Torre del Mangia and City Hall, within which are the actual frescoes.[35]

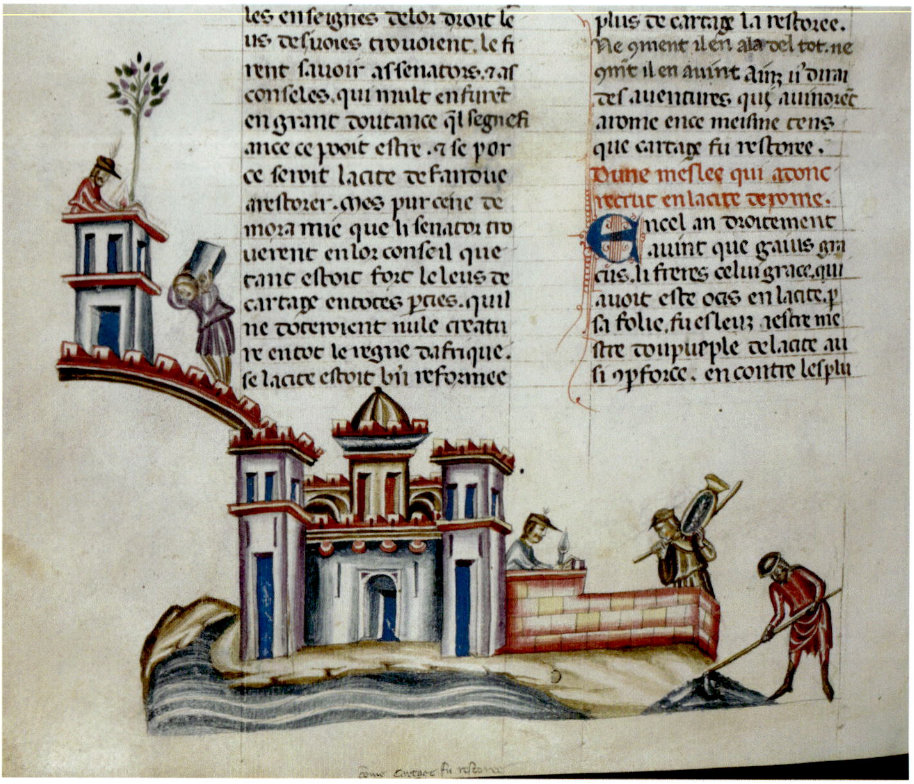

Figure 12: Italian, Naples, 1325-1350, *Building of Carthage*, from *Histoire Ancienne Jusqu'à César*, © British Library Board, London, Royal MS 20 DI, f. 323v.

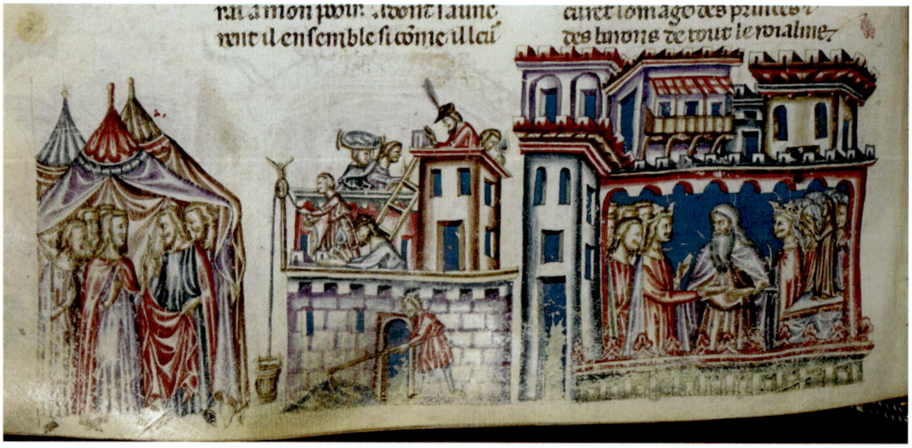

Figure 13: Italian, Naples, 1325-1350, *Rebuilding of Troy*, from *Histoire Ancienne Jusqu'à César*, © British Library Board, London, Royal MS 20 DI, f. 192v.

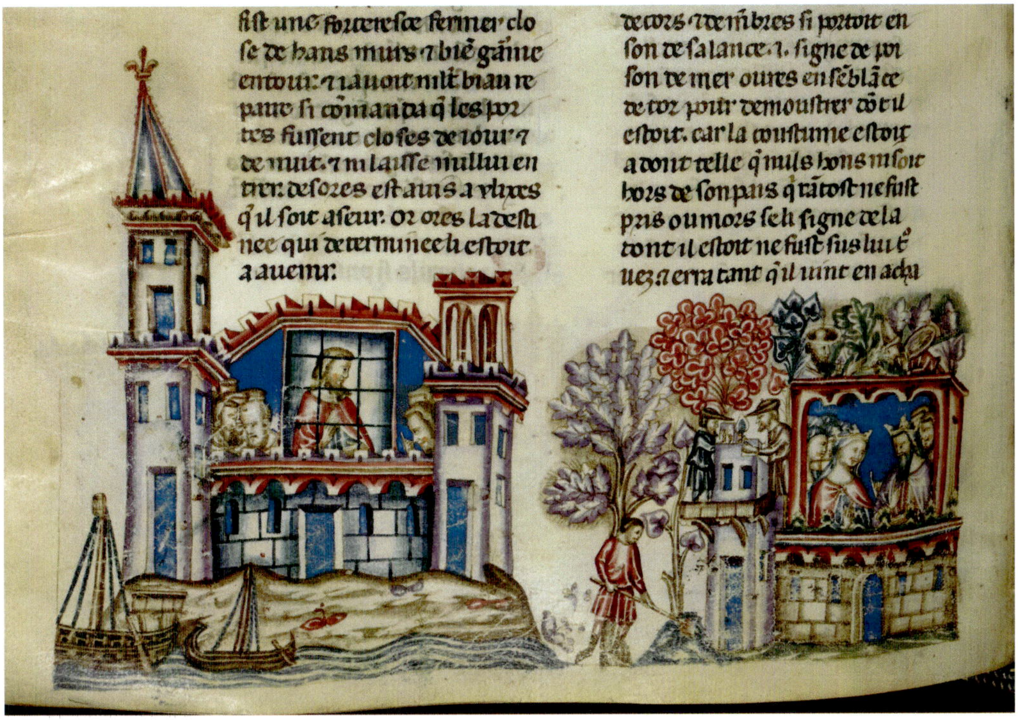

Figure 14: Italian, Naples, 1325-1350, *Ulysses Building a Castle*, from *Histoire Ancienne Jusqu'à César*, © British Library Board, London, Royal MS 20 DI, f. 189v.

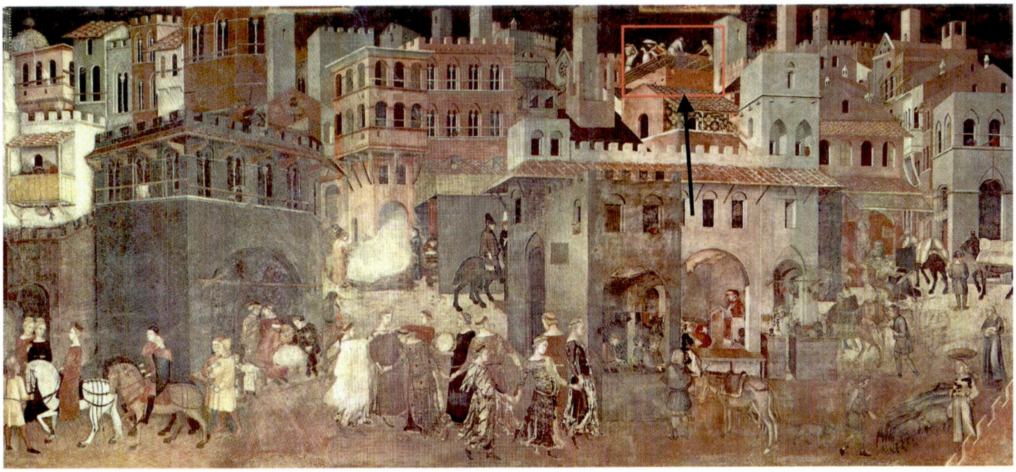

Figure 15a: Ambrogio Lorenzetti, Italian, active 1319-†1348/9, Detail *The Effects of Good Government in the City*, 1338/9, Fresco, Palazzo Pubblico, Siena, Italy. (Jim Forest Creative Commons)

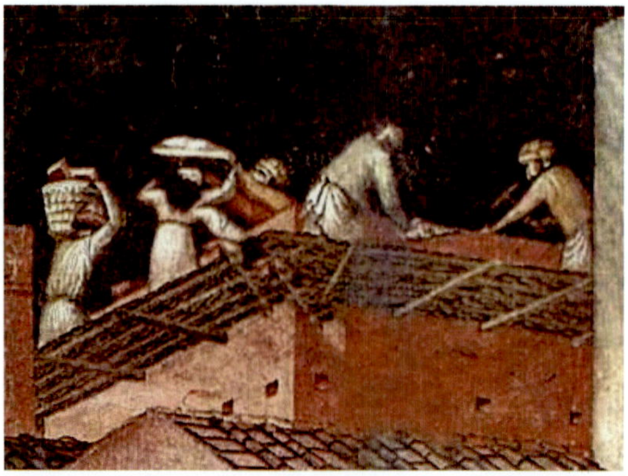

Figure 15b: Ambrogio Lorenzetti, Italian, active 1319-†1348/9, *The Effects of Good Government in the City*, 1338/9, Fresco, Palazzo Pubblico, Siena, Italy. (Jim Forest Creative Commons)

A similar scaffold structure, without standards, although much earlier by almost a century, is to be seen in a French *Bible Historiale* of *c*,1250. (Fig. 16) This is the Tower of Babel. The scaffold is only seen in section on the flank walls of what appears to be a square building. Three storeys have been completed, the upper two set back forming balconies. There are putlogs, but no distinction between these and any platform, the putlogs are additionally supported with brackets. The scaffold does not run across the front of the building and must therefore be assumed to be present as two of the workers each have one foot in mid-air on this imaginary platform; or are they using the top of the balcony parapet for their foothold? The structure of the scaffold is a little dubious, however we must not loose sight that these are fairly primitive images. The putlogs abut the building at the top of the first balcony and would therefore not have any cantilever action, unless they penetrated through into the third storey. A ladder is propped against the building from the ground to the second storey leaning against a pier between two arched openings. A worker climbs this with a piece of stone on his shoulder. The crane is discussed below on page 143.

A charming image, a sketch only in comparison, and later at 1300-1325 is in another *Bible Historiale*. (Fig. 17) A labourer climbs a ladder with a hod of material, it could be mortar or stone; the only other activity is a mason on the ground working a piece of stone with a hammer. There is no scaffold and the only other operatives are shown head and shoulders, one looking out of a window and three others on the battlements of what appears to be a complete building; one of these watches the labourer on the ladder. This is certainly a scene of construction but is of inferior sophistication in its observation of building work and the capability of the draughtsmanship in rendering this. It tells us that the Tower of Babel is being

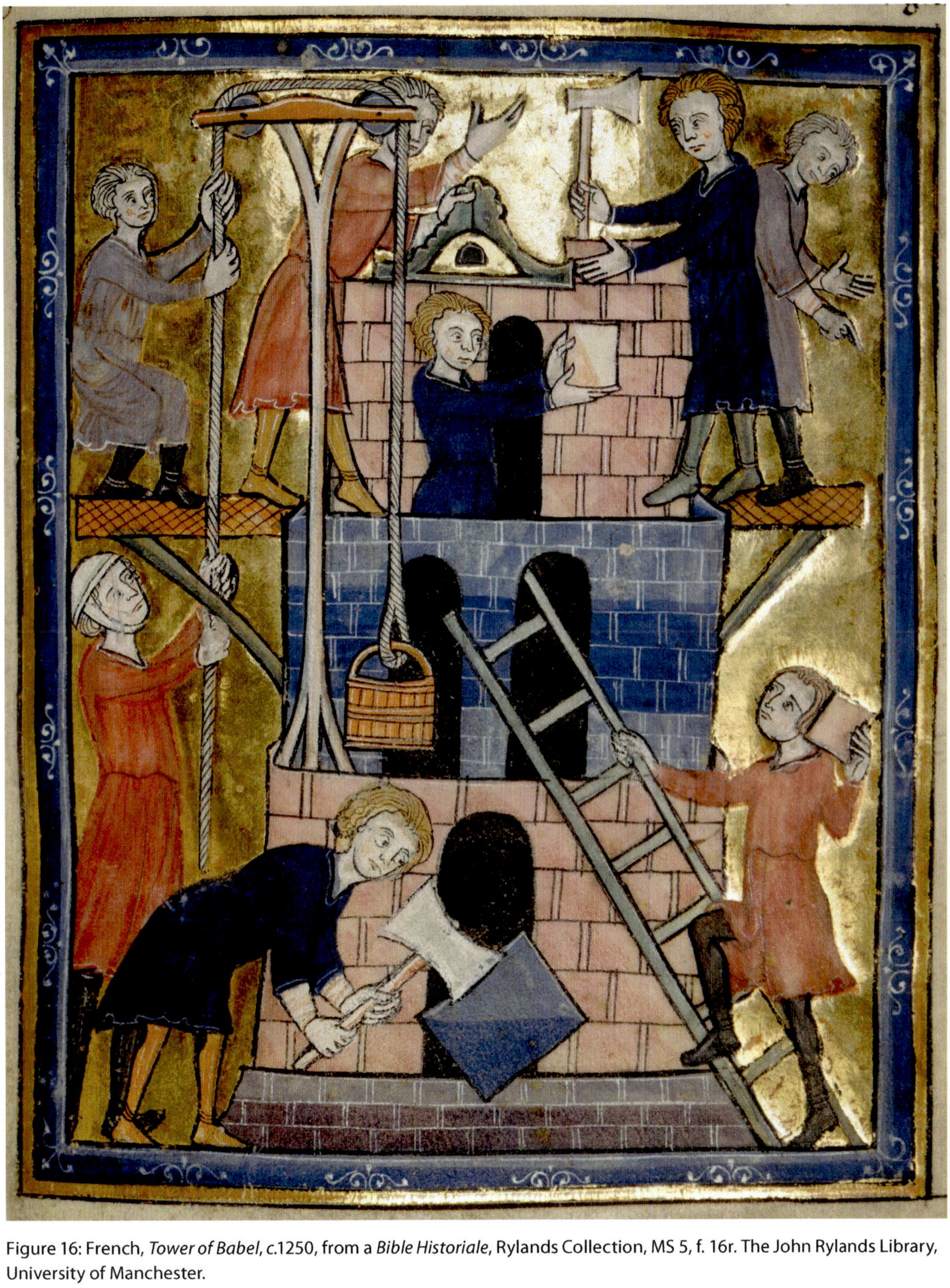

Figure 16: French, *Tower of Babel*, c.1250, from a *Bible Historiale*, Rylands Collection, MS 5, f. 16r. The John Rylands Library, University of Manchester.

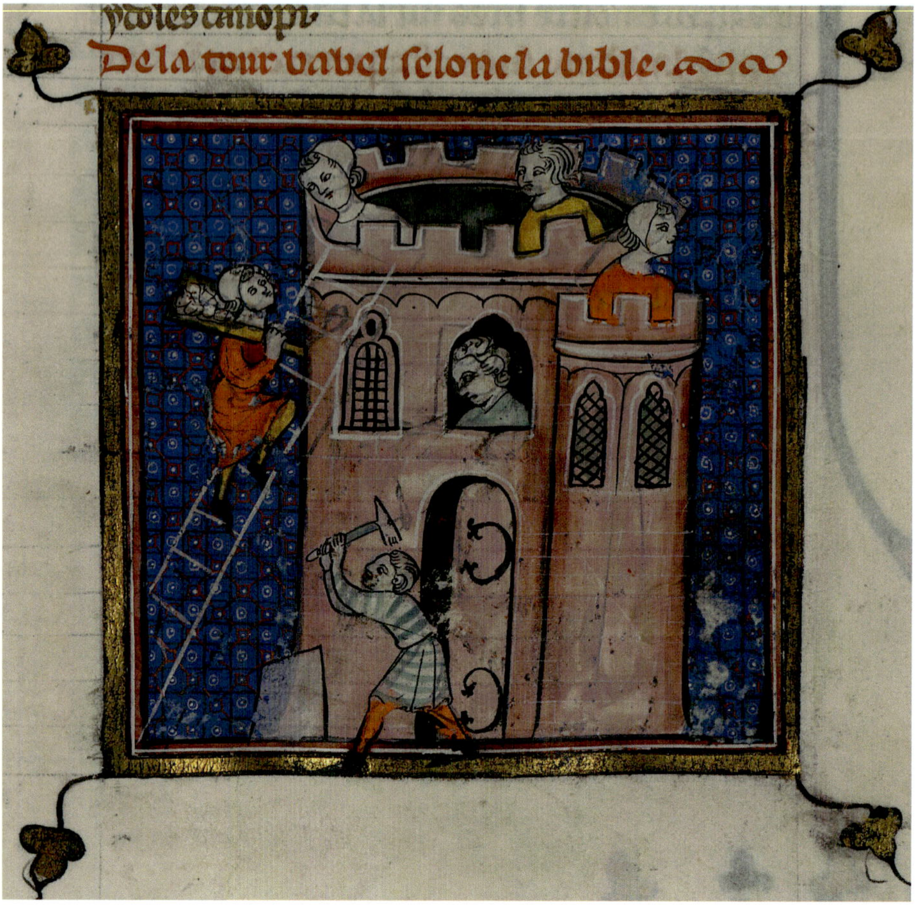

Figure 17: French, *Construction of the Tower of Babel*, 1300-1325, *Bible Historiale*, Bibliotheque de l'abbaye de Clairvaux, ms. 0059, f. 017.

built, but there are more watchers than builders in this strange and charming scene.

A king, we don't exactly know which king, supervises the building of a church in *Estoire del Saint Graal*, a French manuscript of the same date range as Figure 17. (Fig. 18) The miniature shows the king supervising four workmen. One workman is at the top of a ladder; it is not clear what he is doing as the painting is rubbed and indistinct in places, but he may have carried up a piece of stone, placing it on top of a pile of similar pieces stacked on a scaffold lift which the artist again has not included. The mason on the right must also be on this invisible scaffold, otherwise he floats in mid-air. There is also a mason behind the window working overhand. He is standing on a wattlework scaffold platform, clearly seen through the window. Of particular interest, although the scaffold to the exterior is not shown,

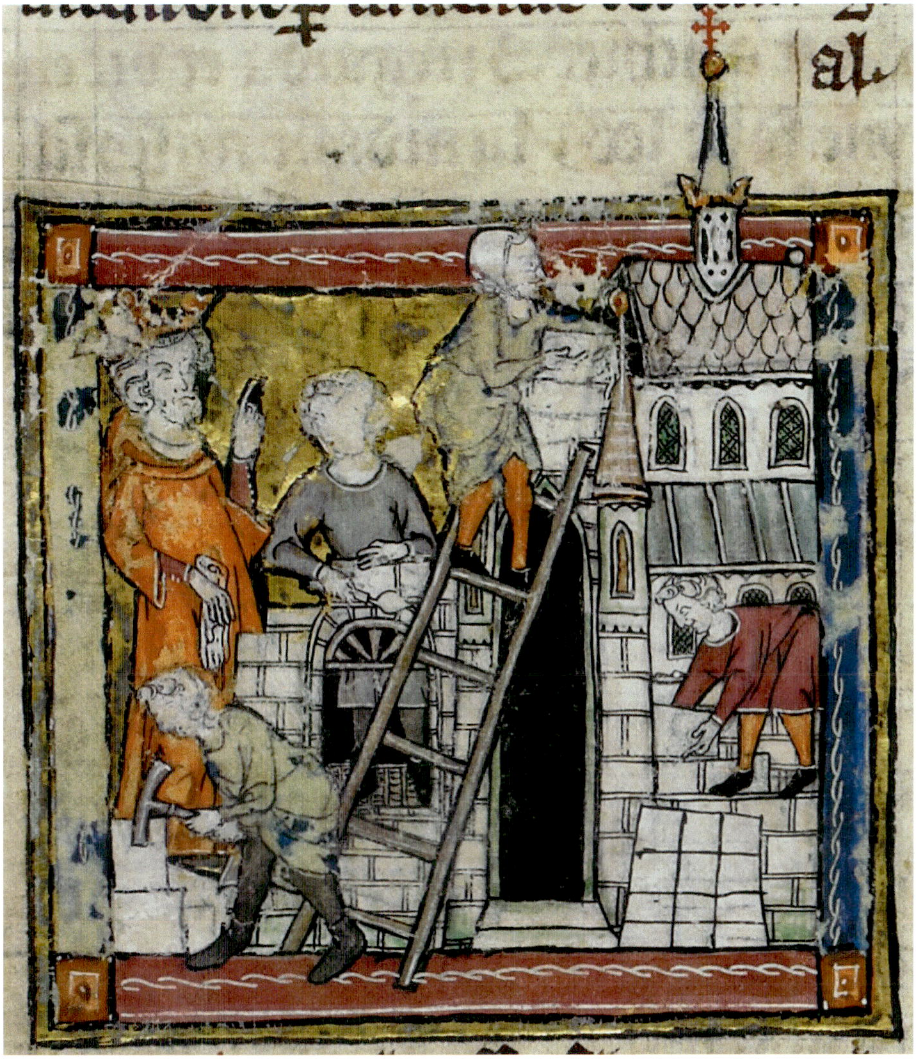

Figure 18: French, 1300-1325, *A King supervising the building of a church*, from *Estoire del Saint Graal*, part of Royal MS 14 E III, f. 85v. London © The British Library Board.

the artist has taken the trouble to show the timber formwork to the window arch; this is supporting the masonry of the actual arch, as well as above, that the mason is working on. Here is important detail in what could otherwise be dismissed as a simple and perhaps naive building scene.

Another monarch, this time ordering the construction of a church, is Charlemagne. This is almost certainly the royal chapel attached to his palace at Aachen, or Aix-la-Chapelle for this is also a French manuscript and of similar date, 1300-1350. The palace complex was begun *c.*790 and completed in the first decade of the ninth

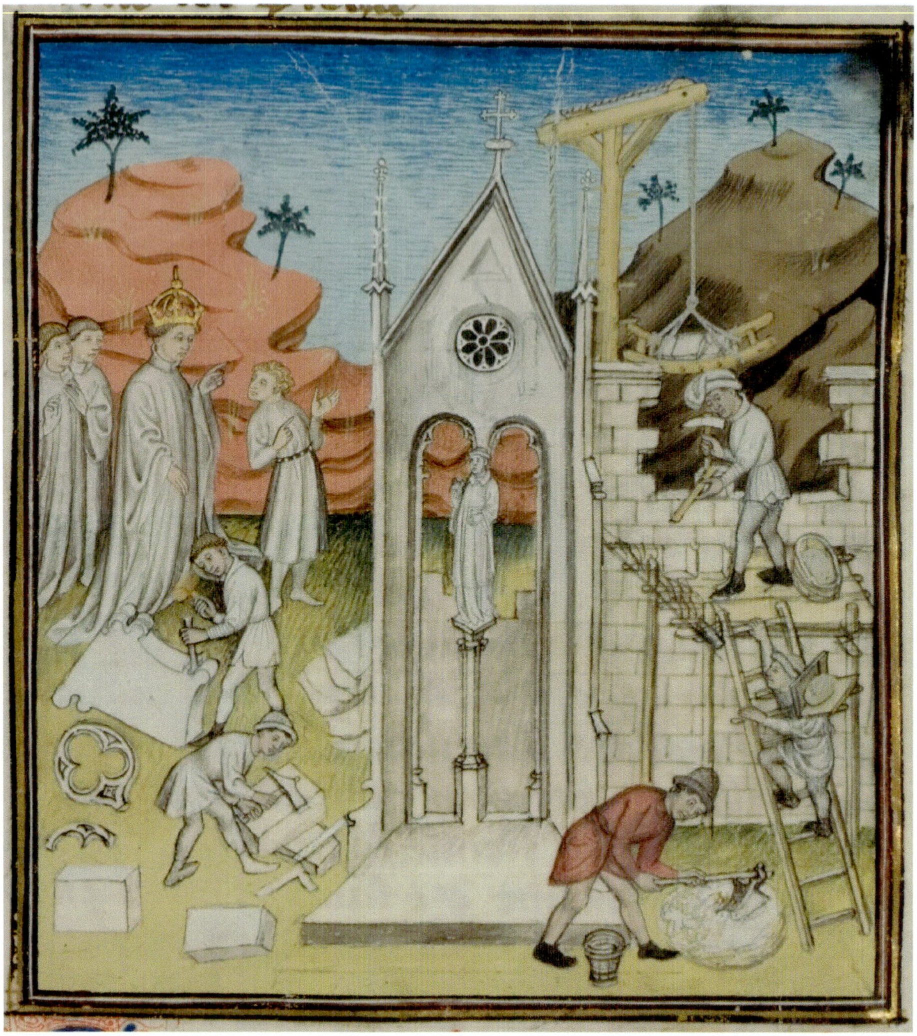

Figure 19: French, first half of the fourteenth century, *Charlemagne orders the construction of a church (Aix-la-Chapelle)*, from the *Grande Chroniques de Saint Denis*, Bibliotheque Municipale de Toulouse, ms. 0512, f. 096r.

century. The building we see under construction (Fig. 19) bears no resemblance to the actual chapel, which is Carolingian and 'Pre-Romanesque' in style. It has a sixteen sided exterior with an octagonal interior with round headed arches to the windows, arcades and upper storeys. It would have been largely intact in the fourteenth century and remains so now, although it has been much restored.[36] If this were a northern French manuscript we would perhaps expect the artist to depict the actual building, but what we have is a Gothic building of the artist's time and produced from his own personal observations in his locality. He may of course have seen later contemporary additions being built; it must therefore, to some degree, be considered a capriccio, as so many of these images are. The scaffold is a simple single lift with standards tied to putlogs, there are also inner standards close to the building. The ties or withies can be seen at the top of the standards binding these to the putlogs. The platform consists of three planks; on this is a mason fixer laying stone, and at his feet are two further pieces of stone and a container of mortar. A labourer climbs a ladder with a hod of mortar. Depicted correctly, he has one hand, his left, for his load, and one for the ladder. Other masons work nearby squaring blocks and carving mouldings. In a later fifteenth-century miniature of a similar scene, *Charlemagne Builds Churches* in the *Grandes Chroniques de France*, the artist Jean Fouquet shows one of the buildings under construction in the Carolingian style.[37] So Fouquet has sought a degree of authenticity for the architecture but at the expense of anything particularly elucidating regarding the process of building. A further miniature *Charlemagne visiting the construction works of Aix la Chapelle*, (Fig. 24) also shows a Carolingian or Pre-Romanesque style building.

The renowned *Très Riches Heures* of Jean, Duke of Berry, created in the early years of the fourteenth century and illustrated by the Limbourg brothers, depicts several scenes featuring King David. In two of these he is shown standing before building sites with very robustly built scaffolding (Figs. 20 & 21). The scenes are of the Limbourgs' time rather than David's. The buildings under construction are unashamedly late Medieval Gothic and bear no resemblance to what David, who reigned *c.*1010-970 BCE, would have understood or recognised some two millennia earlier. We may suppose that these buildings represent the building of the Temple of Jerusalem, which David had planned and assembled materials for. The Bible tells us, however, that God forbade David from building the Temple (1 Chronicles 21 & 22: 1-10), and it was his son Solomon who was the builder. David therefore is here only contemplating the construction of what he had planned. Other depictions and portraits of buildings in the manuscript are of royal palaces and views of cities. These, many of which were sketched and drawn from the actual buildings on site, are also contemporary.

The scaffolds are constructed of what appear to be squared timbers using carpentry techniques almost of the sophistication for a timber framed building. This would have been unduly expensive, except of course, and this is plausible, if such timbers were available in excess of requirements. It is nevertheless an unlikely scenario and perhaps the Linbourgs are over indulging their fantasy of what a scaffold really looked like. In one of the scenes a scaffolder or carpenter is adjusting the scaffold and the platform he is standing on is a woven wattlework clay. (Fig. 21) It is a tragedy that all three of the brothers and their patron died young, at less than 30 years of age, probably of the plague. The work was left unfinished but the illustrations completed *c.*1485 by Jean Colombe, with other mid 15th century additions.

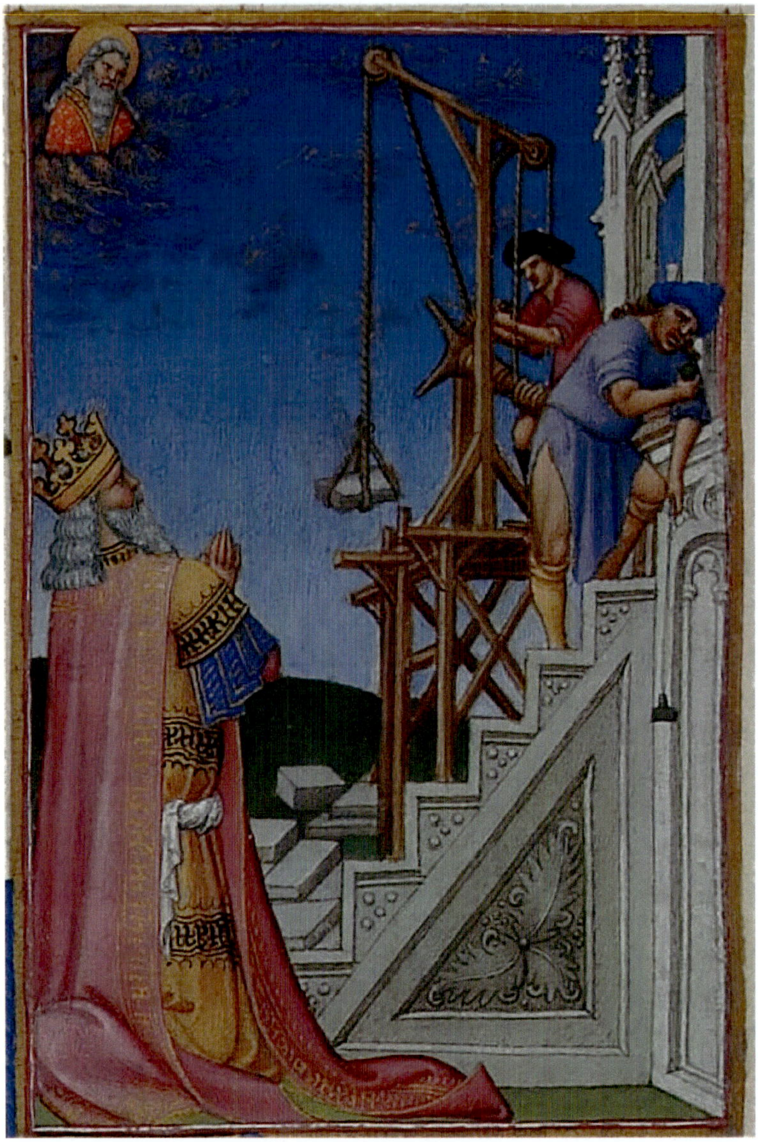

Figure 20: French, first quarter fifteenth century (pre-1416), Limbourg Paul, Jean and Hermann, *Building in Jerusalem*, from *Très Riches Heures du Duc de Berry*, ms.65, f. 49v, Musee Conde, Chantilly, France.

The Temple of Jerusalem stood for approximately 374 years from *c*.960 to 586BCE when it was completely destroyed by the Babylonians. The Persian King Cyrus who ruled 539 to 530 BCE allowed the Jews to rebuild the temple. The foundation was laid in 536 BCE and the edict of Cyrus '*In anno primo Cyris regis* [...]' from the first year of his reign features in many thirteenth-century

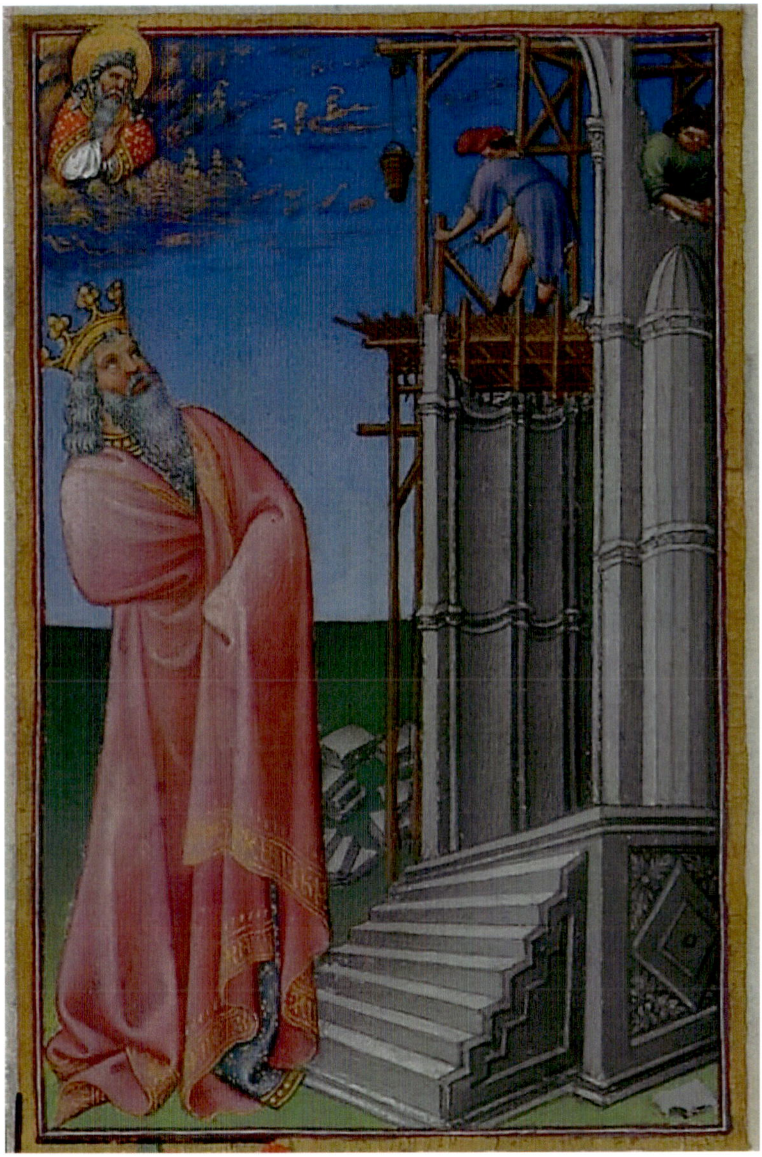

Figure 21: French, first quarter fifteenth century (pre-1416), Limbourg Paul, Jean and Hermann, *The Building of the Jerusalem Temple*, from *Très Riches Heures du Duc de Berry*, ms.65, f. 35v, Musee Conde, Chantilly, France.

bibles as an elongated historiated 'I'. (Fig. 22) Scaffolds are not shown in these examples but ladders and ramps are, as in 1, 3, 4, 5, 7 (an invisible ramp), 10 and 12. Cyrus is either at the base or top commanding the work above or below. The ramps are shown at excruciatingly steep angles in order to squeeze them into the picture.

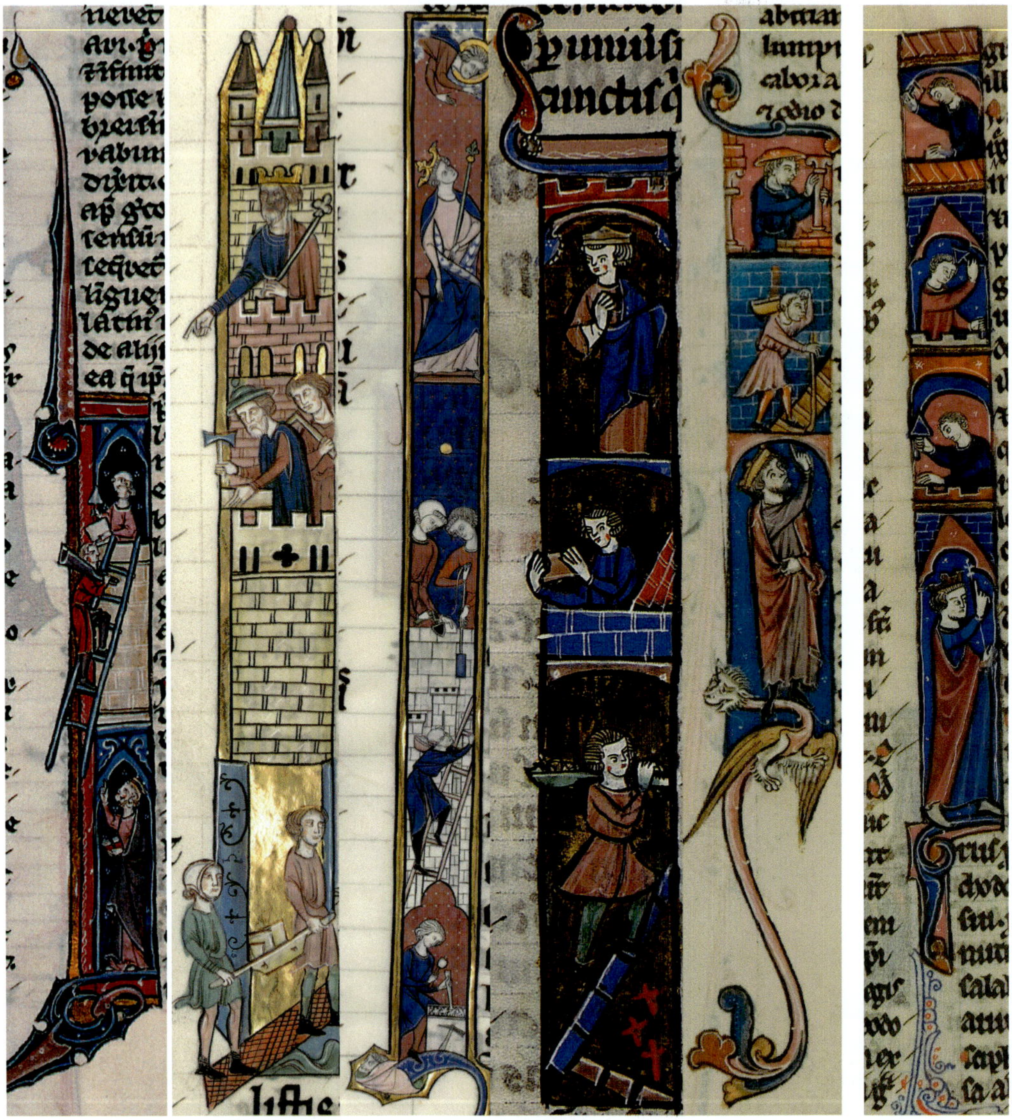

1 2 3 4 5 6

Figure 22: The 13th century bible: Cyrus orders the rebuilding of the temple of Jerusalem, Historiated initials 'I', from French municipal libraries (except 12) as follows:

22.01: 1250-1275, MS. 0001, f.195v. Bibliotheque Municipale de Arles.
22.02: *c*.1220, MS. 0036, f.70v. Reims Bibliotheque et Mediatheque de Reims.
22.03: Late 13th century, MS. 0040, f.163v. Reims, Bibliotheque et Mediatheque de Reims.
22.04: 1225-1250, MS, 0033, f. 86v. Médiathèque du Grand Troyes (L'abbaye de Clairvaux).
22.05: 13th century, MS. 0008, f. 176v. Bibliotheque Municipale de Toulouse.
22.06: 1250-1300, MS. 0005, f. 150v. Bibliotheque Municipale de Toulouse.

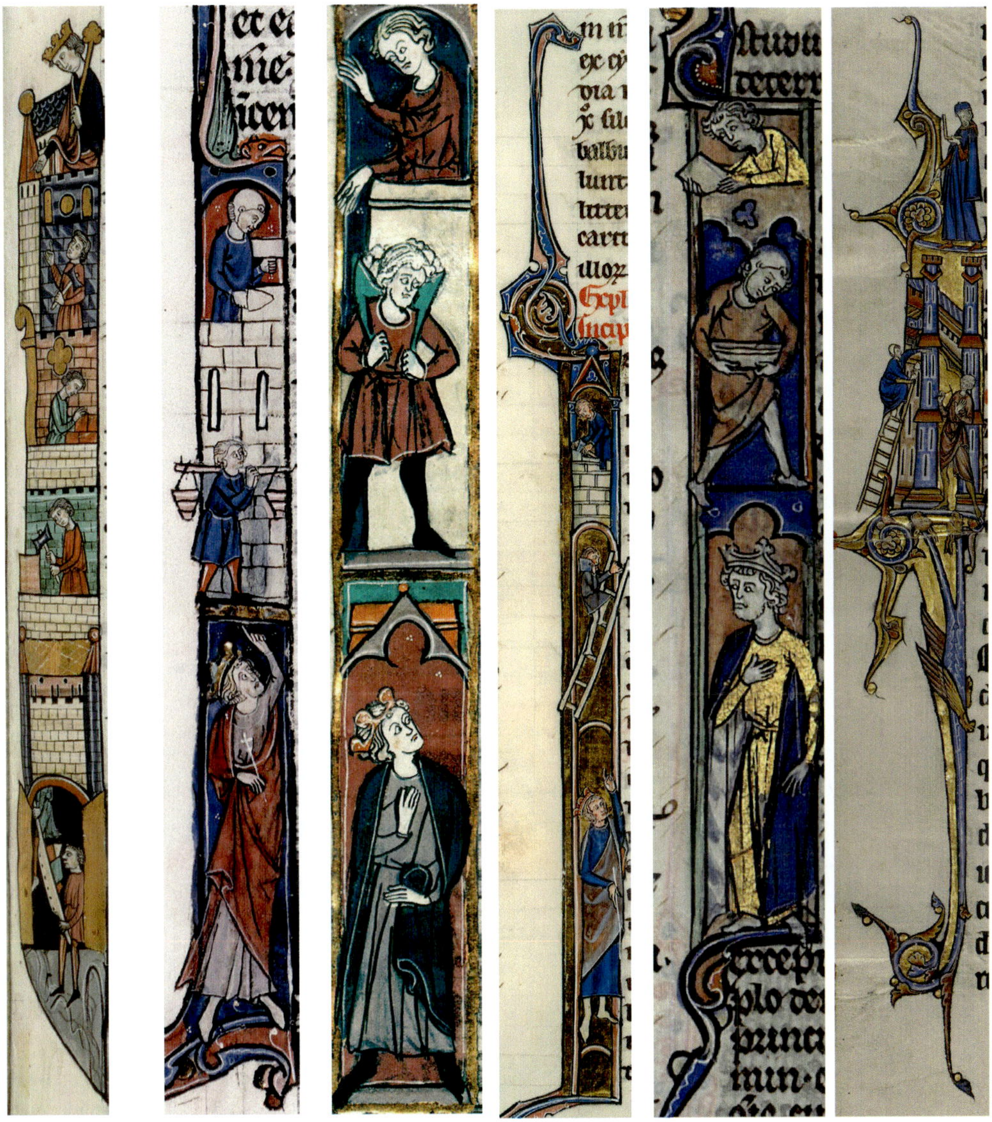

7 8 9 10 11 12

Figure 22: The 13th century bible: Cyrus orders the rebuilding of the temple of Jerusalem, Historiated initials 'I', from French municipal libraries (except 12) as follows:

22.07: Early 13th century, MS. 0023, f. 116. Bibliotheque d'Amiens Metropole.
22.08: 1270-1280, MS. 0146A, f. 188v. Bibliotheque de Autun.
22.09: 1270-1280, MS. 0023, f. 153v. Bibliothèque Gaspard Monge, Ville de Beaune.
22.10: 13th century, MS. 0345, f. 267v. Médiathèque d'Agglomération de Cambrai.
22.11: 1250-1275, MS. 0007, f. 240. Bibliotheque Municpale de Orleans.
22.12: *c.*1270, MS.Ludwig 1 8, v3, f. 256v. Digital image courtesy of the Getty's Open Content Programe.

Loyset Liédet, a Flemish miniaturist, *c.*1420 to 1479, spent his later years illuminating manuscripts for Charles the Bold. In his painting of the *Construction of Alexandria*, (Fig. 23) in the extreme left of the composition, he shows a scaffold with standards and putlogs of two lifts the second boarded with a mason at work. What stands out however is the massive ramp in the left foreground, which rises some two storeys to the top of the wall being worked on. This is stoutly built with uprights, cross-pieces, braces and a handrail all bound together with withies. A labourer walks up the ramp with a hod of mortar and it is easy to see that the wheelbarrow with a load of stone can be pushed up without too much effort. At the top a clerk of works or an overseer exhorts the workers, one of whom drops his load of mortar on top of the wall; should there also be a scaffold here? Meanwhile the architect raises his hat receiving instructions, or is he also being admonished by Alexander on horseback? To the right two labourers are sieving aggregate and mixing mortar.

An image in a hand illuminated early printed book or incunabulum, from the late fifteenth to early sixteenth century (Fig. 24) shows again whole trees used as standards, this was not unusual. Where it suited circumstances nearby trees would simply be felled and used green, as they would have been used as components in the actual building:

> The medieval builder had no hesitation about using green timber. When building operations were undertaken the carpenters were sent to the nearest available wood […] to fell suitable trees, lop the branches and rough them up, to be carried whole or sawn, to the framing place.[38]

For scaffolding purposes they were obviously carried direct to the site. This particular scaffold is made entirely of rough whole trees used whole or cut for the smaller components. Additionally it has braces, the diagonal members used to prevent racking. The standards are set into holes dug in the ground with wedges driven in to steady them. The cranes are described below on page 139; there are no ladders or ramps visible and they are therefore, on this visible elevation, the sole means of bringing up materials.

Jacopo Bellini, *c.*1400-*c.*1470, in one of his albums of drawings, gives us a compelling glimpse of a mid fifteenth century scaffold in Italy. (Fig. 25 a & b) The scaffold is supported both by standards and brackets off string courses on the building. The access is for repairs to the render and possibly cracks in the building as it is decidedly not a new construction. The brackets support putlogs, both of which are squared timber. These, in turn support solid planked lifts. The standards are extremely long without joins; these could be slender pine or fir trunks. There is a ladder from the first to the second lift, where

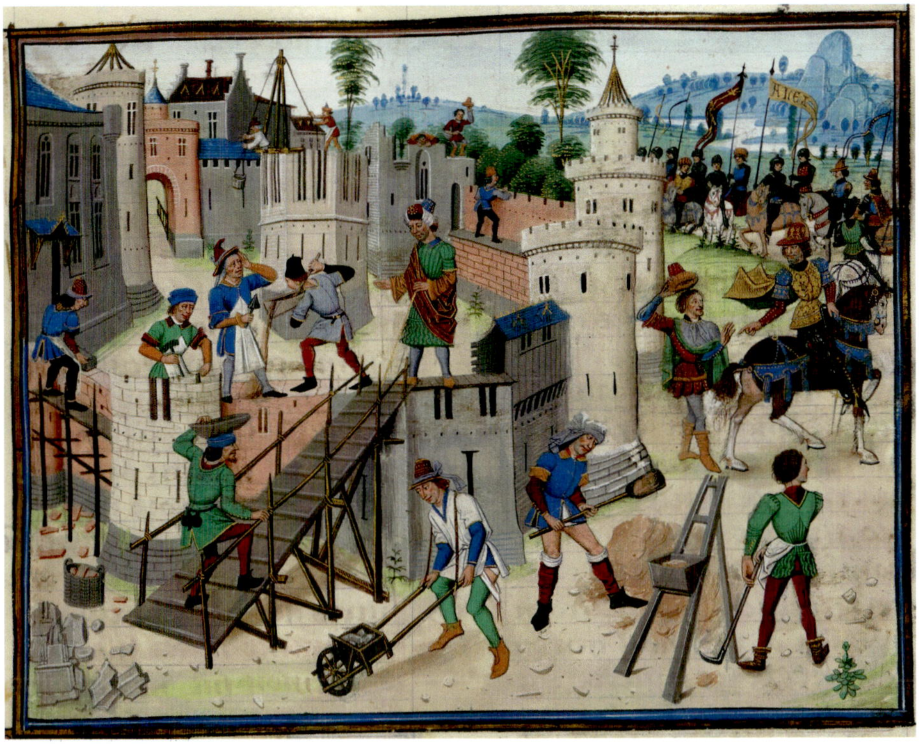

Figure 23: Flemish, c.1480, Loyset Liédet, *Construction of Alexandria*, from *History of Alexander the Great*, Paris, Bibliotheque National ms. Fr 2247, f. 76r.

there is a trap opening to allow access. The scaffold is clearly incomplete at the top; putlogs and a ledger are in place for a third lift and there appears to be a bundle of ties or withies draped over the ledger. There is a lazy laid-back feel to these operations, but not so in the square below. Of the four workers on the scaffold, one is carrying mortar or tools on his shoulder, one is climbing the ladder whilst the other two, one a boy sitting, observe what is happening in the square below. Here, immediately below, a stone carver applies finishing touches to a figure of a saint. This is possibly to be erected on the column and base that lie either side of him, as are others in the background beyond the wall. Here despite the poignant subject of the drawing, working life carries on albeit at a solumn pace. This is a minutely detailed drawing by a master of early Renaissance art.

Another detailed work of similar date this time by Domenico di Bartolo, is a fresco in the Pellegrinaio, the hall for pilgrims at the hospital of Santa Maria della Scala, Siena. (Fig. 26) Here an extension to the hospital is under construction. The bishop, who has commissioned the work, has come to inspect it along with his

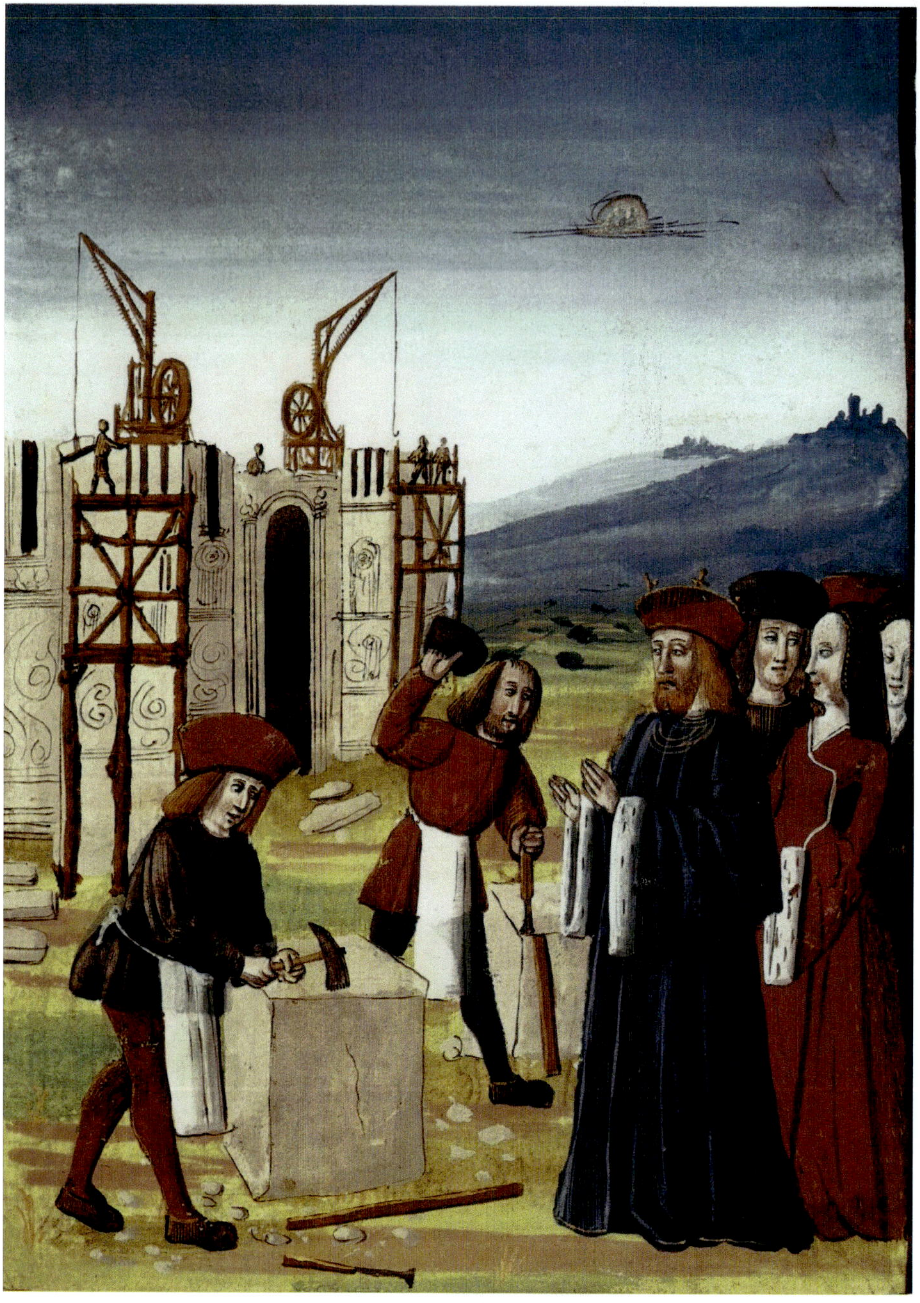

Figure 24: French late fifteenth to early sixteenth century, *Charlemagne visiting the construction of Aix la Chapelle in 796*, illustration from *Grandes Chroniques de France*, edition by Antoine Vérard, hand-painted print in a incunabulum. (Biblioteca Nazionale Universitaria di Torino XV. I. 47, f.133)

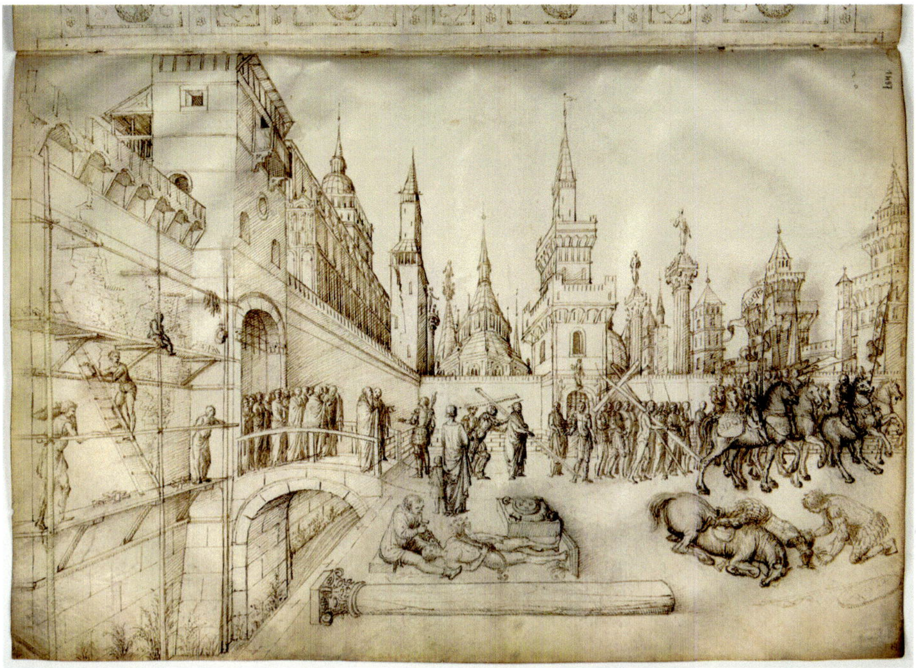

Figure 25a: Jacopo Bellini, *c.*1400-1470, Italian, *The Bearing of the Cross*, *c.*1450, drawing, silver point, pen and brown ink on parchment. (Musée du Louvre, Paris)

canons and nobles on horseback. The bishop has dismounted, he stands talking to the architect or master mason who gestures towards the work. The building work is all to the right-hand of the composition. Here again the standards are whole straight tree trunks, the putlogs are squared timber supported on brackets nailed into the standards. At least one putlog at the building end is supported on a bracket rather than being inserted into the masonry. There are no ledgers as such, the working platform is supported by poles spanning the putlogs. There is formwork on the first lift for arches, which are in the process of construction. The second lift is bracketed out to allow the winch clear working space. This is a busy crowded site with a lot going on. Apart from the mason in the foreground measuring a piece of stone and the overseer or clerk of works, who doffs his hat, there is no pause for the important visitors. The ladder is made by simply nailing or doweling the rungs to the stiles. This is a less sophisticated ladder than that shown in Figure 6 some four centuries earlier, where the rungs are mortised through the stiles.

Considering other fifteenth century ladders: The glazier fitting a window in a miniature of the *Construction of the Abbey of St Denis* stands on a well made ladder with mortised rungs. (Fig. 27) This is an otherwise primitive work by an

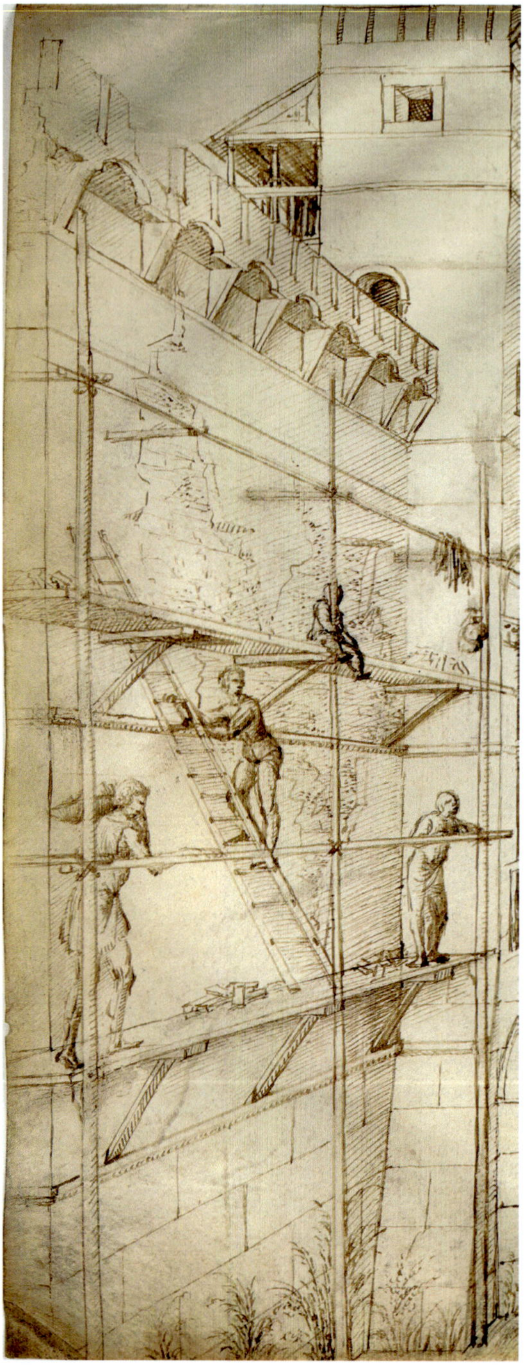

Figure 25b: Jacopo Bellini, *c.*1400-1470, Italian, detail, *The Bearing of the Cross*, *c.*1450, drawing, silver point, pen and brown ink on parchment. (Musée du Louvre, Paris)

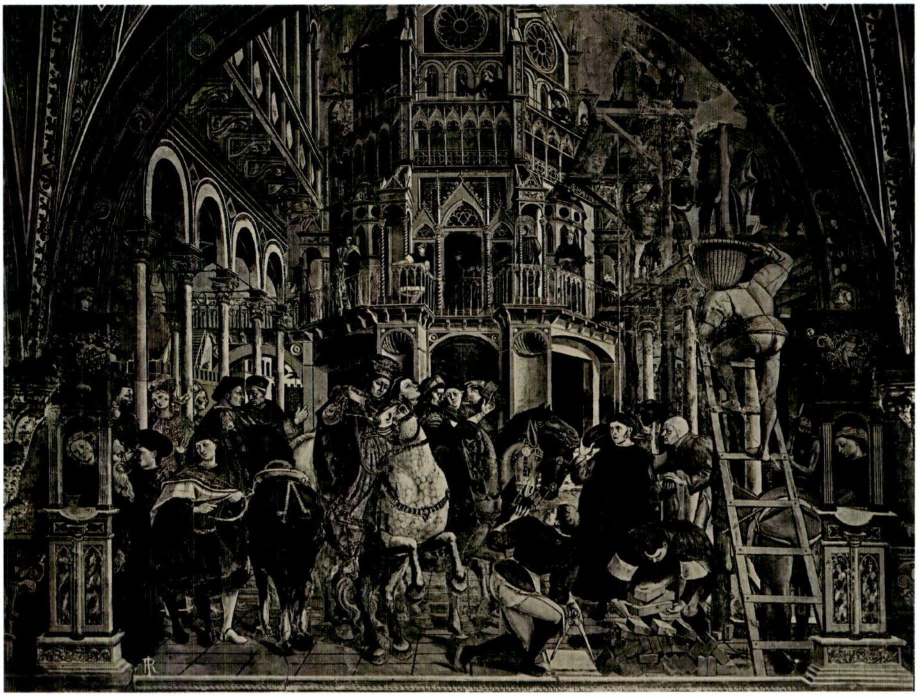

Figure 26: 1443, Domenico di Bartolo, Italian, *c.*1400-*c.*1445, *La Limosina del Vescovo*, fresco, Spedale di Santa Maria della Scala. Siena, Italy. Archive photograph, Fondazione Zeri, Universita di Bologna, Italy.

artist of limited ability. Another miniature *St Bathilda and the construction of a church*, shows a ladder with angled stiles of the type used in orchards. (Fig. 28) A labourer climbs this with a hod of mortar, whilst the mason above him checks the level of his work. They are building a church with exquisite flamboyant tracery. A similar ladder is seen in a Flemish illumination of *The Reconstruction of Rome*, again a hod carrier climbs with mortar. (Fig. 29) In this image there is a female worker with a large basket on her back, although she is not necessarily a construction worker. She is to the extreme left of the composition and appears to be holding a broom or rake. In the painting by Robert Campin, or a follower, of *The Virgin and Child before a Firescreen*, (Fig. 30 a & b) at the top left of the composition is a window through which is visible a town scene. Such a scene, through a window, is a common feature of early Netherlandish paining. The painting is small, some 63 × 48 cm, and although it is immediately clear that this is an urban scene with buildings including a large church, the detail is not readily discernible to the naked eye from where the painting hangs in the National Gallery, London. On closer inspection (Fig. 30b) we see a row of shops and houses with people in the street going about their business. Against one

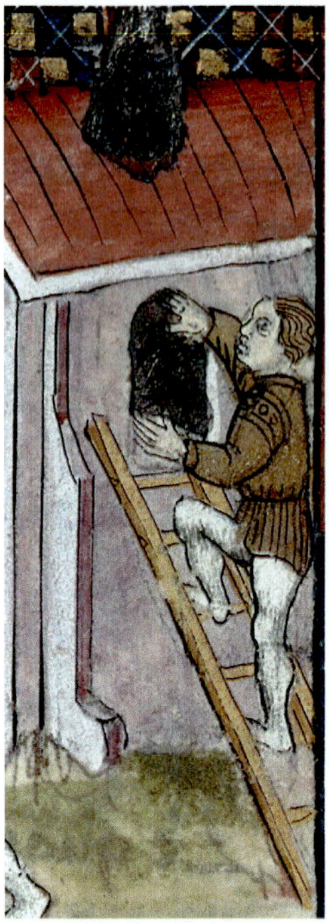

Figure 27: Early fifteenth Century, French, detail, *Construction of the Abbey of St Denis*, illumination from the *Grandes Chroniques de France*, ms. 0863, f. 056v. Bibliotheque Municipale de Besançon, France.

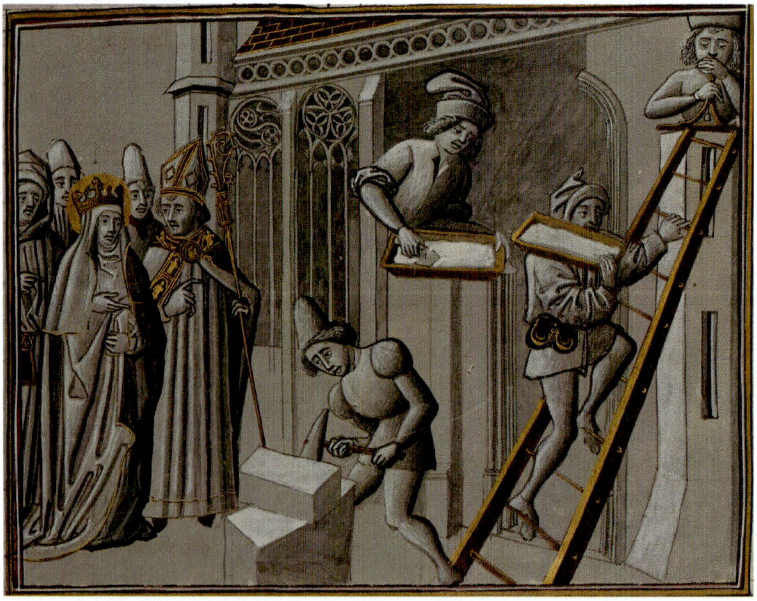

Figure 28: 1444-1461, Flemish (Bruges), *Saint Bathilde et la construction d'une eglise*, illumination from *La Legende Doreé*, ms. 0003, f. 145. Bibliotheque Municipale de Mâcon, France.

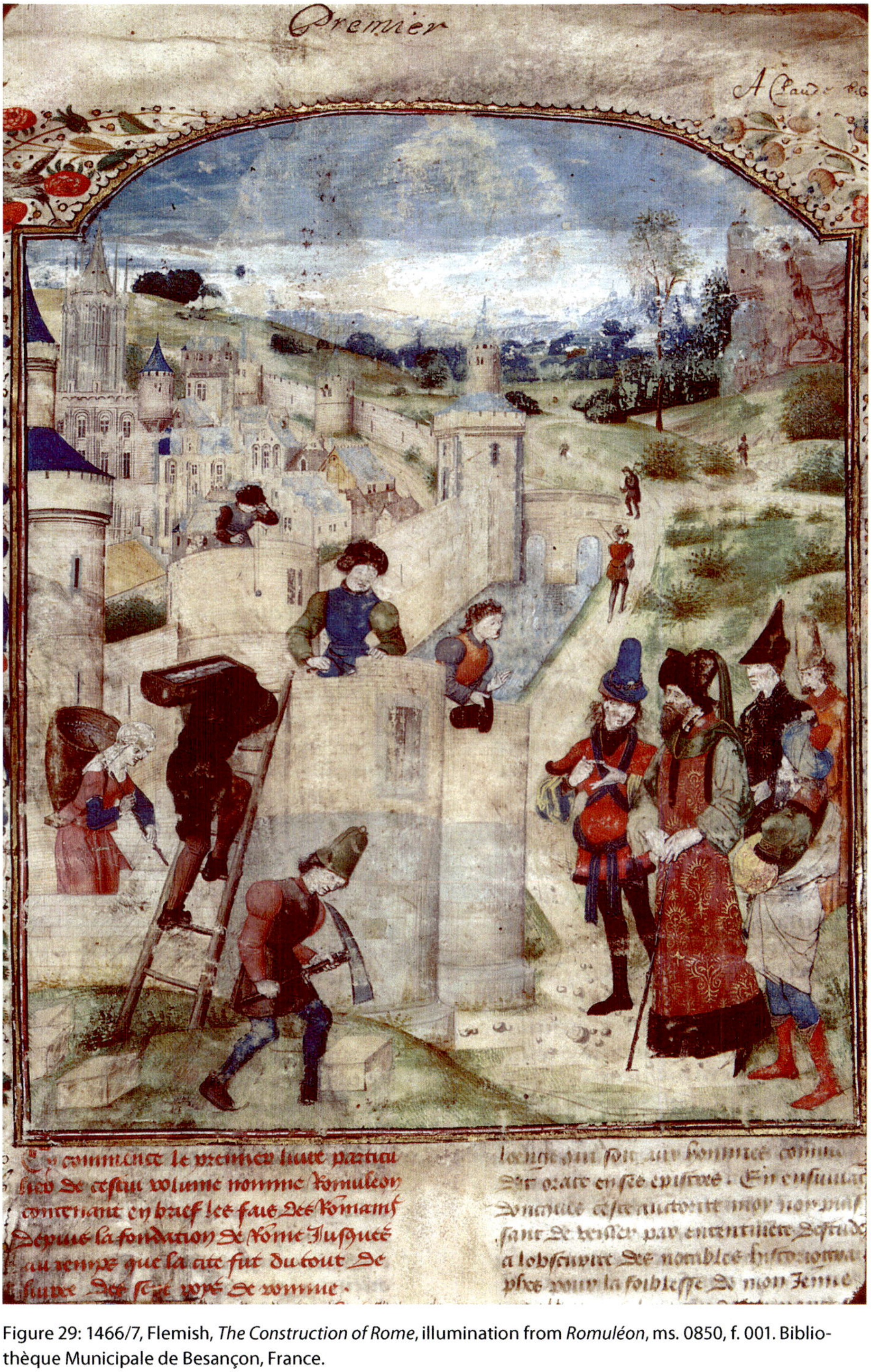

Figure 29: 1466/7, Flemish, *The Construction of Rome*, illumination from *Romuléon*, ms. 0850, f. 001. Bibliothèque Municipale de Besançon, France.

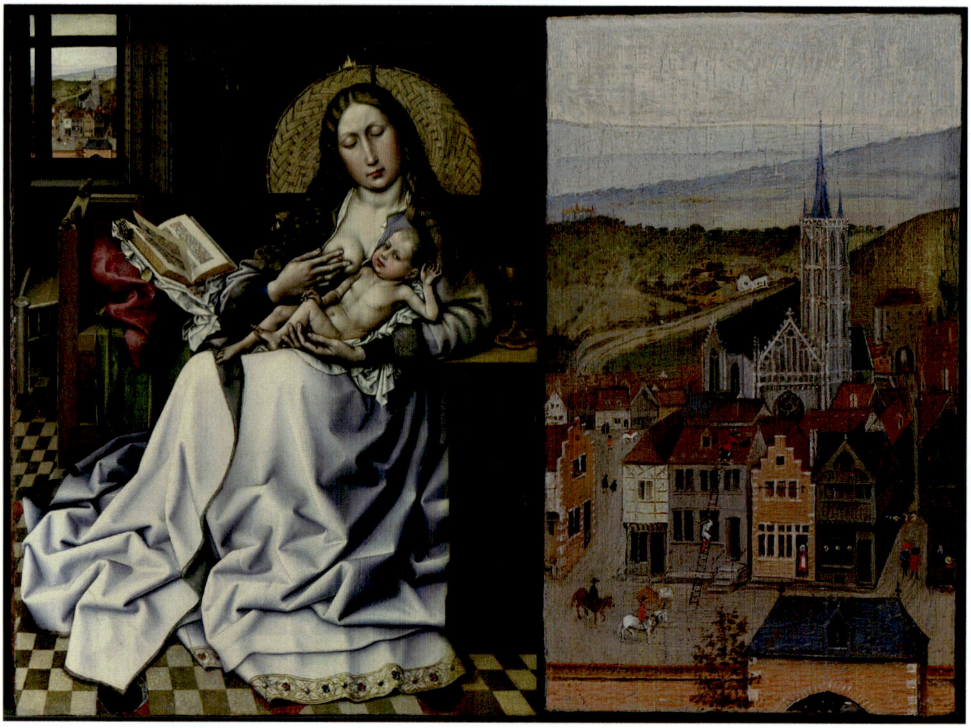

Figure 30a and detail 30b: *c.*1440, Follower of Robert Campin, Flemish, *The Virgin and Child before a Firescreen*, 63.4 × 48.5 cm, oil with egg tempera on oak panel, The National Gallery, London.

of the houses is a very long ladder reaching to the roof, at the edge of the roof is a roofer repairing it whilst on the ladder; another climbs carrying tiles on his shoulders, and piles of sand and mortar are on the ground. This is not a building site as such, but a scene of building maintenance and repair, as is Jacopo Bellini's in Figure 25. The catalogue entry for the painting erroneously tells us that these men are putting out a fire. However 'there is no smoke without fire' and there is certainly no smoke; furthermore the man on the ladder is not carrying a bucket of water, which he would be if they were extinguishing a fire. There are flashes of a pinkish colour on the roof above, immediately above where the roofer stands and these could be mistaken at a glance for flames. But they appear elsewhere on other roofs in the same pink or white and surely represent chimneystacks. What the roofer is addressing is an area of slipped and missing tiles.

Benozzo Gozzoli's fresco, *Building the Tower of Babel*, in the Camposanto Monumentale, Pisa, (Figs. 31 a & b) shows a thoroughly robust scaffold structure and an extremely busy building site with many onlookers. The latter, Nimrod

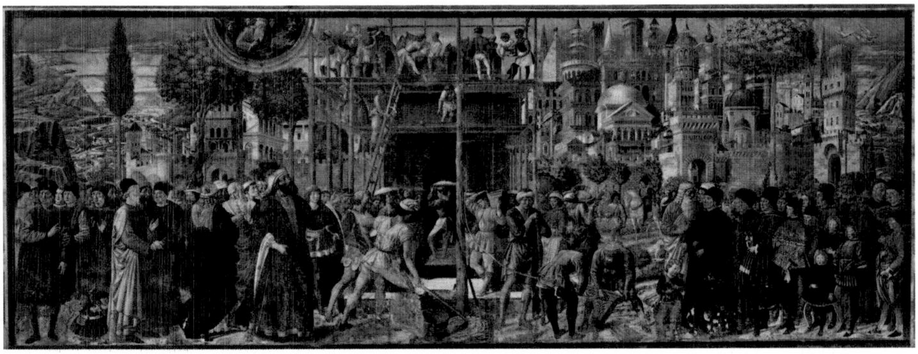

Figure 31a: *c.*1468, Italian, Benozzo Gozzoli, *Building the Tower of Babel*, fresco, Camposanto Monumentale, Pisa, Italy. Archive photograph, 1903, by Giacomo Brogi, Fondazione Zeri, Universita di Bologna, Italy.

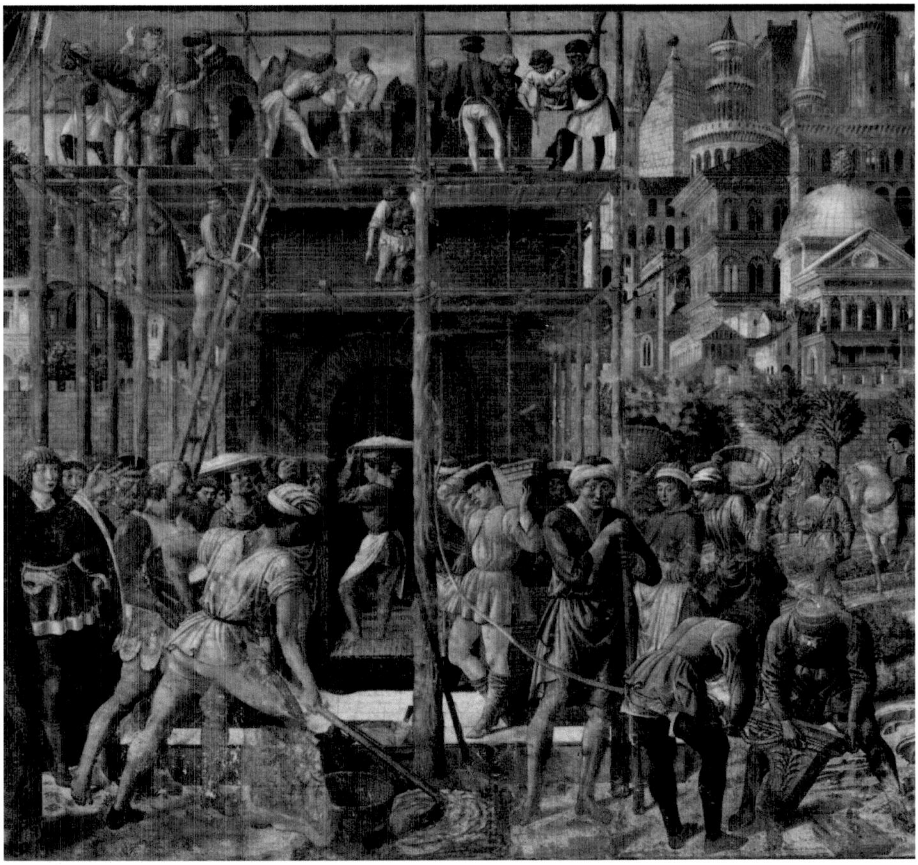

Figure 31b: Detail of figure 31a.

and his retinue, are discussing the merits of what they see before them. The draughtsmanship and observation are compelling, with little left to the imagination here. This, we can be sure, is what a scaffold and building site was really like in Pisa in the 1460s, apart perhaps from the audience, and the worrying lack of standards across the front of the tower. These Gozzoli has presumably, and hopefully for it would be dangerous without them, left out in order to give us a view of the building behind. Judging by the number along the sides, at least three standards are missing, excluding the corner ones. The scaffold standards appear to be whole young trees, probably fir or pine, as are the ledgers, the longitudinal members connecting the standards. The putlogs are a combination of squared timber and round poles and are inserted into the building, as there are no inner standards. Working platforms comprise boards unusually laid across the scaffold, from ledger to ledger, rather than along it, across the putlogs. An internal scaffold is visible through the arched entrance, and elements of this can be seen above the workers at the very top. The ladder is well made and serves both lifts, it is set at the correct angle, and at the top it partly rests against a small cantilevered platform. The standards running back, along the right flank, have brackets supporting the ledgers as well as ties or withies. These withies can be clearly seen across the front, particularly to the main standard, which is just off centre. There is a problem with this standard: at the levels of the two platform lifts it is on the same plane as the other standards, indicated by the ledgers that span across, at ground level it is set further forward away from the building. This gives the composition a slightly Escheresque feel compounded by the fantastic and capricious depiction of Babylon in the background.[39] It is interesting that, although sited on a hill, many of the buildings are far bigger and far higher than the tower of Babel under construction. However, this large standard may well be a 'star' or *stella* lifting device, discussed below on page 154, although if it is, most of the rigging is missing. Two men can be seen attaching a stone bracket to a rope which rises up beside the post. The frescos in the Camposanto and the building itself were seriously damaged in 1944, when Allied bombing raids, although there was no a direct hit, caused the roof to catch fire and it was completely destroyed, leading to widespread damage internally. The illustration here (Fig 31) is an archive photograph of 1903, which therefore predates the damage. The frescoes have since been restored.

The legend of St Barbara, recorded in the *Golden Legend* or *Legenda Aurea* compiled by Jacobus da Varagine in the mid-thirteenth century and of great popularity in medieval Europe.[40] This tells us that Barbara's father, a rich pagan, kept her locked in a tower to protect her modesty. In her isolation and unknown to him she converted to Christianity, and on his discovery of her conversion her father killed her. This all took place in the third century and she subsequently became a martyr and a saint. The legend became a popular subject for illumination and paintings throughout the Late Medieval period. In the third quarter of the fifteenth century an unidentified painter, now known as the Master of the Joseph Sequence, from the South Netherlands, produced an altarpiece dedicated to St Barbara, dated between 1470 and 1500. The church for which it was made is unknown and all that remains is one panel, probably a wing. The painting depicts St Barbara at a key moment in her legend directing operations to provide a third window in the tower in which she was kept. In this case it is being incorporated into a second storey which is part built. This third window reflects Barbara's conversion to Christianity and her belief in the Holy Trinity. The scene is particularly relevant here for its scaffolding. (Figs. 32 & 33) When existing buildings were added to, maintained or repaired, to save the expense of erecting a scaffold from the ground they were often rigged part way up by utilising existing openings and previous putlog holes. This could be done on ladders, or out of windows if conveniently placed, as is possible here. The lower putlogs are rigged through the existing windows with assumed counter-weights inside the building. From these were erected the scaffold as seen, the upper putlogs would have to be tightly inserted and packed into the putlog holes to prevent the whole structure tipping over under imposed loads, although the counter-weights would assist. It is not clear what these masons, completely out of scale, are doing, because the parapet behind which they are working is complete and was presumably part of the original building. Above them is the extra storey with its third window under construction. The scaffold here is simple and easily understood; the putlogs are supported by brackets directly from the wall, as opposed to upright timbers below, and go through the wall, with the one on the left still not built around. The scaffold is made of squared timbers and planks, and the withies are particularly well depicted and appear to be of rope. St Barbara became the patron saint of those in danger of sudden death, such as soldiers and fire fighters; to this we can well add builders, particularly those working at height.

Another altarpiece, this time complete and still in the church it was created for is the St Wolfgang Altarpiece in the church of St Wolfgang im Salzkammergut in central Austria. This is an exquisite and complex work both carved and painted by Michael Pacher, possibly with the assistance of his brother Friedrich. Michael Pacher was a Tyrolean sculptor and painter active from *c.*1462. The altarpiece is immense, approximately twelve meters high and six meters wide. It is a polyptych or Wandelaltar, with two pairs of movable wings and offers three aspects: closed for weekdays, open for Sundays, and fully open for feast days. In the closed aspect the two lower panels depict scenes of building construction (Figs. 34 and 156). That on the right is *St Wolfgang Healing the*

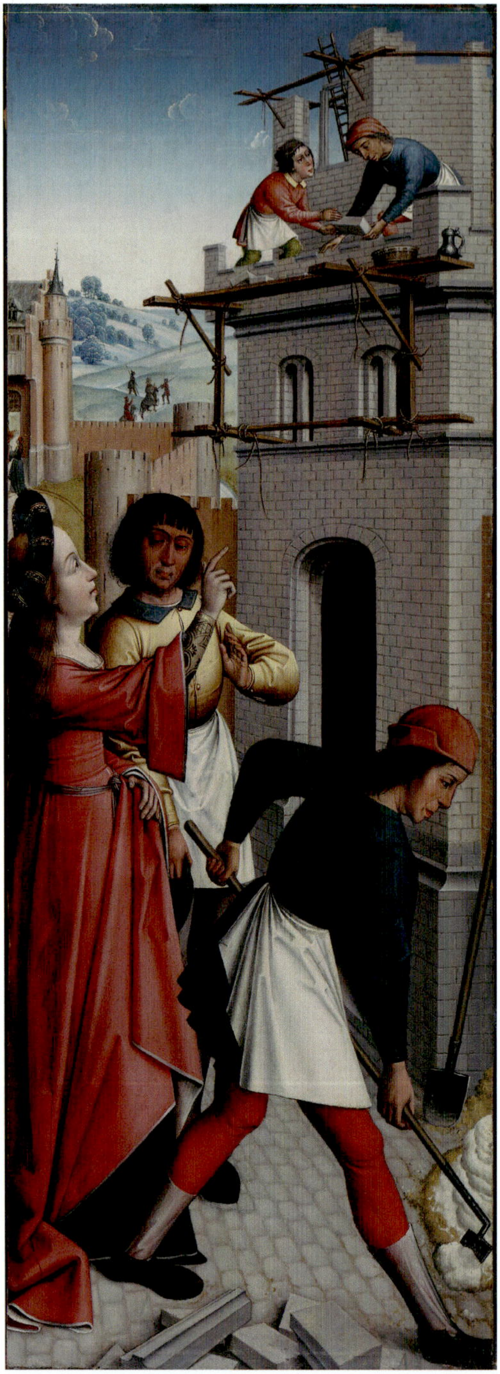

Figure 32: South Netherlandish, Master of the Joseph Sequence, *St Barbara Directing the Construction of a Third Window in Her Tower*, 98 × 35.4 cm, oil on panel. Walters Art Museum, Baltimore, Maryland, USA.

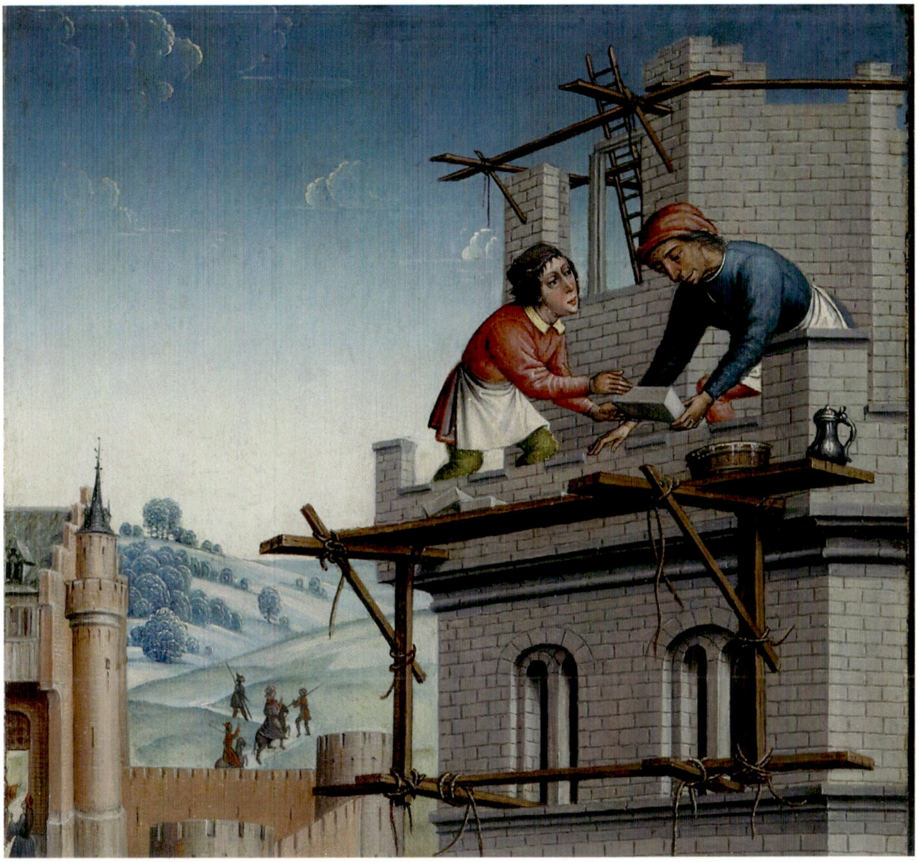

Figure 33: (Detail) South Netherlandish, Master of the Joseph Sequence, *St Barbara Directing the Construction of a Third Window in Her Tower*, 98 × 35.4 cm, oil on panel. Walters Art Museum, Baltimore, Maryland, USA.

Possessed Woman. Against a backdrop of a church under construction we see the west end of a Gothic church with three lifts of scaffold entirely supported by putlogs. The scaffold continues back along both flanks of the building in near perfect perspective. The putlogs are good-sized poles of varying diameter and length. They are assuredly the product of woodland management as outlined above. The lifts consist of planks, which overlap at the putlogs on the middle lift where two men are at work. The one on the left in black appears to have just lifted a plank into place on the lift above him. There is a striking air of precise observation and authenticity in this scaffold, a secondary element only in the overall scene. The second panel is discussed in chapter 2.

In a Flemish Book of Hours of *c*.1478 (Fig. 35) St Barbara features again. Here the whole tower appears to be under renovation as well as having an extra storey being added to it's construction. The new window constructed to approximately

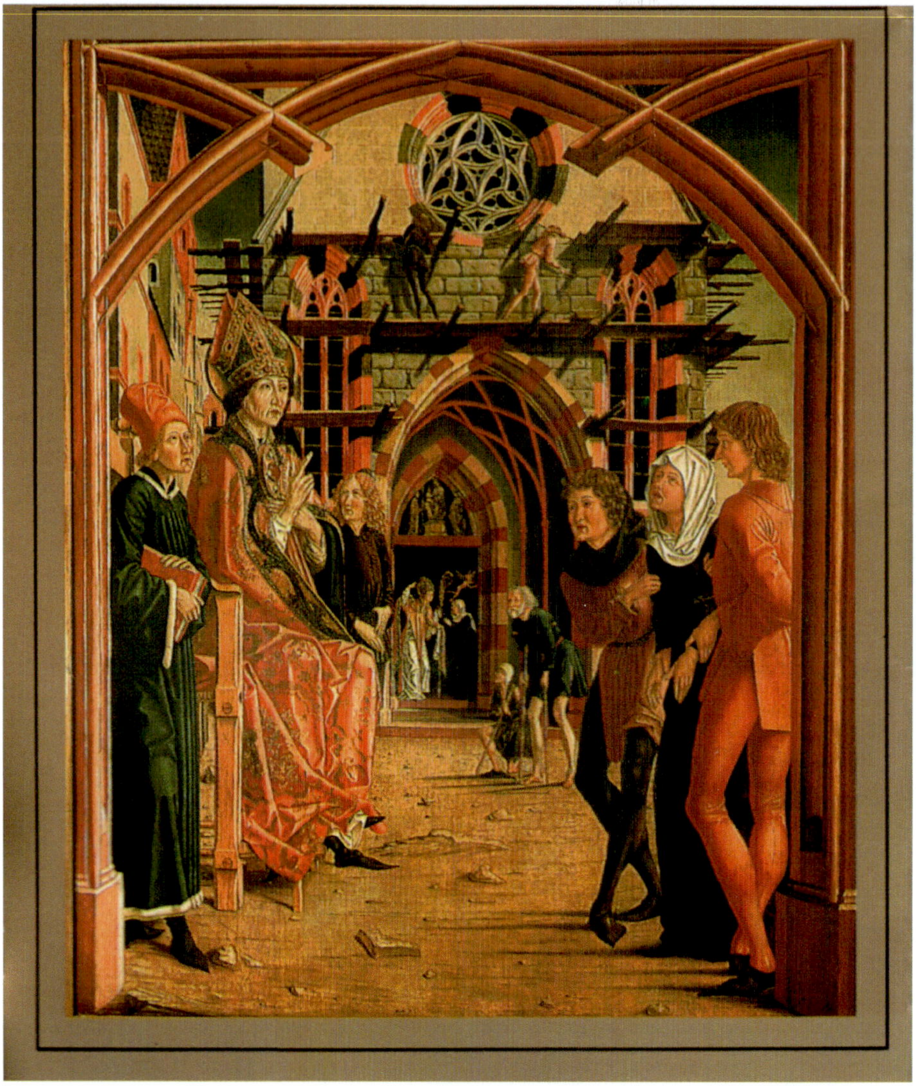

Figure 34: 1471-1481, Austrian, Michael Pacher, *c.*1435-1498, *St Wolfgang Heals the Possessed Woman*, Lower right-hand panel of the workday (closed) aspect of the *St Wolfgang Altarpiece*, 173 × 140 cm, oil on panel. St Wolfgang Church, Salzkammergut, Austria. Flickr, Jaime Antonio Alverez, with consent.

half height is immediately above the gabled tall entrance porch. A full scaffold has been erected comprising standards, ledgers, putlogs and diagonal braces, the latter for stability to prevent the scaffold racking either under loading or in the wind. The withies are sketched in here and there. There are only very few boards for working on, the roofer at the base of the pitched roof on the porch sits on a single board,

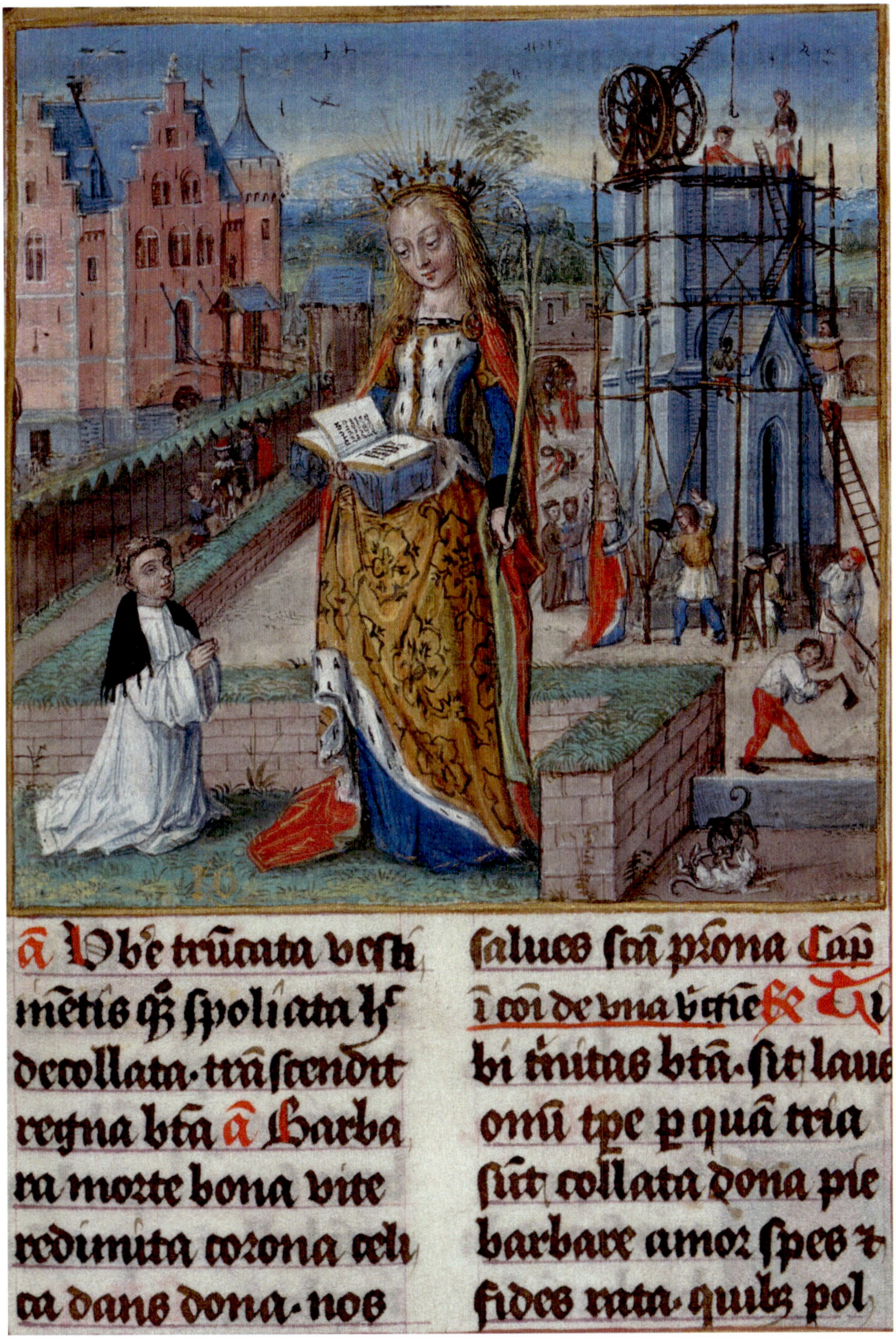

Figure 35: Flemish, *c.*1478, *St Barbara*, illumination from a Book of Hours, Latin ms. 39, f. 151r. Rylands Collection, University of Manchester.

he may be a slater using a slater's hammer. Ladders connect the four lifts except the first to the second, which may be hidden behind the porch roof. The crane is discussed below on page 139.

A domestic scene of house building, albeit a grand house, nears completion in an illumination from a French manuscript of *c.*1480. (Fig. 36) This is the translation of a Treatise written *c.*1310 by Petrus de Crescentiis a Bolognese lawyer, on agriculture, the *Ruralia Commodorum*. This is in multiple parts and covers several different themes, one of which is the best location and arrangement for a manor, villa or farm. The house under construction is set within a rural scene and appears to be part of an estate. It is being built adjacent to a chapel or church and in addition to the building operations it is juxtaposed with other rural activities. The scaffold is tall, of four lifts although only the top has a working platform of hurdles or narrow planks, supported on putlogs. The bottom lift still has putlogs clearly inserted into the building, but on the intervening two lifts they have been removed and the holes filled in. The standards are very long slender poles as are the ledgers; here and there brackets are visible attached to the standards which support the ledgers. A worker starts to climb the exaggeratedly long ladder with a pitcher of beer to slake the thirst of the bricklayers above. Is this a ladder with angled stiles, or is it perspective, which is well observed in some elements but less so in others? For further commentary on this illumination see p. 122 below.

Scaffolders at work can be seen in a miniature of *c.*1500, the *London Rothschild Hours*. (Fig. 37) This manuscript is not English but originates from the Southern Netherlands and is also known as the *Hours of Joanna I of Castile*. This is a scene depicting the building of the Tower of Babel but is otherwise what we must accept as, in the artist's eyes, a completely accurate rendering of a typically large building project on a fortress, cathedral or other large building.

Two scaffolders can be seen erecting the scaffold at the very top in the background. One is at the top of a ladder fixing a ledger, whilst the other appears to be driving in a wedge or nailing on a bracket, similar to that on the standard at the left hand extremity just above the figure who appears to be steadying the stone being lifted by the crane. Other than these significant elements, there is nothing new depicted. Everything is constructed with poles bound together with withies, but these not always visible. Interestingly, there are several braces; one is in the foreground above the arch, nailed to the standards, while the others appear to be bound with withies.

A strange and unusual scaffold platform is depicted in one of the frescoes of the *Life of St Benedict* by Giovanni Antonio Bazzi, at the Abbey of Monte Oliveto Maggiore in Asciano, Italy. Here a platform is supported on poles or ledgers laid horizontally and resting on the abaci of the capitals, or that part that juts out beyond the springing of the vault, of the colonnade. (Fig 38) On the right side the ledger is a

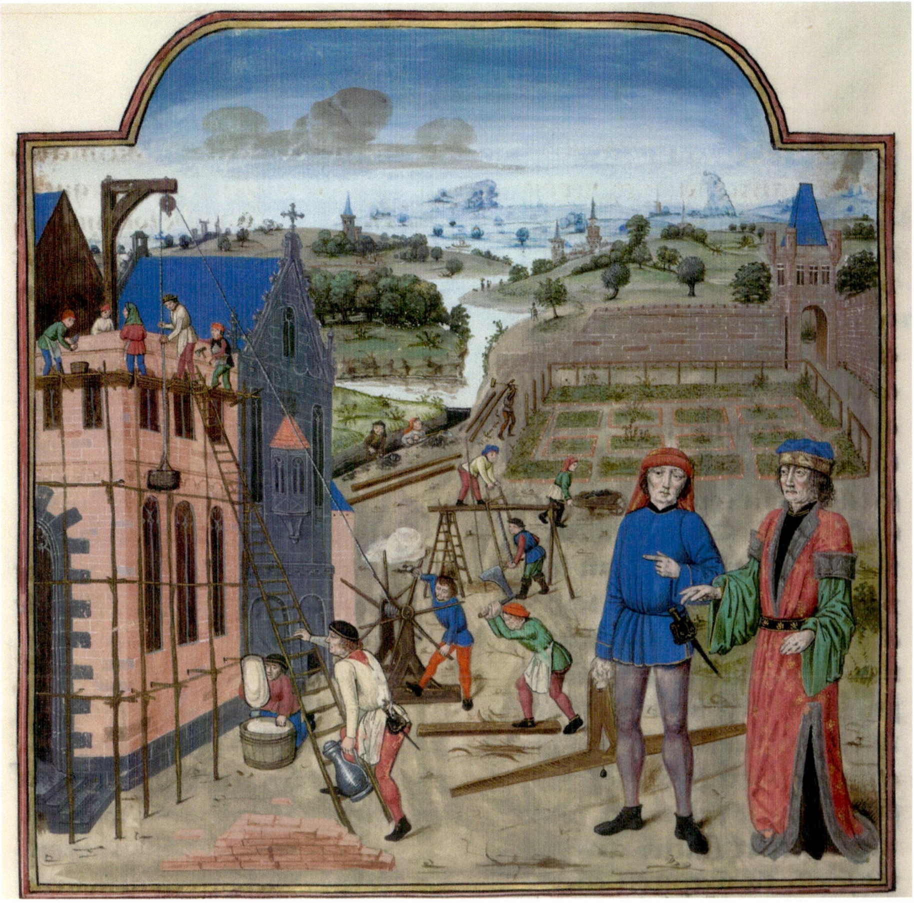

Figure 36: Bruges Master of 1482, Flemish, French manuscript *c.*1480, *Building a house*, from Petrus de Crescentiis, *Ruralia Commodorum*, additional MS 19720, f. 27. © The British Library Board, London.

pole whilst on the left it is constructed with squared section timber. Both of these 'ledgers' are also suspended by rope slung from somewhere above but out of view. This is a rudimentary affair and the workers have placed a great deal of trust in the strength of the stone from which the capitals are carved. The scene is entitled, in a painted plaque below the fresco *Come Benedetto Compie la Edificazione di Dodice Monasteri* ('The foundation or building of twelve monasteries'). Here St Benedict is supervising the building of one of these monasteries and is perhaps alarmed at the danger above.

Raphael's *Disputa*, in the Stanza della Segnatura in the Vatican Palace, (Fig. 39 a & b) contains a landscape between the Church Triumphant in the upper sphere of the fresco, and the Church Militant below, where a building is under construction on a hillside to the left. A great deal has been written on the iconology of the *Disputa* but

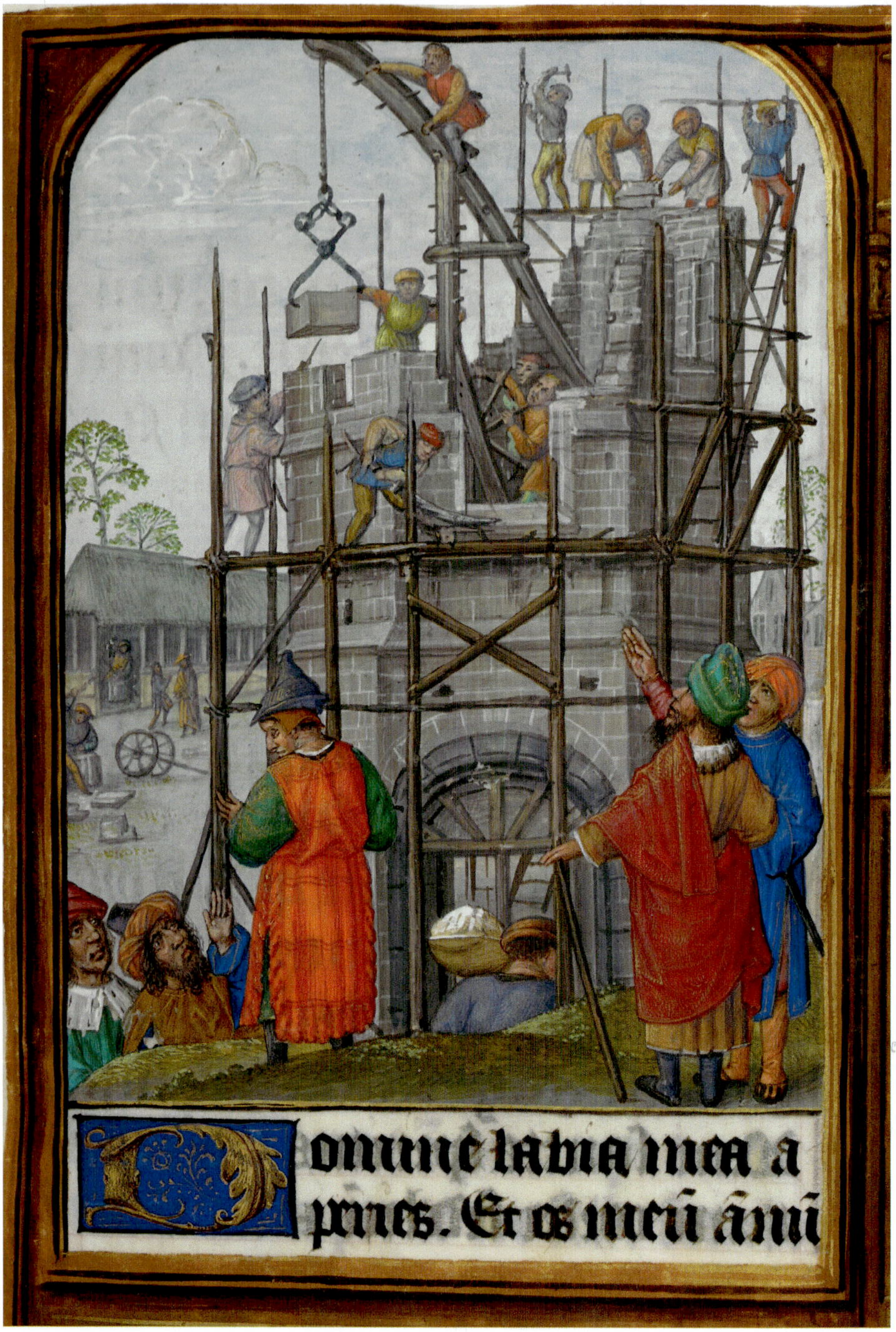

Figure 37: Southern Netherlands (Ghent?), *c.*1500, *The Building of the Tower of Babel*, from a Book of Hours, the 'London Rothschild Hours' or the 'Hours of Joanna 1 of Castile', f. 34r, © The British Library Board, Add MS 35313.

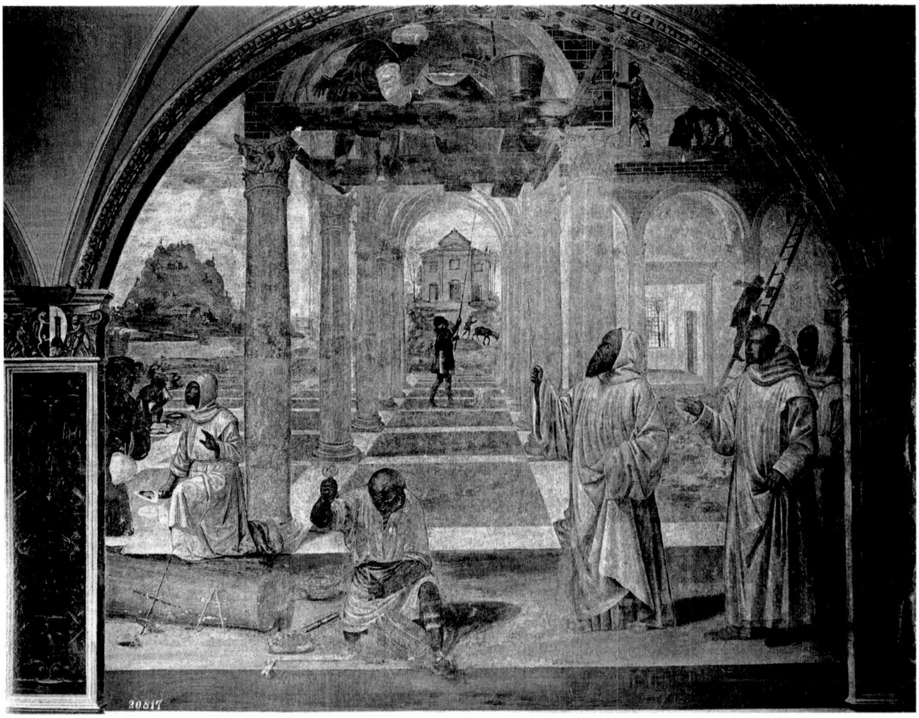

Figure 38: 1505-1508, Giovanni Antonio Bazzi, called Il Sodoma, Italian, *Benedict Founds Twelve Monastries,* from a series of frescoes depicting the Life of St Benedict, Abbey of Monte Olieveto Maggiore, Asciano, Italy. Archive photograph *c.*1895, Fratelli Alinari. Fondazione Zeri, Universita di Bologna, Italy.

what is this building site doing here and what is its meaning? The detail is distant and rural, it is possibly a farm or domestic building, but it is surely too vernacular to be associated with Donato Bramante, the architect of St Peter's at the time, and distantly related to Raphael, who was partly responsible for bringing Raphael to Rome from Urbino. What does have meaning however, is the enormous and apparently incongruous slab like structure to the right, occupying the foreground of the same landscape. This is said to represent the first of four massive piers, part of Bramante's design for St Peter's, and under which was laid the foundation stone by Pope Julius II on 18 April 1506.[41] This therefore is also a building site, although not as immediately obvious as that which is opposite. Bramante is said to be depicted to the left leaning on a balustrade and holding an open book; he occupies the left foreground of the detail below the domestic building site.[42] (Fig. 39b) A scaffolding, part way around the building to the right, is visible, and there is also a ramp to a working platform at the level at which the building is currently under construction. On the left is a separate scaffold to another part of the building. Judging by the actions of the two people working on it, this scaffold appears to be in the process of being erected.

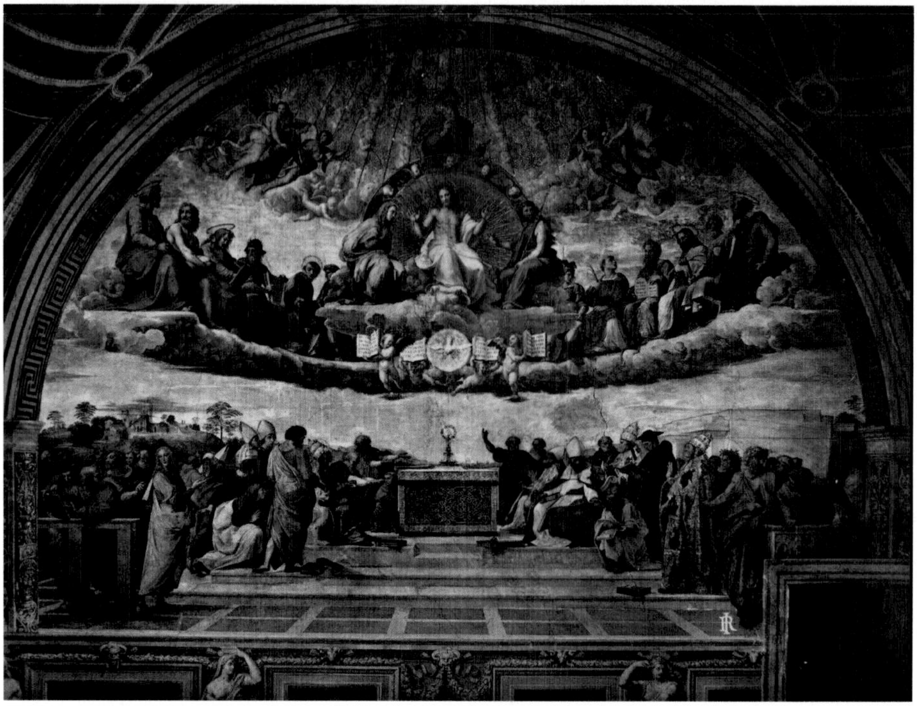

Figure 39a: 1509-1510, Raffaello Sanzio da Urbino, known as Raphael, 1483-1520, *The Disputa*, fresco, Stanza della Segnatura, Vatican Palace, Rome. Archive photograph, *c.*1920, by Giacomo Brogi, Fondazione Zeri, Universita di Bologna, Italy.

In the *Spinola Hours*, from about 1510-1520, illustrated by the Master of James IV of Scotland, or possibly the Flemish master Gerard Horenbout, with whom the Master is often associated, is an illumination showing *Elijah Begging for Fire from Heaven*. The illustration is part framed, at the base and the right-hand side by a robust and wonderful scene of construction. (Fig. 40) The second lift, upon which a labourer is unloading a hod of mortar into a wooden tub, is bracketed off the building as well as supported on the ledgers and putlogs. All the other lifts are supported off ledgers and putlogs. The ledgers rely on brackets fixed to the standards as well as also being bound together. There are various diagonal braces nailed to the standards to give stability and some of these appear to be quite random but have doubtless been put in place to deal with deflections as they arise. The second standard in from the front corner does not run to the ground, but is supported off a beam or large putlog, which is also supported on brackets. It is unclear why this should be so. The treadwheel crane at the very top of the main part of the building rests partly on the top lift of scaffold and partly inside the building. A realistic estimate, by comparison with the figures at work, of the diameter of the standards is between 10 to 12 inches (25-30cm) which is commensurate with the

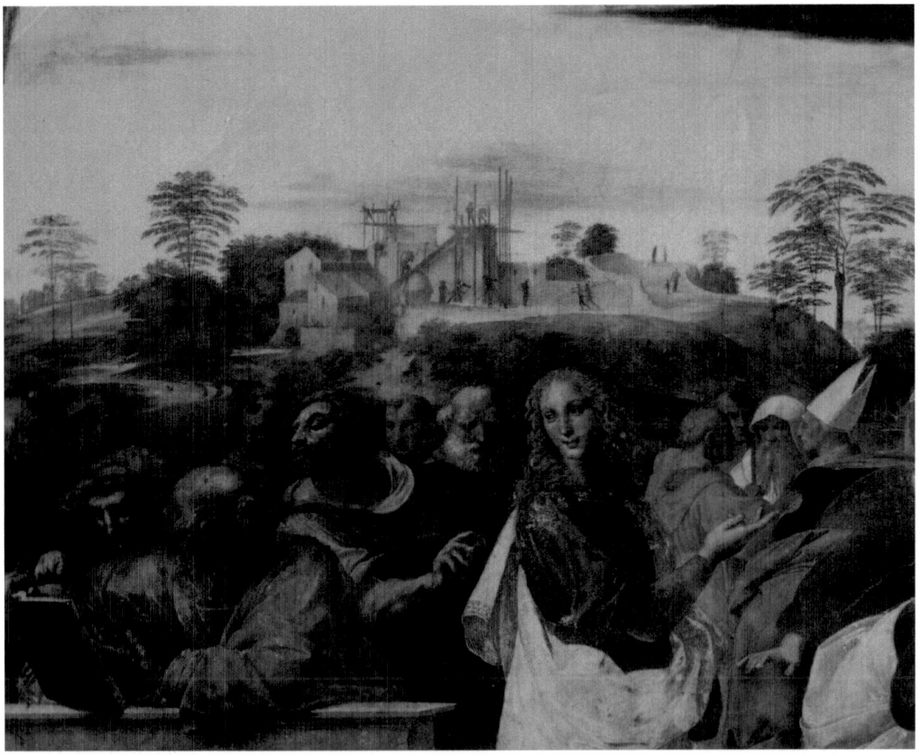

Figure 39b: 1509-1510, Raffaello Sanzio da Urbino, known as Raphael, 1483-1520, detail, *The Disputa*, fresco, Stanza della Segnatura, Vatican Palace, Rome. Archive photograph, *c.*1920, by Giacomo Brogi, Fondazione Zeri, Universita di Bologna, Italy.

need to support such a large piece of machinery. A ladder runs from the fourth to the fifth lift, on which a man climbs carrying a pole. On the extreme right a hod carrier disappears into the building, indicating that there is also access within. On the fourth lift, a carpenter is sawing a piece of wood, a pole possibly, and he may be a scaffolder. The carpenters were in charge of organizing, erecting and striking (dismantling) the scaffolds. In the Proctors's accounts for the rebuilding of Christ Church Cathedral, Dublin, 1564-1565, it is clear that the master carpenter John Brenaghe is also in charge of scaffolding and cranes. There are numerous entries relating to his wages in relation to all three functions.[43]

In a wonderfully detailed Netherlandish painting of 1520-1525, *A Sermon on Charity*, a partly built church has been left by the builders while a service is in progress within. (Fig. 41 a & b) The painting is allegorical: A nobleman hears Mass and is then seen distributing alms to the poor; the church is left open beyond the chancel so that we can see inside. This may possibly be the conversion of St Anthony, who upon hearing the words 'If thou wilt be perfect, go sell what

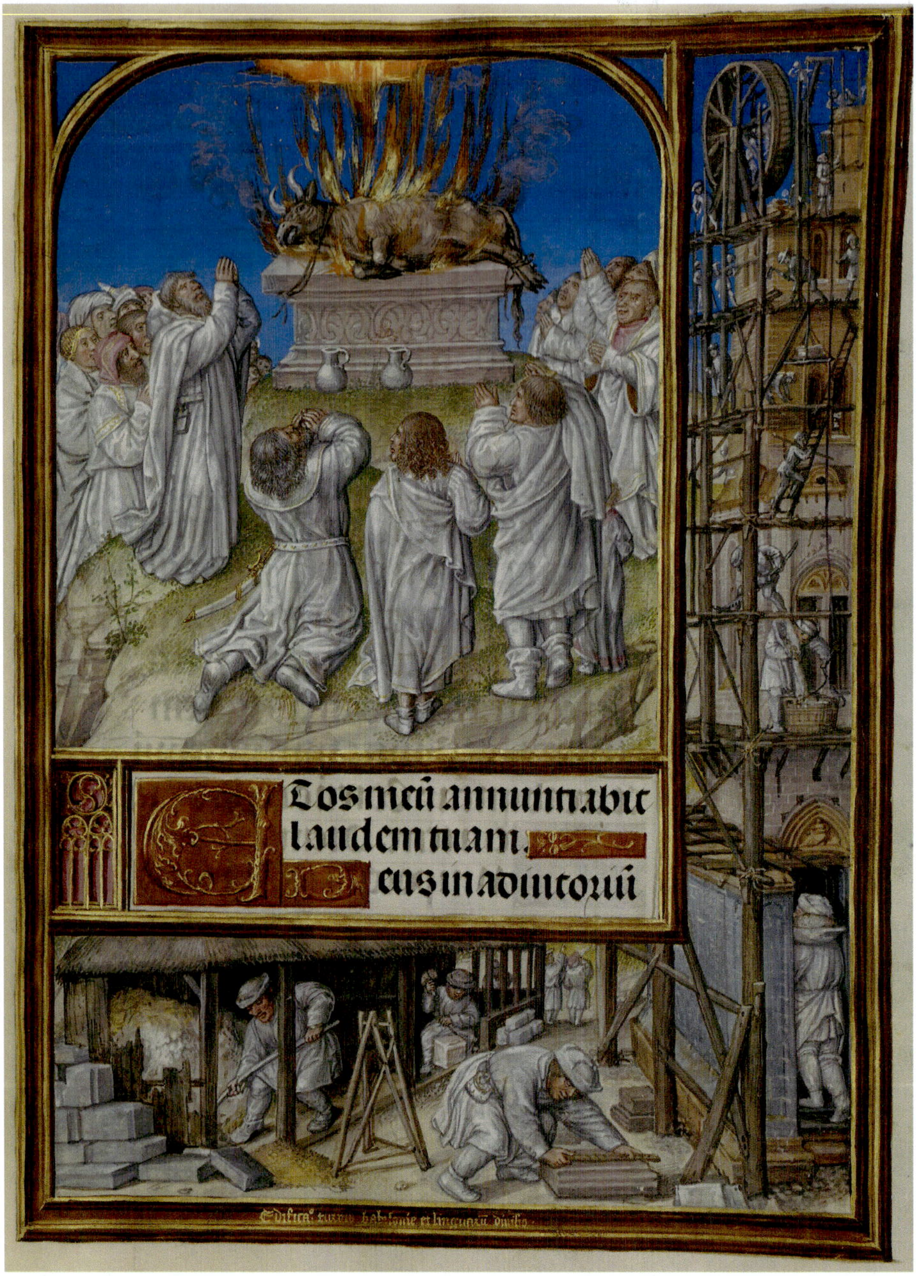

Figure 40: Master of James IV of Scotland or Gerard Horenbout, *Elijah Begging for Fire from Heaven*, 1510-1520, Flemish, Ghent or Bruges, from the Spinola Hours, Ms Ludwig IX 18, f. 32. (The J Paul Getty Museum, digital image courtesy of the Getty's Open Content Program)

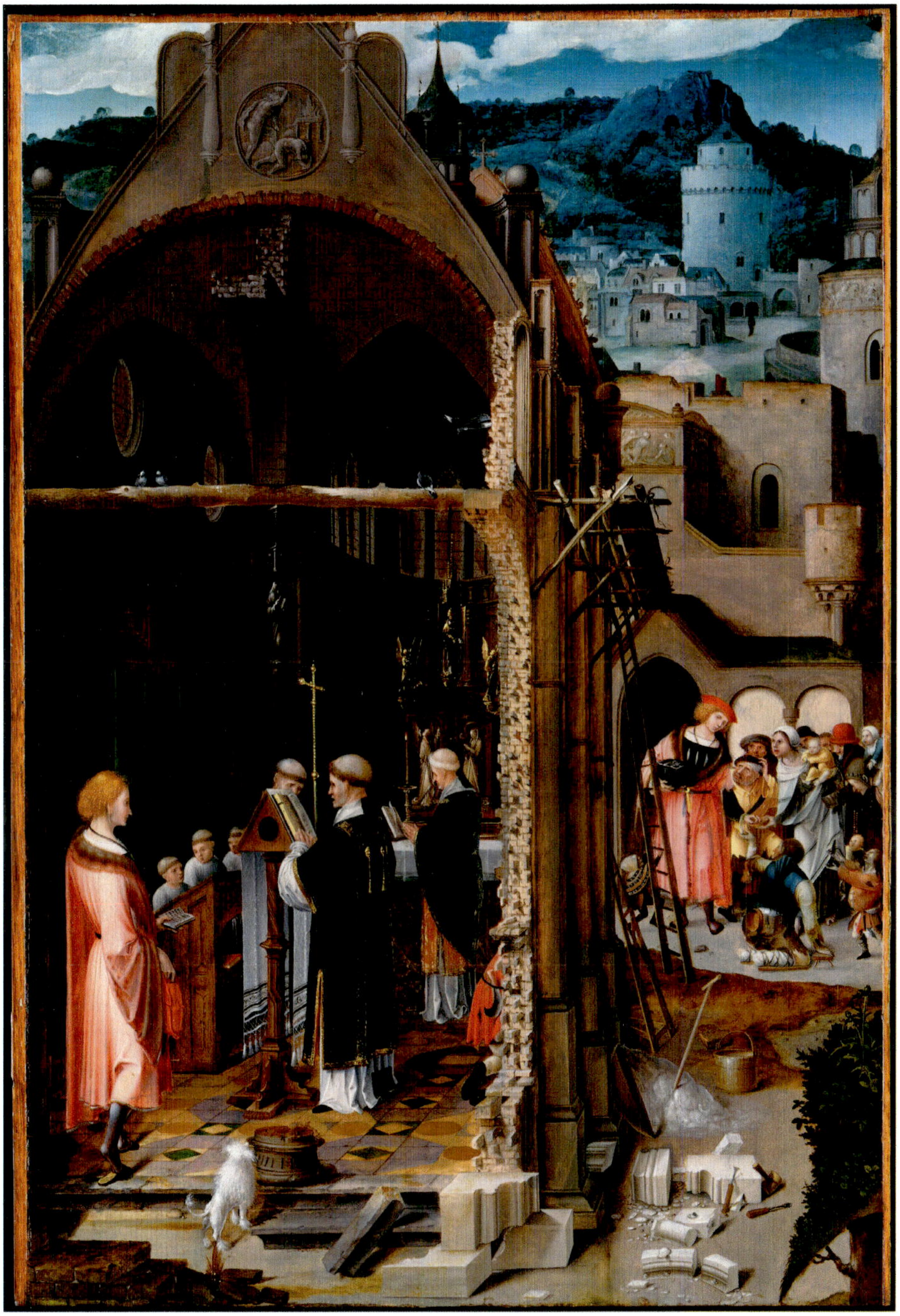

Figure 41a: 1520-1525, Netherlandish (Antwerp Mannerist) Painter, *A Sermon on Charity (possibly the Conversion of St Anthony)*, 85.1 × 58.4 cm, oil on panel. Metropolitan Museum of Art, New York, OASC.

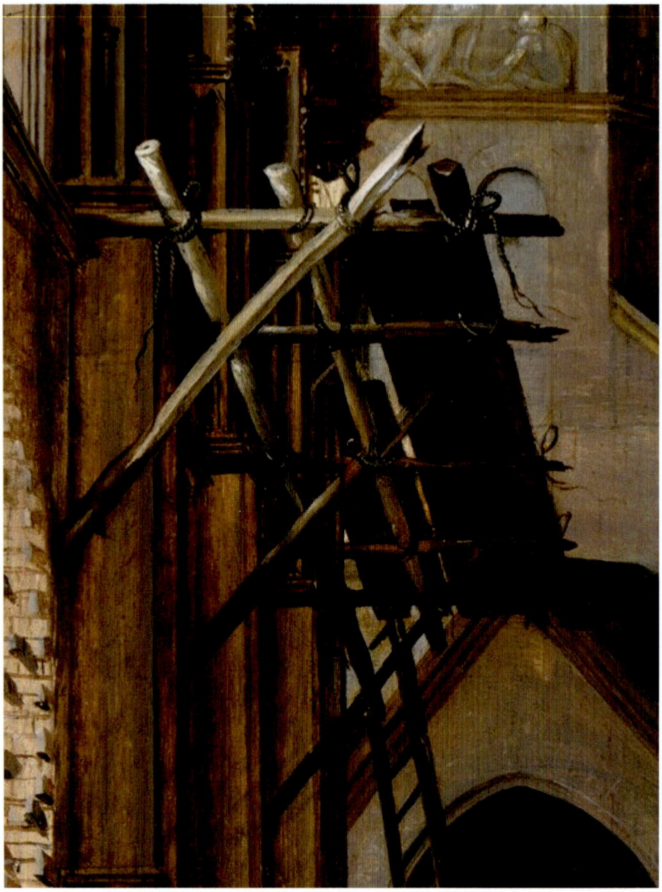

Figure 41b: detail of scaffolding.

thou hast and give to the poor'. This he did before he then became a hermit. The open aspect of the church with the jagged masonry is not unusual but in reality would have been boarded up at this east end. Churches and cathedrals were often constructed chancel first, as liturgically this was the most important part of the church, with the nave and crossing following later. The jagged edge to the masonry is in fact toothing to allow the continuation of the brickwork to be properly bonded in. On the outside is a scaffold supported on brackets and putlogs set into the building. This is high up at the level of the clerestory windows, and a ladder is leaning at the far end. On close examination (Fig. 41b) the details are so precise as to be almost photographic. This is without doubt something the artist has drawn and recorded from a building site he has studied. It is the perfect example of the use of ordinary poles simply tied together with withies and two or three planks laid on top for a working platform. The eye for detail and accuracy is also reflected in the other items of equipment and tools left lying around below this scaffold.

The Dutch painter Abraham Bloemaert painted landscapes, mythological, and religious scenes. He was the son of an architect and sculptor and his abilities at drawing reflect some of what he must have learnt from his father. His skill in drawing and painting vernacular rural buildings in considerable detail is evident. Although we cannot judge the buildings and scenes against what he produced there is a sense of compelling truthfulness in these works. In his *Tobias and the Angel*, (Figs. 42 a and b) setting aside the scene from the apocryphal Book of Tobit, the rural scene in the background with a man climbing a ladder to a rickety scaffold tells us a lot about domestic construction and maintenance in the rural area where Bloemaert lived. There is a sense of a degree of poverty, making do and patching up. Let us assume that the man is arranging the boards to give a better foothold. He is then perhaps going to do one of two things: Attend to the large crack just above the level of the scaffold; or insert two further putlogs into the holes higher up to eventually gain access to the roof. The roof is clearly undergoing some form of work; the pole at the ridge is too high to be the ridge itself, so the roof is not laid bare to the weather as could easily be assumed at a glance. There are other poles laid across, and at least one other, at right angles to these, running up the roof; is this some form of access structure? What we certainly see in this painting is a wonderful and precise depiction of the function of a putlog; all that supports the working platform is the cantilever action of the two putlogs, as is often seen in Medieval and later scaffolds. The building in the middle ground is clearly a workshop with a stack of bricks piled up in front, evidence that this estate is more than just a farmyard. The very same building at a slightly different angle; is the subject of another work by Bloemaert, a drawing, now in The Cleveland Museum of Art, although here it is a peasant's cottage. The main building beyond is also there albeit not quite as depicted in this painting.[44] One work is clearly derived from the other, with the drawing possibly being an initial study. The figure on the ladder appears, almost identical, in an engraving after Bloemeart by Jan Saenredam; although here he is repairing a thatched roof and another figure is below him on the ladder.[45]

Formwork

The other craft of the carpenter or wright was formwork, which is a comparative rarity in medieval manuscript illuminations. This is temporary work in timber to support masonry structures whilst they are being built, particularly arches and vaults. The formwork to the arch of the doorway into the building, shown in

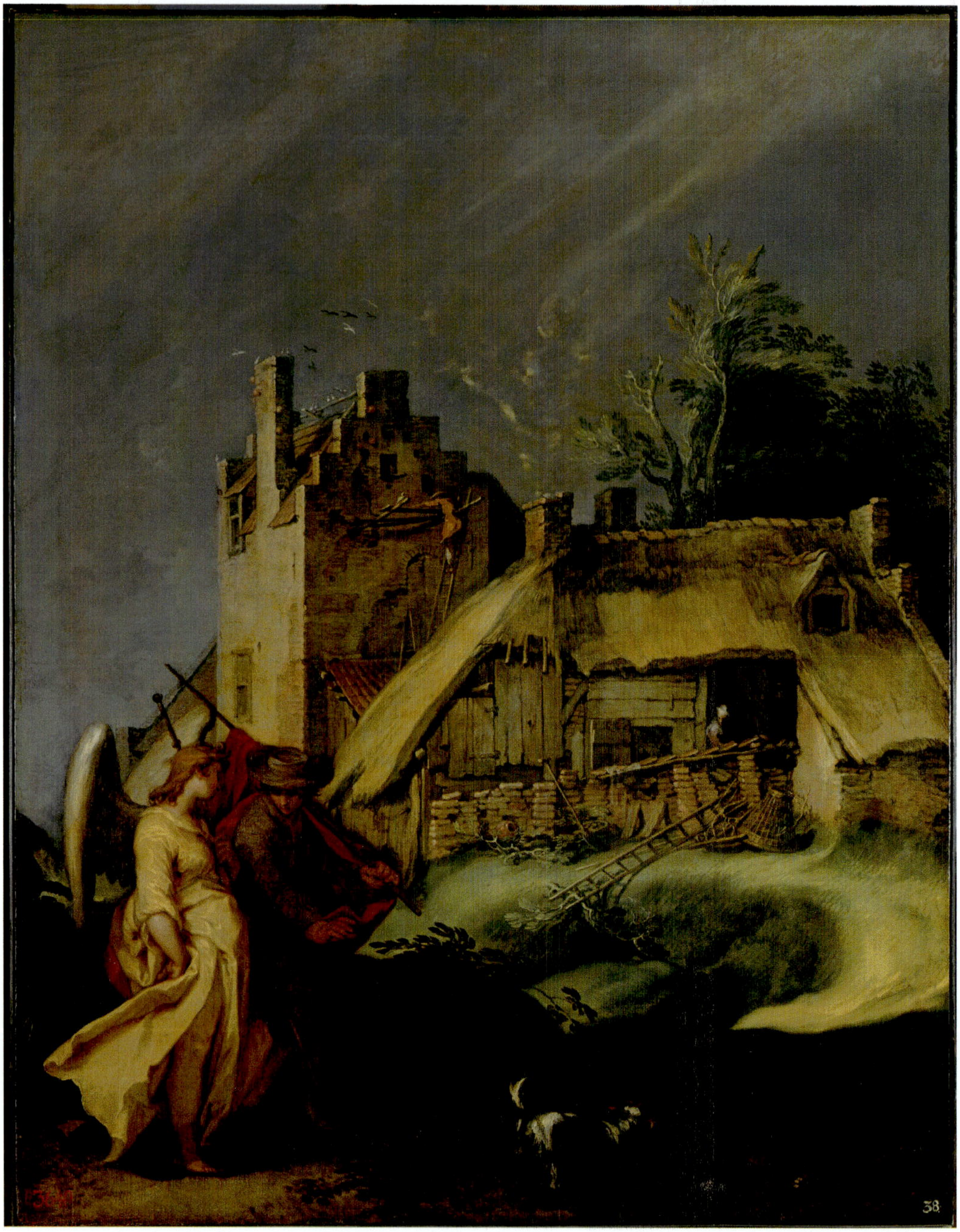

Figure 42a: Abraham Bloemaert, Dutch, 1566-1651, *Tobias and the Angel*, c.1600, oil on canvas. (State Hermitage Museum, St Petersburg. Photograph © The State Hermitage Museum, Photo by Vladimir Terebenin)

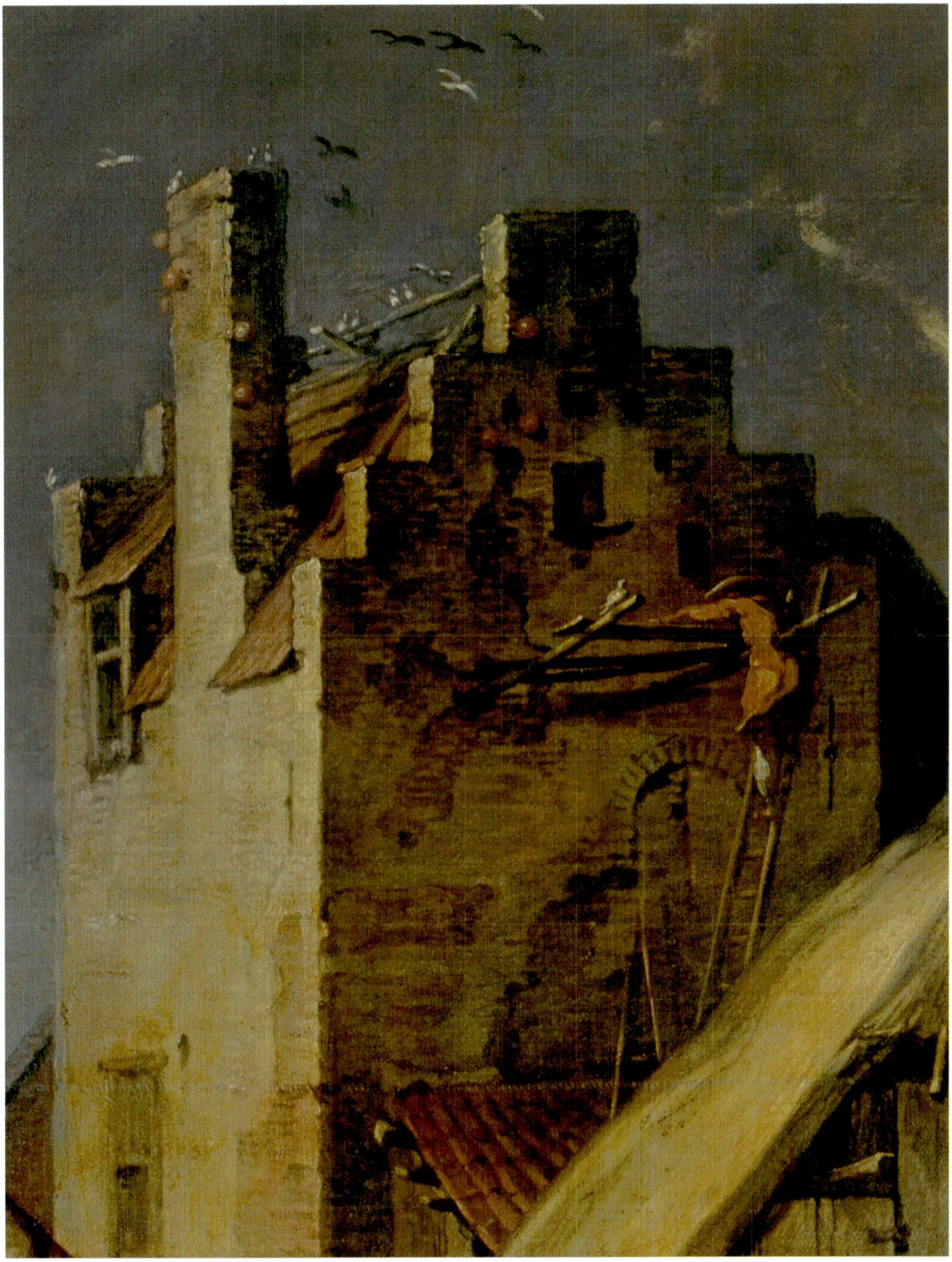

Figure 42b: Abraham Bloemaert, detail of figure 42a.

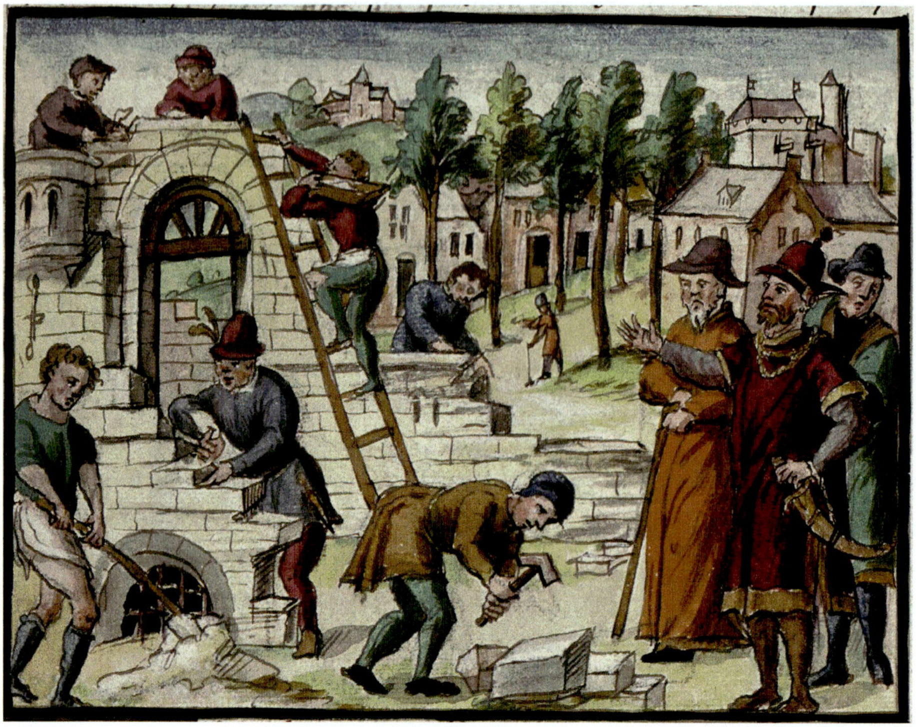

Figure 43: Northern French, second half of the sixteenth century, *Caradoc et la Construction de la Porte Cardon à Valenciennes*, from *Recueil des Antiquites de Valenciennes*, BM de Douai, France, ms 1183, f. 018. © Institut de recherché et d'histoire des textes – CNRS.

Figure 37 is still in place, despite a good deal of masonry having been completed above it. It was normal to leave such formwork in place to allow the mortar to cure and achieve strength. There is no such formwork in place yet for the window openings, but this will surely follow. In the northern French manuscript of the second half of the sixteenth century, *Recueil des Antiquites de Valenciennes*, there is shown the construction of the Porte Cardon gate (demolished 1890-1892) in the fortifications of the city. (Fig. 43) In the main portal of the building is similar formwork supporting the arch. See also figure 18 above.

In Pieter Bruegel the Elder's painting of *The Tower of Babel*, (Fig. 166) most of the carpenters on this vast site are working on formwork. (Fig. 44) In the left hand image scaffolders are erecting formwork for the vault, of what will become an internal passageway, with arched timber pieces that will carry the brickwork. On the level above (right) formwork is left in place in the two open arches, as they are in many other instances throughout the site. These presumably will be partly in-filled to match those to the left. In Figure 44 right, masons are laying bricks on the formwork of the left-hand arch whilst a carpenter finishes the timberwork for that on the

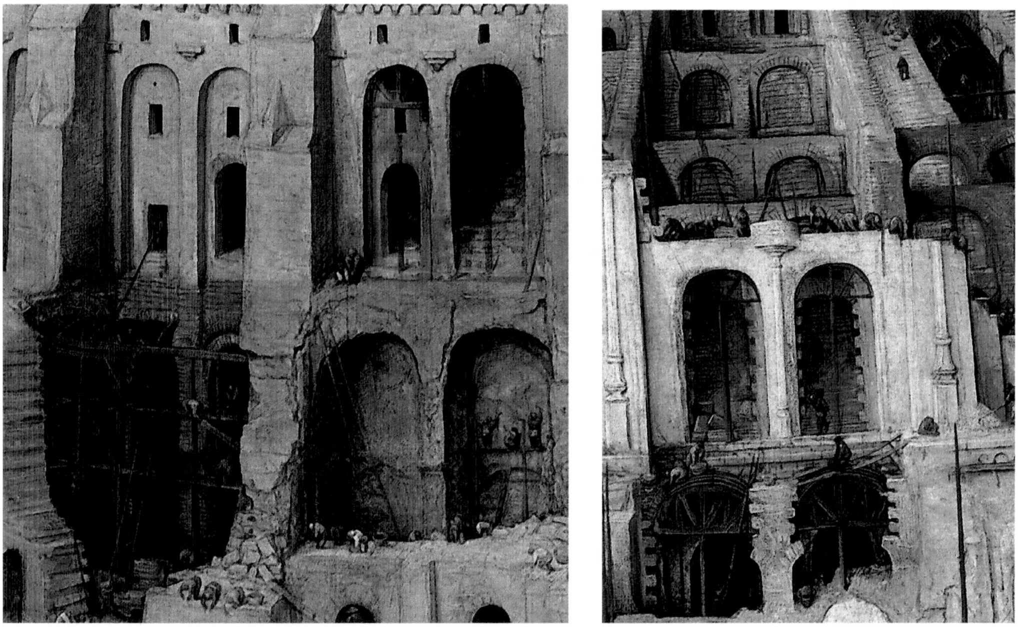

Figures 44a and 44b: 1563, details of Figure 166: Pieter Bruegel the Elder, Flemish, 1525-1569, *The Tower of Babel*, showing some of the formwork on this huge site. Oil on Oak panel. Kunsthistorisches Museum, Vienna: KHM-Museumsverband.

right. At the top right-hand of the detail is the complicated formwork for a sloping barrel vault under construction. On the dockside are various timber stores and on the water is a raft formed of timbers or logs lashed together which is approaching the dock. Where river transport was readily available such raw materials, for use as timber, which will float, was favoured rather than road or track. Poles, on the other hand would tend to be transported by barge or overland. (Fig. 166)

In Eustache le Sueur's painting of the construction of a monastery (Fig. 102) there is robust formwork in the arch immediately behind St Bruno and his architect, and also in the square window on the flank of the building below the scaffold. The first floor joists are partly in place, over which has been laid the flimsy looking platform for the crane.

Scaffolding in the Seventeenth Century and Beyond

Anton Möller's huge painting *Rebuilding of a Temple by King Joash*, dated 1602, is a scene of unabated building frenzy. (Fig. 45) Although this is a biblical scene, Joash was King of Judah who reigned *c.*836-796 BCE and had the Temple repaired after

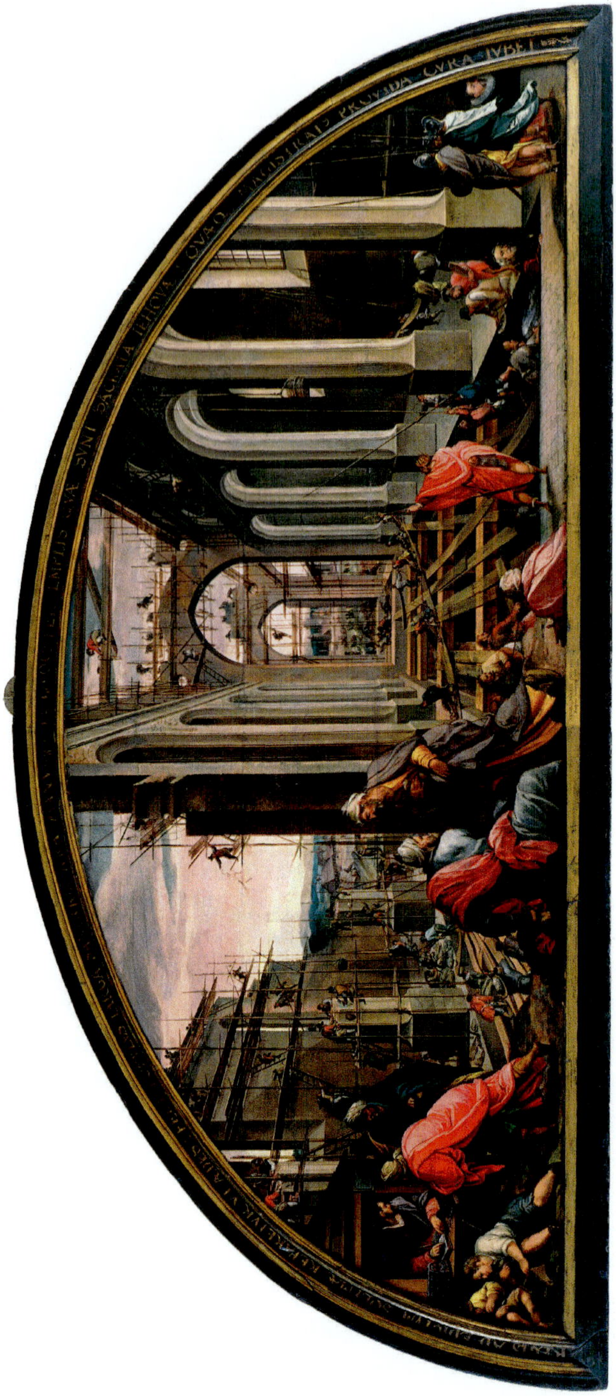

Figure 45: Anton Möller, *Rebuilding of a Temple by King Joash*, 1602, Gdansk, oil on board, 129 × 326 cm, inv. No.: MNG/SD/494/M © Muzeum Narodowe, Gdansk, Poland.

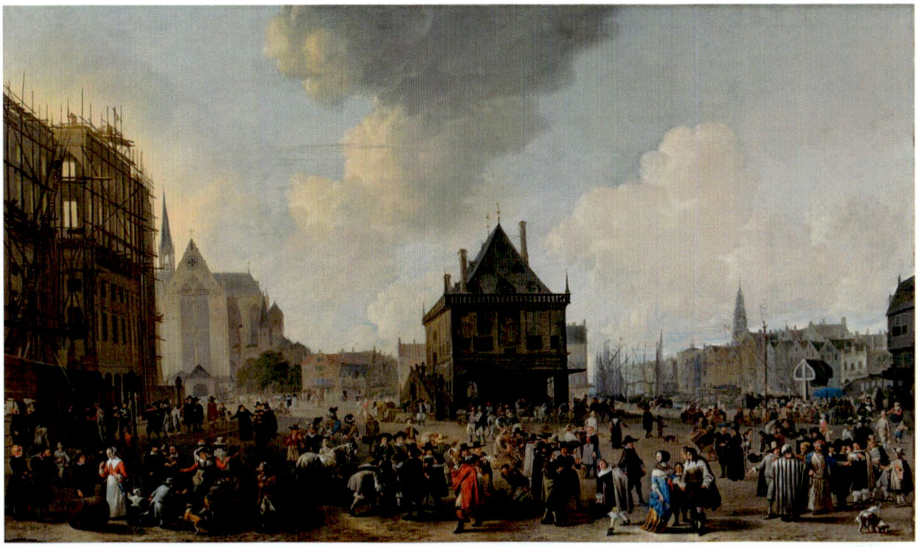

Figure 46: Johannes Lingelbach, Dutch, *Dam Square, Amsterdam, with the new City Hall under construction*, *c.*1650, oil on canvas, 122.5 × 206 cm, inv. No.: SA 3044, Amsterdam Museum, Netherlands.

it had been destroyed (Chronicles 23-24); Möller depicts the construction of a vast late Gothic/Renaissance church of cathedral like proportions. He specialised in large ceiling and wall paintings for public and church buildings and this work must have graced the interior of a large hall or room being some 3.26 metres wide by 1.3 high; its semicircular shape could have fitted an arch or lunette. The church has two naves with aisles either side, with the right-hand side being more advanced. Looking into the painting we appear to be facing east, or ecclesiastical east, with at the end of the right-hand nave a massive crossing. The scaffolding here is complex and has been rigged across the span of the nave and around the first arch independent of support from the ground. The second or furthest crossing arch has scaffold erected from the ground, as does the lateral arch between the two. The scaffolds are essentially putlog scaffolds with no inner standards. Here we see for the first time the scaffolder's pulley and rope, known today as a 'gin wheel and rope' and almost universal on British building sites. They were rigged up easily to haul up components for the scaffold and as an aid for the builders for lighter loads of materials. Three can be seen; the closest, to the right opposite the second bay of the nave, where two labourers are hauling on a rope with a bucket suspended at the other end, the pulley above them out of sight. The second is on the left in the aisle between the second and third piers under construction; again there are two men hauling on the rope whilst another is aloft steadying the rope and waiting to drag the bucket or basket onto the platform. The third, not in use, is in the centre above a worker who appears to be throwing a short length of rope down to someone below him.

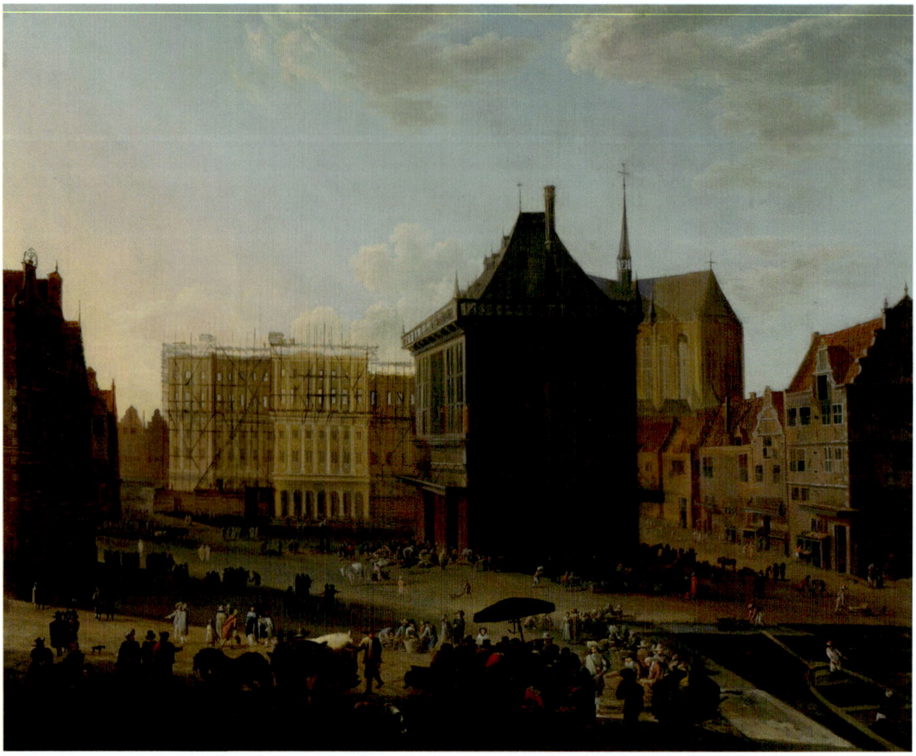

Figure 47: Attributed to Jacob van der Ulft, previously attributed to Johannes Lingelbach, *Dam Square with the City Hall under construction*, c.1650, oil on canvas, 81 × 100 cm, inv. No.: SB 1175, Amsterdam Museum, on loan Rijksmuseum Amsterdam.

There are scaffolders at work here and there; on the left on the first boarded lift, one receives, or passes, a pole through a tall window. At the crossing arch the man on the angled platform is possibly fixing the brace he is holding with his left hand and above him to the right another handles a pole. Lower on the right at floor level of the nave and adjacent to the two men hauling on the rope is another scaffolder handling a long pole. Behind him in the aisle can be seen a stack of poles leaning against the wall with a ladder next to them. Many ladders are in use around the site. King Joash is in the centre of the nave viewing the building's progress; his guards and the other figures in the foreground are in 'biblical' dress, whilst the workforce appear to be in contemporary garb for Möller's time. In the right foreground are the head and shoulders of two figures, also in contemporary clothing, looking out of the painting at the observer. Are these protraits of real people? In the left foreground is a counting house; Chronicles 24:8 tells us: 'At the King's command, a chest was made and placed outside, at the gate of the temple' which was used to collect a tax to pay for the work of rebuilding (Chronicles 24:6-15).

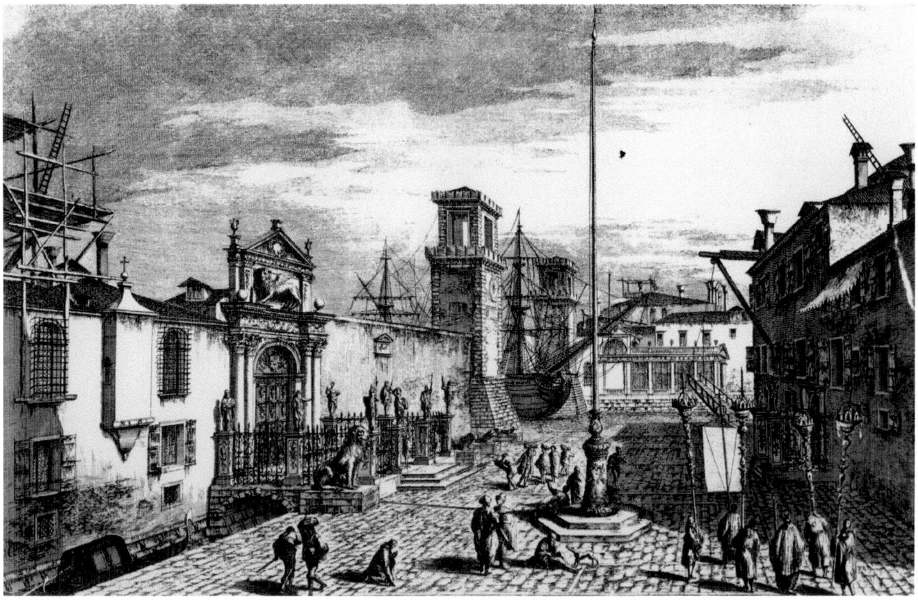

Figure 48: Michele Marieschi, Italian, 1710-1744, *Entrance to the Arsenale, Venice*, 1741, engraving, sheet 311 × 470 mm. Fondazione Federico Zeri, Bologna, Italy.

When the city of Amsterdam were building a new city hall in the mid-seventeenth century, completed in 1655, the scene was captured on canvas by Johannes Lingelbach and Jacob van der Ulft (attributed). (Figs. 46 and 47) Both these scenes of Dam Square, in the heart of the city, show the bustling life of the city in full flow. Lingelbach's painting looks north to the Nieuwe Kerk with the city hall to the left and the *Waag* or weighing house on the right. The main front of the hall, the set forward bays, is completed at ground level and two storeys above, with the arcade visible. Above the third storey is a scaffold protection fan to stop debris falling onto pedestrians below. Above the fan is a further scaffold to where work is proceeding on what will be a false pediment. To the left, closer to the viewer, the scaffold is still based upon the ground with a hoarding to protect, and keep out, the public, the first instance of such a safety measure in these depictions. This incorporates a kiosk, which may be a construction or information office. What appears to be a continuous ladder to the highest level of work must be at least four ladders tied together, which is unusual as ladders were normally staggered; it would have been a perilous climb to the top. Beyond the arcade is the third section of the building with again hoarding and scaffold from the ground. Van der Ulft's painting (or Lingelbach's, as it was originally attributed to him) looks to the west façade of the hall to the left of the *Waag*. The detail is almost exactly the same: The hoarding; the kiosk; the exceptionally long stretch of ladders; only the bracing to the central

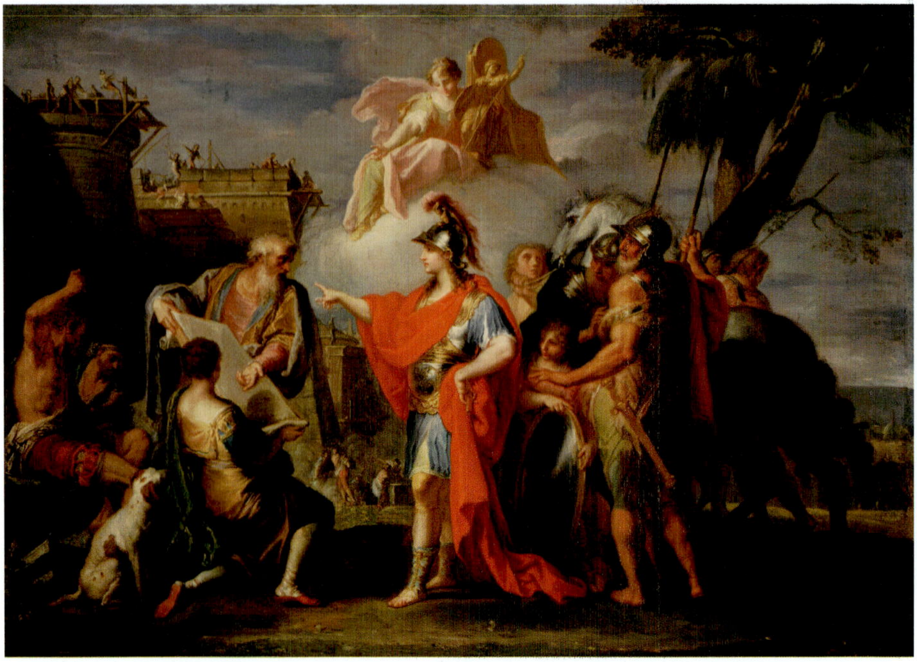

Figure 49: Placido Costanzi, Italian, 1702-1759, *Alexander the Great Founding Alexandria*, oil on canvas, 46.3 × 65 cm. Walters Art Museum, Baltimore, Maryland, USA.

section is different and no appreciable progress has been made on the visible external envelope of the building. The moment at which these scenes of construction were captured can only be days apart. The City Hall became a Royal Palace at the beginning of the nineteenth century and the *Waag* was demolished in 1808.

The Venetian painter of views and capriccio Michele Marieschi (1710-1743), published in 1741 a set of 22 engravings of Venetian views. One of these, *Entrance to the Arsenale*, (Fig. 48) shows the Campo dell l'Arsenale with the Rio de l'Arsenale and the entrance to the 'Arsenale Nuovo', built 1451-1457, beyond. At the extreme left is a scaffold erected over the roof of the administrative offices with two standards supported on a window sill immediately above the small branch of Rio de l'Arsenale. The gate beyond is the main pedestrian entrance to the Arsenal, decorated and embellished, which includes the Venetian Lion, to a degree commensurate with an establishment that was once at the centre of the commercial and military world. Marieschi also painted this scene in oils, a much larger work although it is less detailed.[46] The scaffold in the painting is slightly surreal, although it shows the working platforms are boarded with planks rather than what appears to be wattlework in the engraving.

Another depiction of Alexander the Great founding Alexandria, (which earlier appeared in a Flemish miniature of 1480 in Figure 23) is a painting of *c.*1736 by the Italian artist Placido Costanzi. (Fig. 49) Here Alexander discusses plans with the

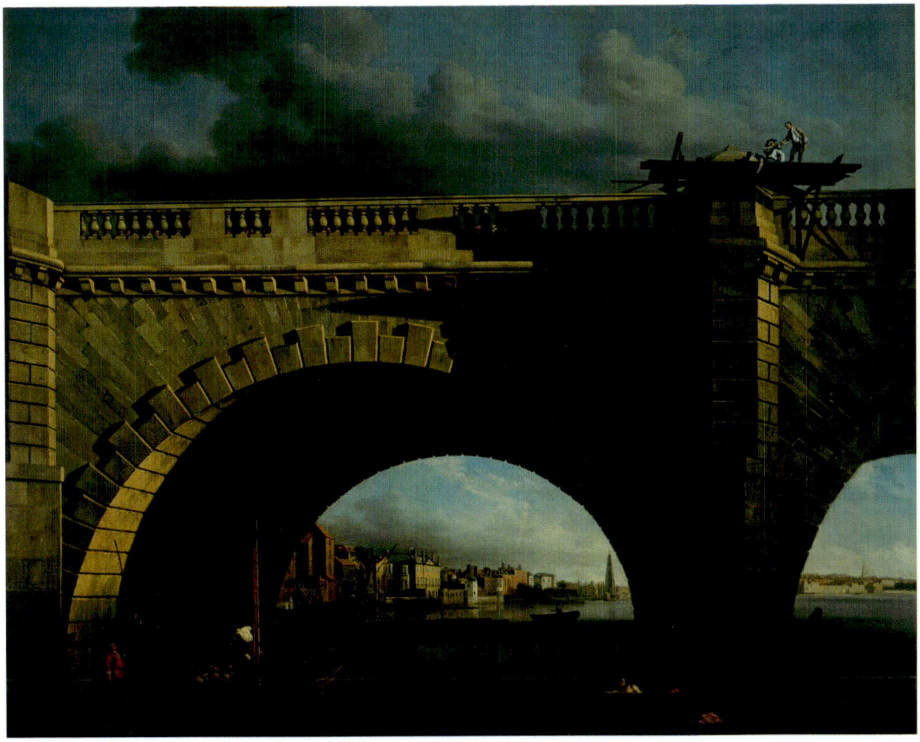

Figure 50: Samuel Scott, English, 1702-1772, *An Arch of Westminster Bridge*, oil on canvas, 135.7 × 163.8 cm, Tate Gallery, London.

Greek architect Dinocrates, although the city would appear to be nearing completion, judging by the buildings behind them and in the middle distance. At the top of massive battered walls is a large platform on top of a rounded tower, this overhangs the wall by means of heavy timber brackets supported from a string-course. But it does not overhang sufficiently for the winch on the platform; anything hoisted from the ground will drag against the tower for the first third or so of the lift. Behind this is another tower with squared corners. This has two putlog scaffolds; the one facing the viewer is supported entirely on putlogs. There are three workers on this platform, with another either climbing off, or lowering himself onto, the lift through a break in the parapet. On the other elevation the second lift is supported above the string course with putlogs and additionally with brackets or braces set into the wall. Between Alexander and Dinocrates can be seen classical buildings complete with sculptures, below these there is a hint of scaffold poles tied together and the mixing of mortar and burning of lime producing smoke which turns into the clouds supporting a deity who blesses this founding of Alexandria. This may be a Baroque representation of Athena the Goddess of war, military strategy and the protectress

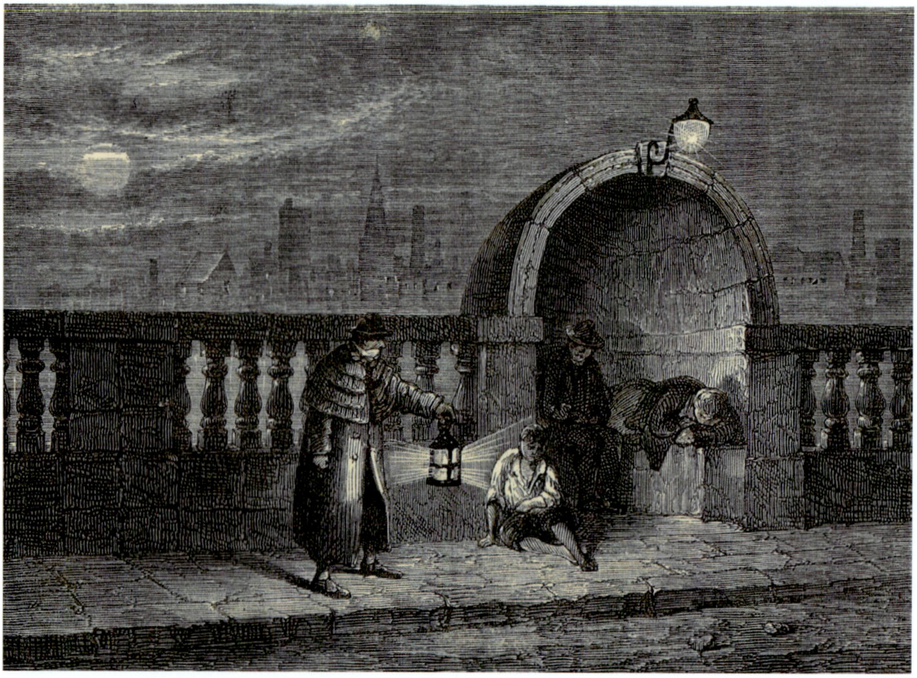

Figure 51: *Alcoves of Old Westminster Bridge*, from Walter Thornbury, 'The River Thames: Part 2 of 3', in *Old and New London: Volume 3*, London, 1878, p.300, engraving. Author's collection.

of towns, forts and harbours (among many other attributes), who although not in her traditional garb, holds a statue of herself, which *is* dressed accordingly. This all fits in well with Alexander; even the empty throne, which awaits him.

Samuel Scott's painting of *An Arch of Westminster Bridge, c.*1750, (Fig. 50) shows a similar scaffold to the nearest and furthest of Constanzi's scaffolds in Figure 49. This is supported off the large bracketed string-course on the river side of the bridge. There are no putlogs as such except for two members either side of the scaffold which support the working platform. One of these passes through the balustrade, and the other, on the left, passes over it and partly performs its function by means of a short length of rope; the precise purpose of this is unclear. These members must be supported by some counter weight structure on the road side of the bridge. Otherwise the platform is supported off of braces and uprights founded on the string-course. The scaffold appears to have been erected to construct or carry out repair works to the stone roof of the alcove behind. One of these alcoves is shown in an unsigned print of 1878 entitled *Alcoves of Old Westminster Bridge*, (Fig. 51) a simple engraving which shows a 'down and out' sleeping rough and being moved on by a night watchman. It is important because the scale of the scene is compelling in its apparent accuracy; it shows that the pavement level is exactly that of the upper level of the exterior string-course and that,

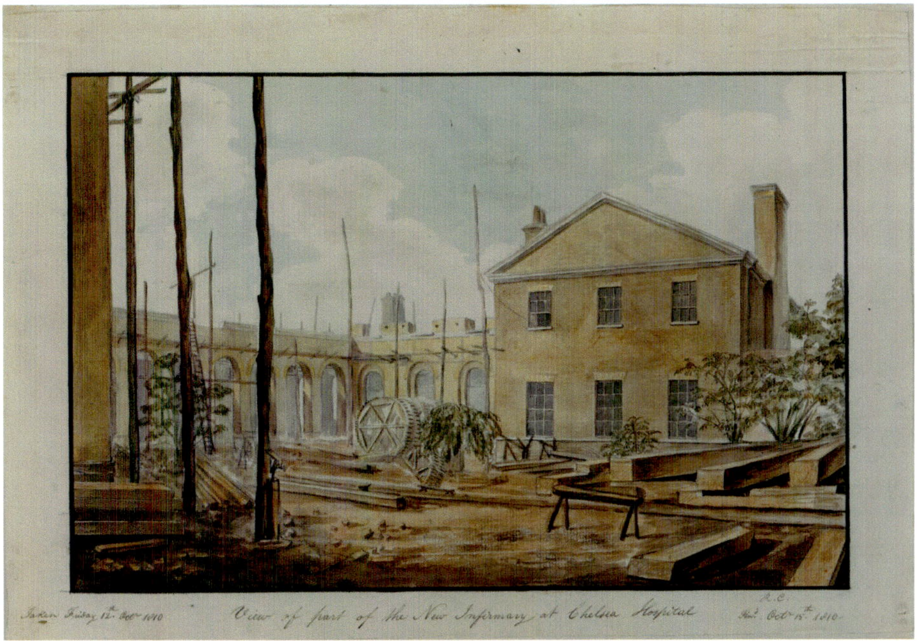

Figure 52: R D Chantrell, *View of part of the New Infirmary at Chelsea Hospital*, October 1810, pen and watercolour. By Courtesy of the Trustees of Sir John Soane's Museum.

let us say, the average height of an adult, the night watchman, who is stooping slightly, is approximately level or just above the height of the parapet. In Scott's painting of the figures, therefore, there is a strange and un-reassuring conflict of scale, and they are far too small. The two men on the scaffold, who are clearly not workmen for their clothes are too good, but more likely weekend revellers, would barely reach the centre point of the balusters, and the faces peeking through the balustrade, which are there but possibly not visible in the illustration, are positively Lilliputian. This is inexplicable and strange indeed for such an excellent draughtsman; Scott made several paintings of Westminster bridge during and after construction based on numerous drawings.[47]

The architect Sir John Soane had site record drawings made of progress during the building of the infirmary at Chelsea Hospital, on the north bank of the river Thames in London. The seven images shown here are watercolour sketches and paintings. (Figs 52-58) They are by Soane's pupils, John Buxton, R.D. Chantrell and Thomas Lee and are all dated October 1810. These paintings are important for their depictions of scaffolding at the turn of the nineteenth century. The building is at a stage where the scaffold is almost finished with; one gets a sense of this because there are not many working lifts and no one is working on them except in the internal view. (Fig. 58) Most of the standards are long straight poles, possibly whole young Fir or Pine trees from plantations, meaning that they were not the product of woodmanship but of

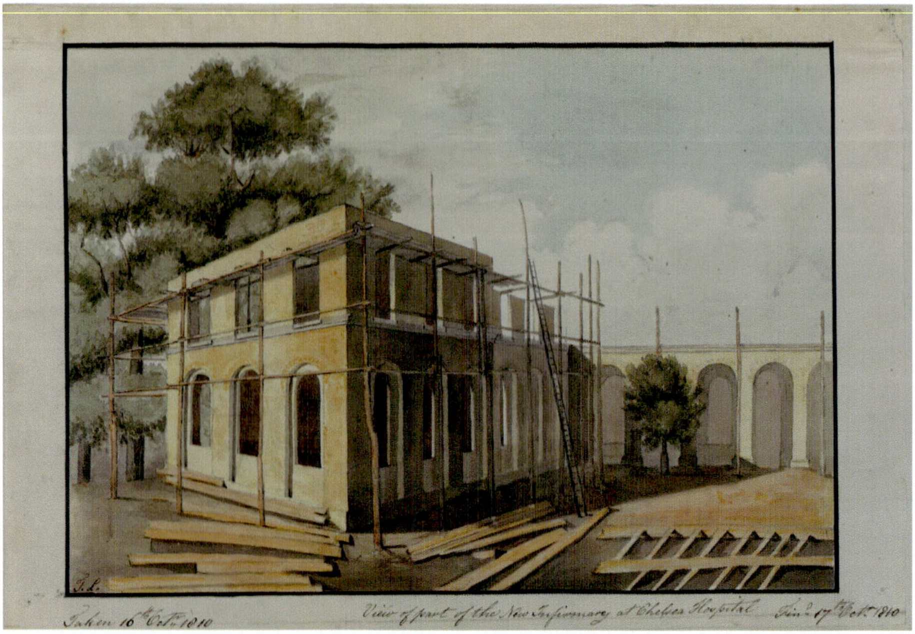

Figure 53: Thomas Lee, *View of part of the New Infirmary at Chelsea Hospital*, October 1810, pen and watercolour. By Courtesy of the Trustees of Sir John Soane's Museum.

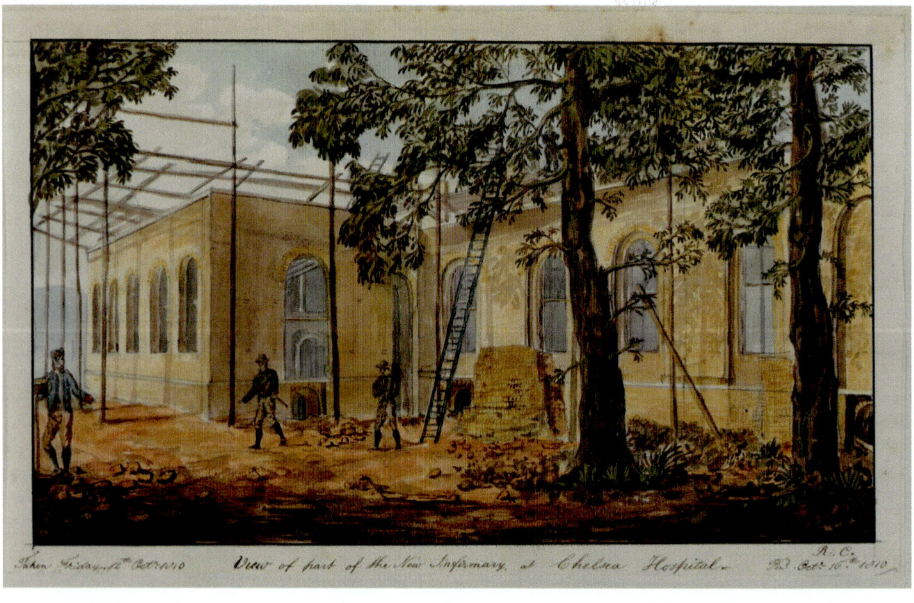

Figure 54 R D Chantrell, *View of part of the New Infirmary at Chelsea Hospital*, October 1810, pen and watercolour. By Courtesy of the Trustees of Sir John Soane's Museum.

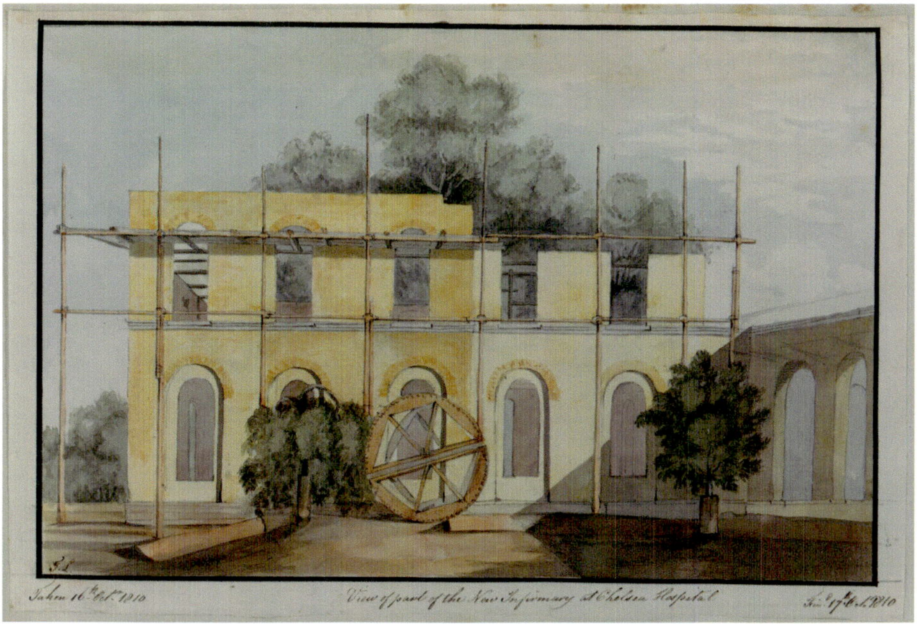

Figure 55: Thomas Lee, *View of part of the New Infirmary at Chelsea Hospital*, October 1810, pen and watercolour. By Courtesy of the Trustees of Sir John Soane's Museum.

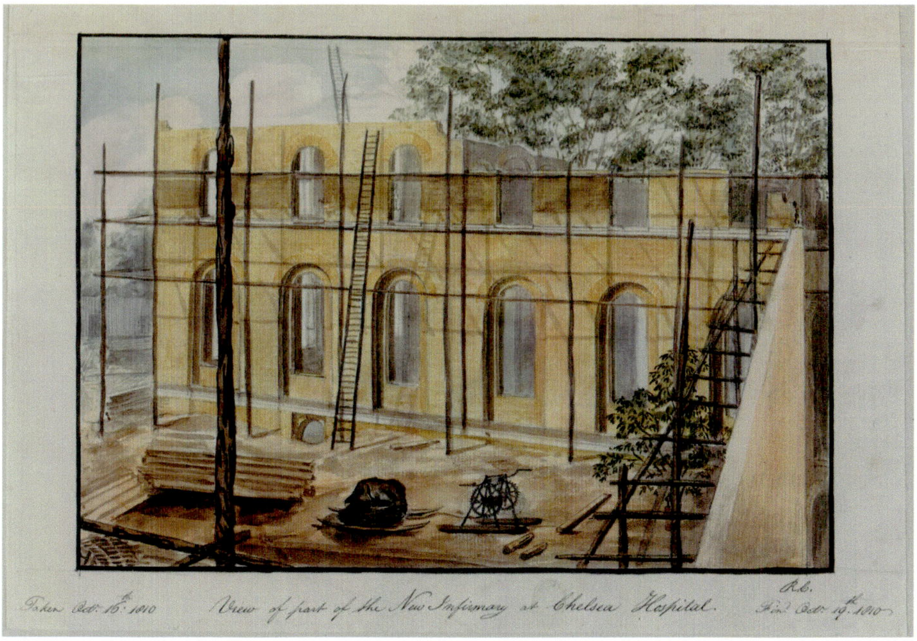

Figure 56: R D Chantrell, *View of part of the New Infirmary at Chelsea Hospital*, October 1810, pen and watercolour. By Courtesy of the Trustees of Sir John Soane's Museum.

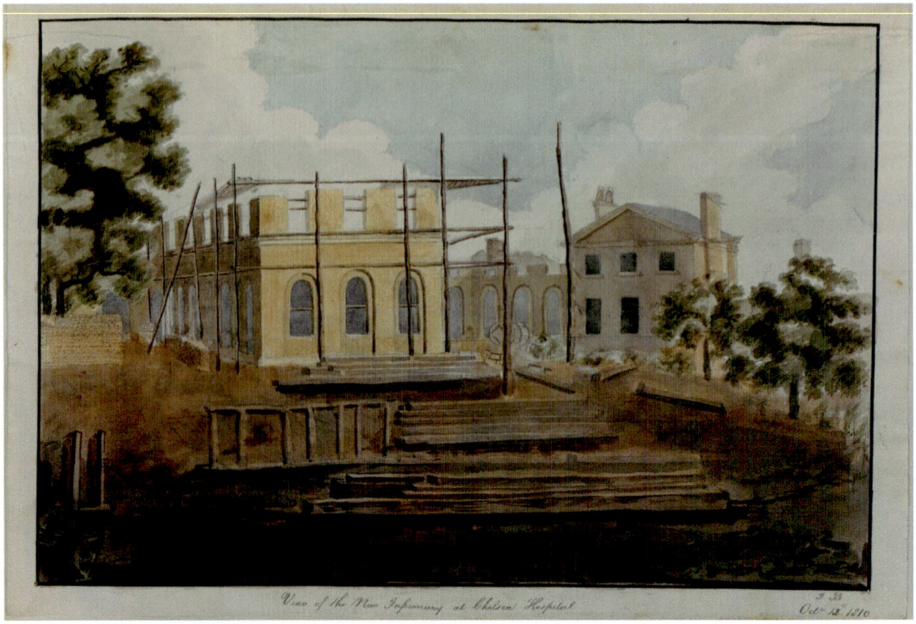

Figure 57: John Buxton, *View of part of the New Infirmary at Chelsea Hospital*, October 1810, pen and watercolour. By Courtesy of the Trustees of Sir John Soane's Museum.

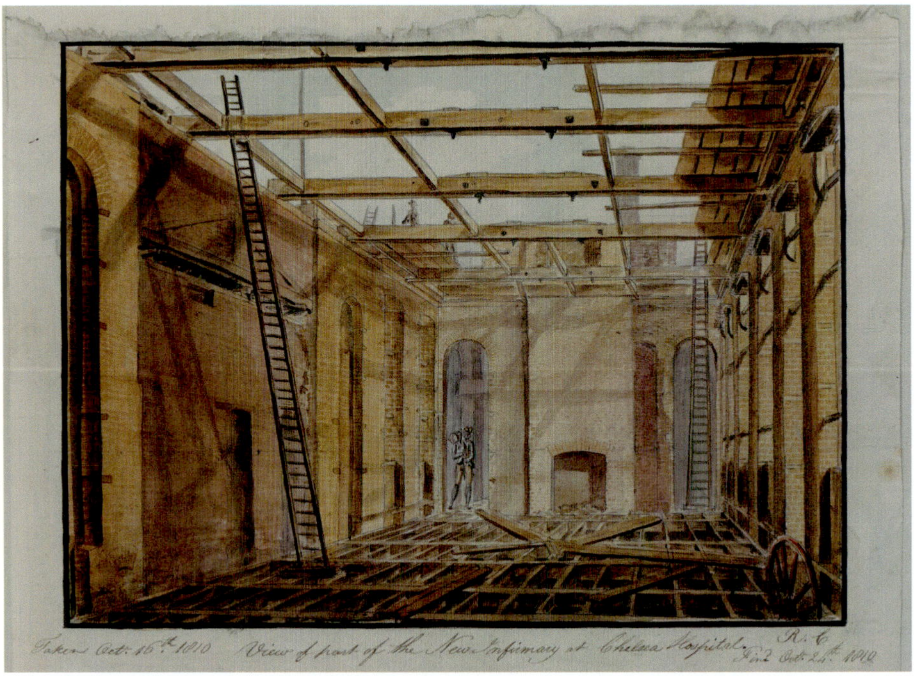

Figure 58: R D Chantrell, *View of part of the New Infirmary at Chelsea Hospital*, October 1810, pen and watercolour. By Courtesy of the Trustees of Sir John Soane's Museum.

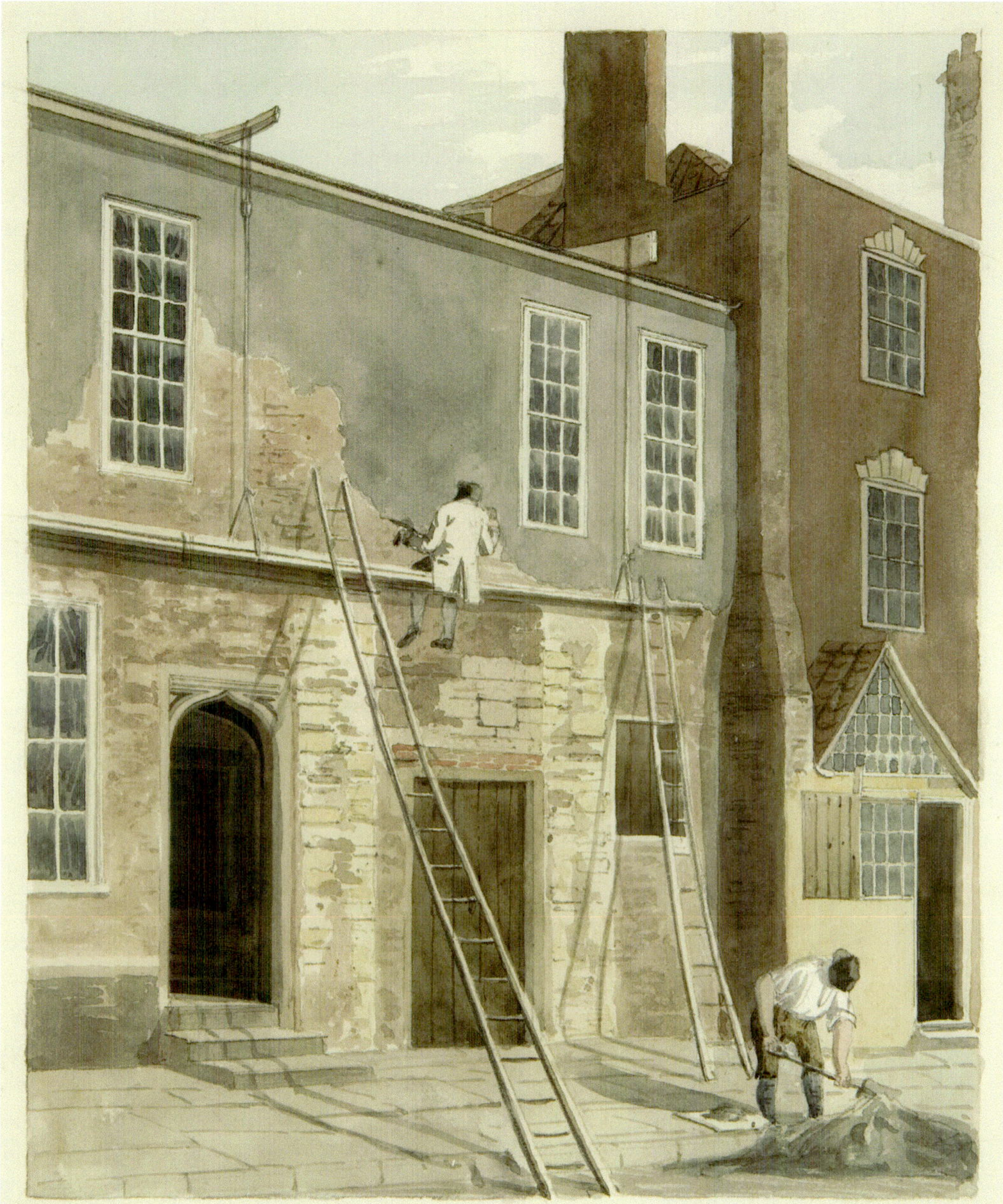

Figure 59: Samuel Jackson, English, 1794-1868, *House in Castle Green*, pencil and watercolour on paper, 214 × 175mm. Bristol City Museum and Art Gallery, Bequest of William J Braikenridge, 1908. Bridgeman Images.

the modern concept of forestry.[48] These standards are very clearly seen, particularly in Figures 52 and 56; the ledgers too are long poles and the putlogs cut from similar stuff. Some standards are joined, visible in Figure 53, the second and third in form the end and in Figure 55, the second from the left, with a great deal of overlap, and the furthest on the right. In most of these views the ties or withies are also clearly visible. The standards are dug into the ground for stability as is seen in the single freestanding one on the right in Figure 57 and those to the left in Figure 50. The interior view, Figure 58, demonstrates the paucity of boarded access across the completed ground floor structure with single boards only and none, except the joist, for access to the ladder on the left. The boarded lifts above rely on the joists of the first floor, along with additional putlogs, for support. The bulk of the carpentry in these later buildings, particularly from the eighteenth century, is hidden once the building is complete. The pen and watercolour drawing by R.D. Chantrell shown in Figure 58 eloquently illustrates the large amount of carpenters work that one would expect in an early nineteenth century building, and there are, as yet, no roof timbers in place.

In many of these images much timber is seen on the ground, including scaffold planks removed from the scaffold. By this time planks would have become standard as sawing became mechanised and timber was cheaper. Also seen are large section timbers in Figure 52 for use in floor structures as seen in Figure 58. Other timber components lie around the site including a treadwheel from a now disused hoist and stacks of scaffold poles and other stocks of timber awaiting use by the carpenters. It is clear when comparing these images that Chantrell was the most accomplished draghtsman.

Samuel Jackson, the Bristol artist (1794-1869), in his *House in Castle Green*, depicts a novel and cheap method of gaining height to re-render the upper part of the front wall. (Fig. 59) Here a board is suspended from overhanging timbers placed over the parapet at roof level. These cantilevered members must be counter-weighted further back behind the parapet out of sight. There are two ladders propped against the board and the plasterer having reached a lower section of the work sits on the board. The whole ensemble must have been extremely mobile and unstable, as the plasterer applied the render, the pressure would have pushed him away and as he released it, he must have used his feet to keep the board from swinging into the face of the building and smashing his knees. Furthermore the worker mixing the mortar when climbing one of the ladders, to replenish his co-worker's spot board, would have produced his own momentum. Jackson was essentially a watercolourist and a leading figure in what has become known as the 'Bristol School' of artists. He painted many views of the Bristol townscape and landscape, and some of these depict scenes associated with building construction such as grading gravel; brick or lime kilns; baulks of timber, road mending (laying cobbles) and, shown below in Figure 169, masons at work sawing large blocks of stone.[49]

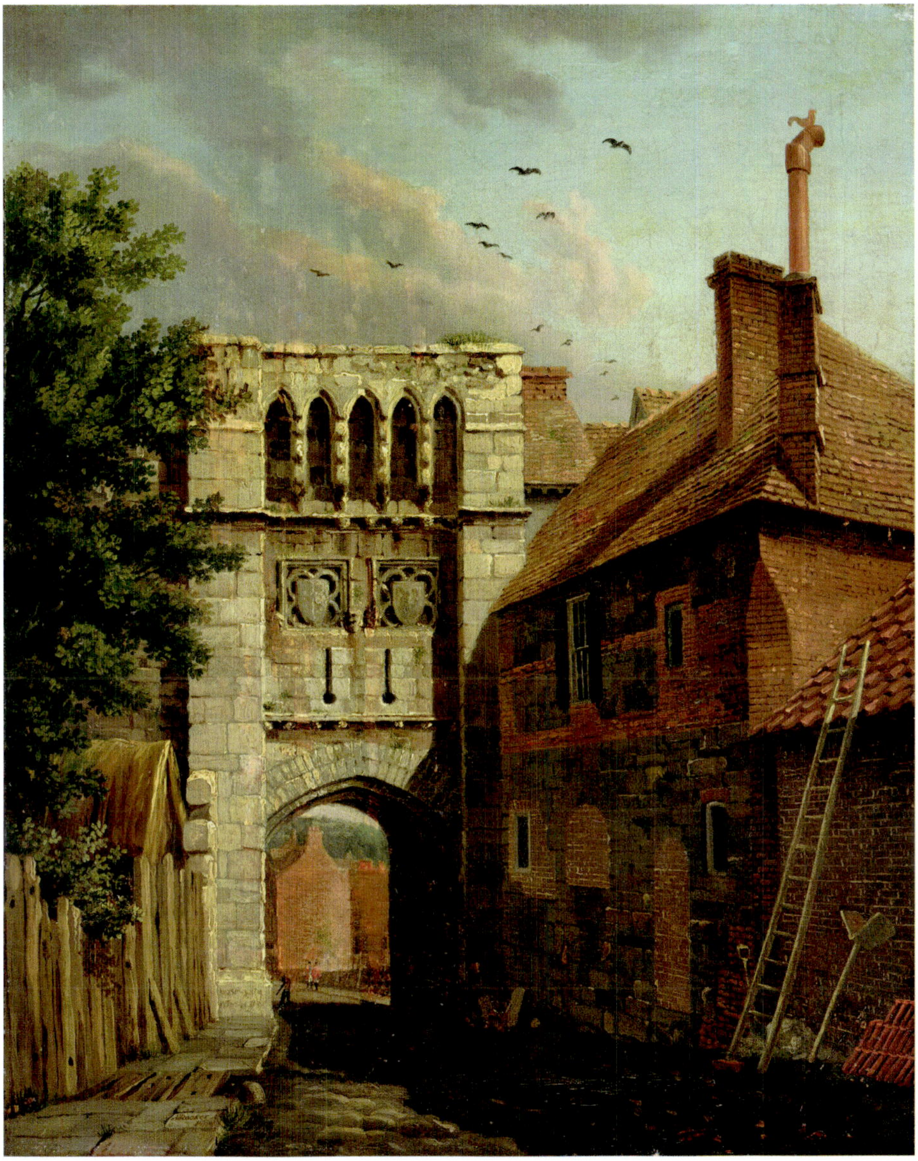

Figure 60: 1779, English, 1743-1801, *Westgate, Winchester*, oil on canvas, Manchester Art Gallery, Bridgeman Images.

Another small scale domestic scene, this time in Winchester and depicting roof maintenance, is an English oil painting of 1779. (Fig 60) Here the builders are missing but their ladder leans against the building as do their materials; a pile of mortar with a shovel and a hod waiting to be loaded from the stack of roof tiles to the right.

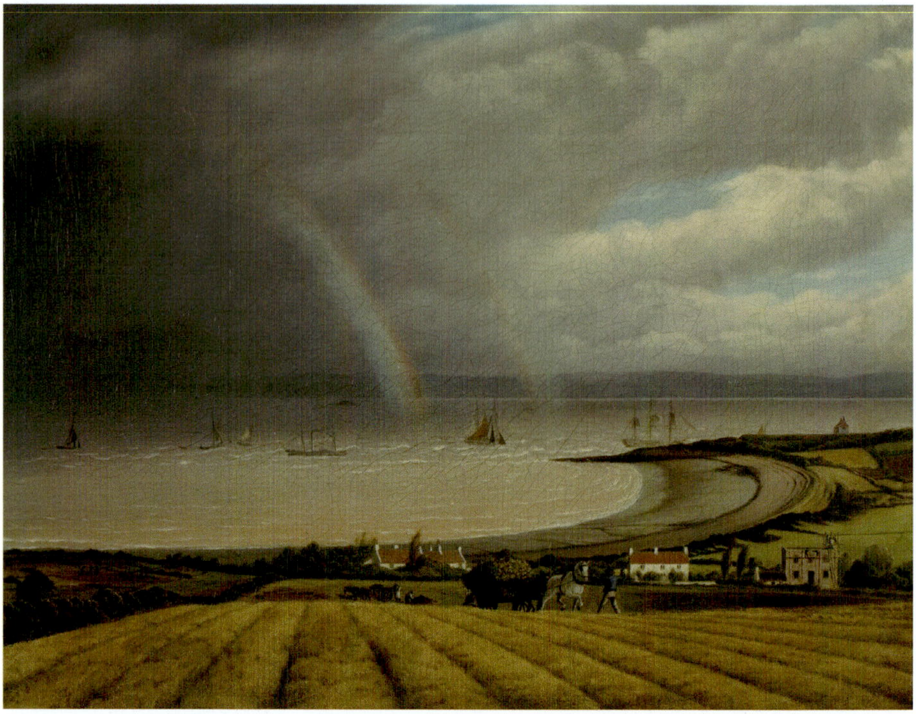

Figure 61: Joseph Walter, English 1783-1856, *A View from Portishead towards Wales*, *c.*1832, oil on canvas, 48.2 × 63.5 cm. Bristol City Museum and Art Gallery, given by James Fuller Eberle, 1921. Bridgeman Images.

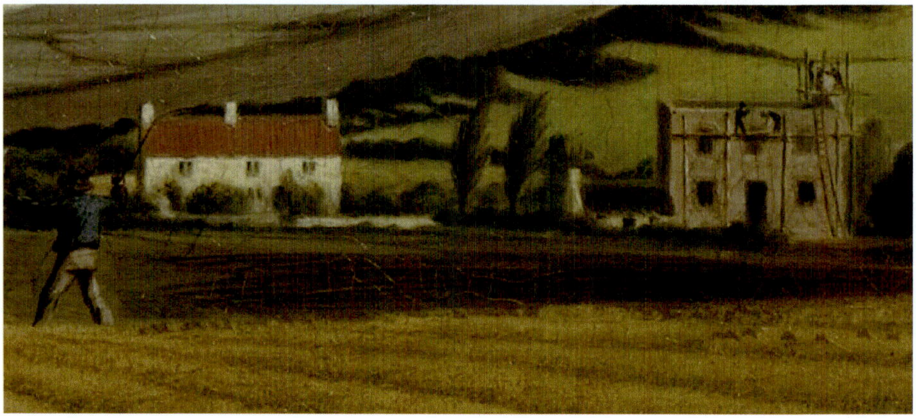

Figure 62: Detail, Joseph Walter, English 1783-1856, *A View from Portishead towards Wales*, *c.*1832, oil on canvas, 48.2 × 63.5 cm. Bristol City Museum and Art Gallery, given by James Fuller Eberle, 1921. Bridgeman Images.

Also in the English west country is Joseph Walter's *A View from Portishead towards Wales*, of *c.*1832, a charming rural landscape scene of harvesting and muck-spreading, on the high ground above the bay which sweeps round to Portishead Point at the

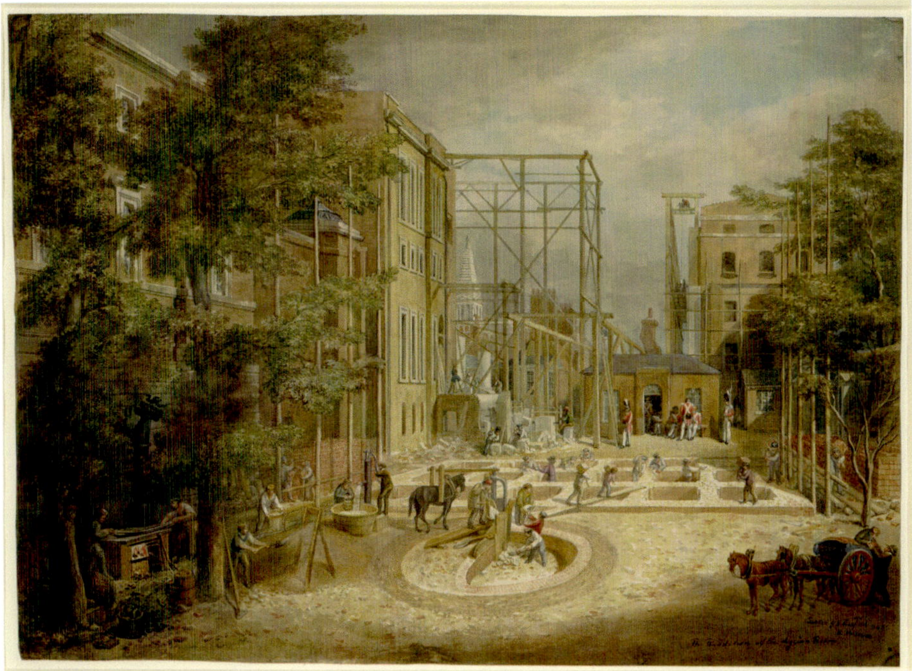

Figure 63: George Scharf, German, 1788-1860, *Laying the Foundations of the Lycian Room, the British Museum*, dated 1845, but drawn earlier, watercolour over graphite on paper. © The Trustees of the British Museum

mouth of the River Avon. (Fig. 61) A storm approaches up the Bristol Channel from the west and the farmer is in a hurry to get the cart loaded with the harvest into shelter before it gets drenched. Another worker is seen behind the cart frantically pushing. To the right of this scene, along a lane further down the hill, is a house under construction. (Fig. 62) Two men work on a rickety scaffold at eaves level. This is a simple single lift scaffold with putlogs set into the building, essential for some degree of stability, as there are no braces. Another two are working on the gable at the end.

George Scharf (1788-1860), a German artist who settled in London, was a prodigious draftsman, illustrator and print-maker. He recorded urban life in London. He produced scientific illustrations and prints for a living; among the subjects that fascinated him were scenes of demolition and construction, the latter both of buildings and civil engineering.[50] Among these are scenes at the British Museum (Figs. 63, 64 & 65), the College of Surgeons, (Figs. 66, 111 and 113) New Covent Garden Market (Fig. 67 & 114) and 'Craig's' house. (Fig. 68) At the British Museum in the watercolour *The Foundation of the Lycian Room*, dated 1845, shown in Figure 63, we see an extensive and busy building site. The view is taken from the north at the Montague Place end, and looks due south. The church behind the large timber framing and scaffold, with the unmistakable stepped spire, is St George Bloomsbury

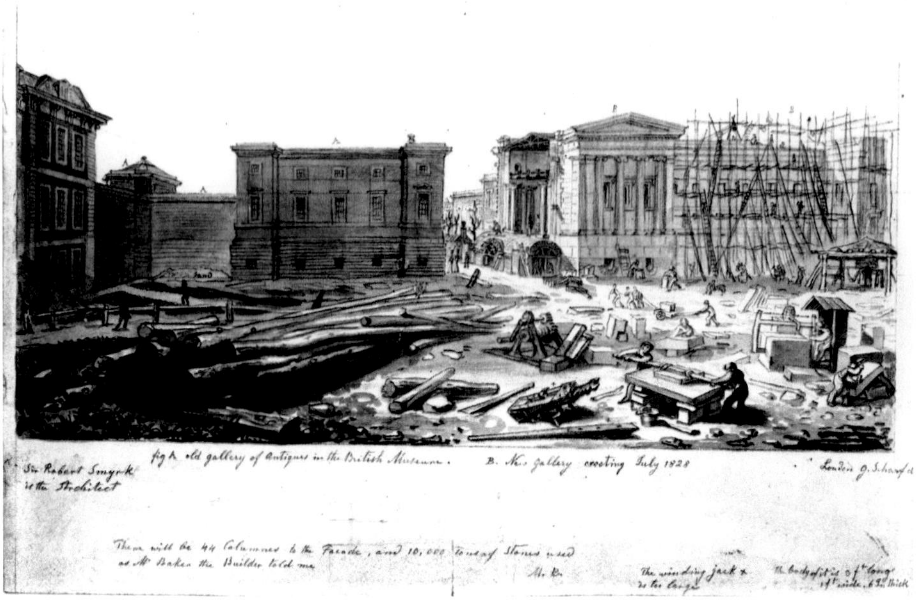

Figure 64: George Scharf, German, 1788-1860, *Building the New Gallery of the British Museum*, dated 1828, graphite (with ink and wash?) on paper. Private Collection. Bridgeman Images.

in Bloomsbury Way. The grid of the foundation is clearly seen in the middle ground in front of the guardhouse, which we imagine is soon to be demolished, judging by the timber framing that spans over it. The Lycian room is now room fifteen but still houses antiquities from Athens and Lycia. Scaffold poles have been set into the ground either side of the foundations, four on the right and five on the left, in readiness for building up the walls. These are long straight Firs from plantations. Both sides have ledgers tied on, at about shoulder or waist height of the men seen working adjacent, and no other elements have yet been rigged. The very large and complex timber-work framing further back, is a combination of scaffold components and timber baulks joined or bolted together. It is so complex that Scharf appears to have lost his way with some of the detail. We can assume that the heavier elements are constructed for the hoisting of the main portico, the partially built columns of which can be seen immediately below the spire of St George, and the continuing colonnade, which runs around the front of the building. This is indicated by the framing, which spans forward over the guardhouse and will then return in front of the large building to the right. This building has a regular wooden scaffold across its rear, comprising standards and ledgers. This is connected to a massive, almost free standing, timber framework along the flank of the building with a single boarded lift approximately half way up. Above this at the very top is a travelling

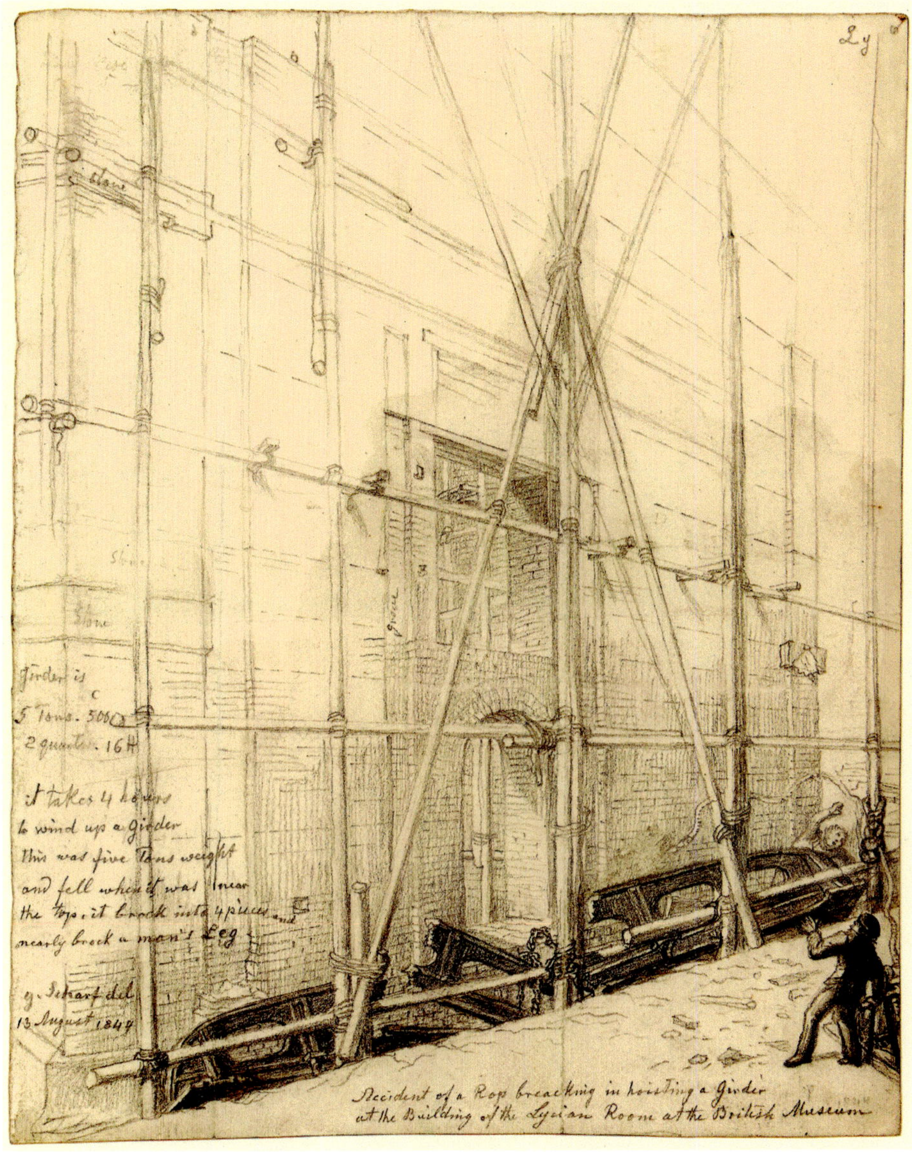

Figure 65: George Scharf, German, 1788-1860, *Accident of a rope breaking when hoisting a girder at the building of the Lycian Room of the British Museum*, 1844, graphite on paper. © The Trustees of the British Museum

gantry with a lifting operation in progress. Quite why this framework is so huge and precisely what its function is, is not clear. The building is part of the original Montague House, the first home of the British Museum, and yet to be demolished, as indeed are the other buildings on the site.

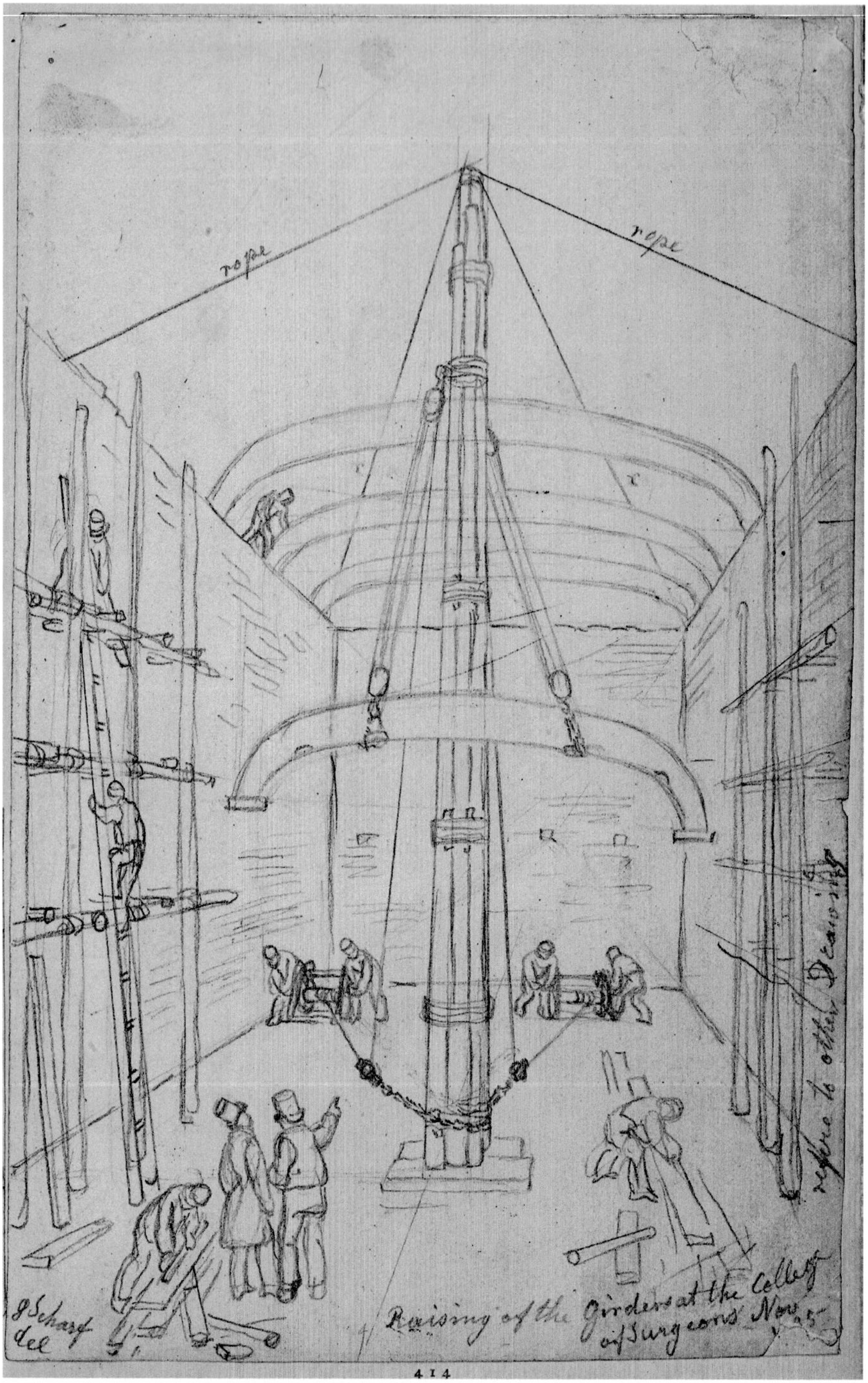

Figure 66: George Scharf, German, 1788-1860, *Raising of the girders at the College of Surgeons*, 1835(?), graphite on paper. © The Trustees of the British Museum

That the building of the new museum was carried out piecemeal around Montague House, is borne out by a drawing Scharf made on site, of works progressing within what became the Great Court of the new building by the architect Sir Robert Smike. (Fig. 64) This was to protect the collection until parts of the new building were ready. What we see in Scharf's drawing, on the right, is part of the west courtyard elevation of the east wing under construction. This is the King's Library, now room one. On the left is the main frontage of Montague House facing Great Russell Street with the east wing still intact. Scharf has clearly annotated this 'A' and 'B': 'Fig. A old galleries of Antiques in the British Museum. B New gallery erecting July 1828.' To the far right of the image the new building is fully scaffolded, albeit very rapidly sketched with four lifts and five or six ladders. In the left foreground is a pile of large scaffolding poles, perhaps just removed and awaiting the next phase. Another annotation reads 'There will be 44 Columns to the Façade, and 10,000 tons of stone used as Mr Baker the Builder told me.' A further annotation concerns the winding jack, for which see below page 164. The view in Figure 63 is taken from between the two buildings looking towards the viewpoint of this image.

The third image depicting work at the British Museum sadly records an accident, but it is nevertheless a good study of the scaffolding inside the Lycian Room. From 1844 it predates Figure 63, which shows only the foundations with no walls built and only standards erected with a single ledger at the base. We can be fairly certain of the dates recorded in Scharf's sketchbooks, in fact this is precisely dated 13 August 1844. It is assumed therefore that the date of the watercolour records when he completed it in his studio and not when he was on site. In this image (Fig. 65) two of the standards are doubled up and braces have been rigged in the form of shear legs, all for the hoisting of the roof girder, which he tells us weighs some five tons. The withies are clearly visible as are the wedges driven in to tighten them. Scharf's annotations give the title to the drawing and on the left describe the girder, how long it took to lift; four hours, that it fell when near the top and broke into four pieces and that it 'nearly broke a man's leg'. What he fails to tell us however, is how the girder passed above the putlogs at the level of the third ledger. These are embedded in the wall and the girder is definitely inside the scaffold. It must have been lifted vertically rather than horizontally and fell flat on impact with the floor.

At the College of Surgeons Scharf recorded, in October and November 1835, the installation of iron roof beams or girders in three notable drawings. In figure 66 we see scaffold being struck or dismantled. As the installation of the girders moves towards the viewer the scaffold needs to be removed in order to allow the hoisting. A scaffolder on the top lift has removed a board and is passing it down for the man on the ladder to take down to ground level, another board leans against the wall and others lay on the ground.

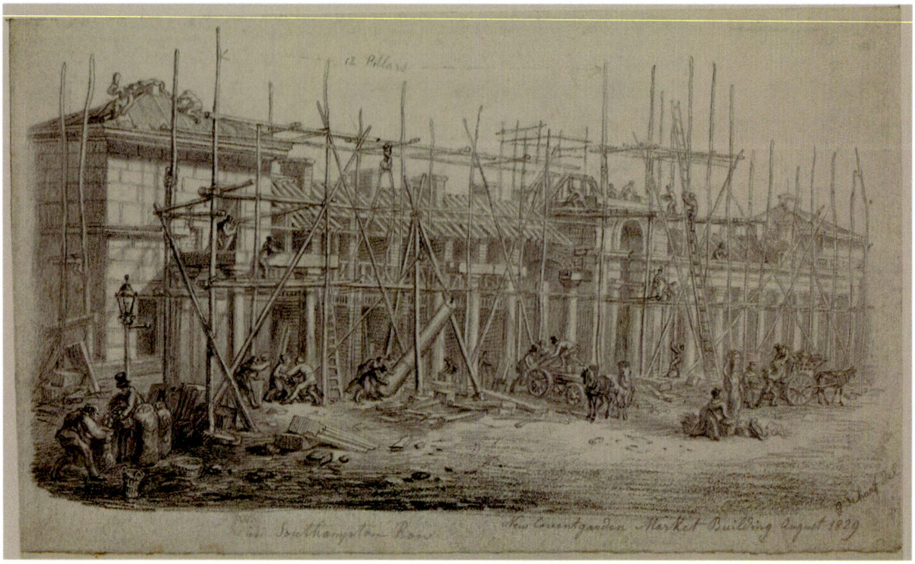

Figure 67: George Scharf, German, 1788-1860, *New Covent Garden Market Building taken from Southampton Row*, 1829, graphite on paper. © The Trustees of the British Museum

New Covent Garden Market is a building with an open frontage with a colonnade and a putlog scaffold does not work, so internal standards are required. We see in Scharf's drawing of 1829, (Fig. 67) therefore, a freestanding scaffold with inner standards and at least one buttress brace on the extreme left. Here the putlogs span between the inner and outer standards, and the scaffold is also heavily cross or wind braced. There are three lifts, although only the bottom one is boarded. To the right in front of the main arched entrance and behind the ladder are three scaffolders attending to a standard. The ladder, up which a worker climbs, is exceptionally long. Compare this to the ladder shown in Figure 58 of the interior of the New Infirmary at Chelsea Hospital of similar date. Three poles have been rigged to form shear legs for lifting the columns and above this is a gin wheel and rope although the latter is not in use, see page 160 for a discussion of this operation. Market activities are continuing along the front of the building site.

One of Scharf's most instructive drawings of scaffold is his *Outside Craig's house in Francis Street, Summer 1841*. (Fig. 68) In his title he indicates that Francis Street is 'now Walker's'. Scharf lived in Francis Street from 1830 to 1848 and Craig must have been a neighbour who was having his house repaired.[51] This is almost exclusively a scene of scaffold erection although there are one or two other things going on. The standard poles are only short and are joined at the level of the second ledger or lift; at ground level they strangely have rounded bases. Starting from the right; the worker on the ladder is not part of the scaffolding team, he is raking out mortar for repointing; then there are three men joining a standard, with one holding the

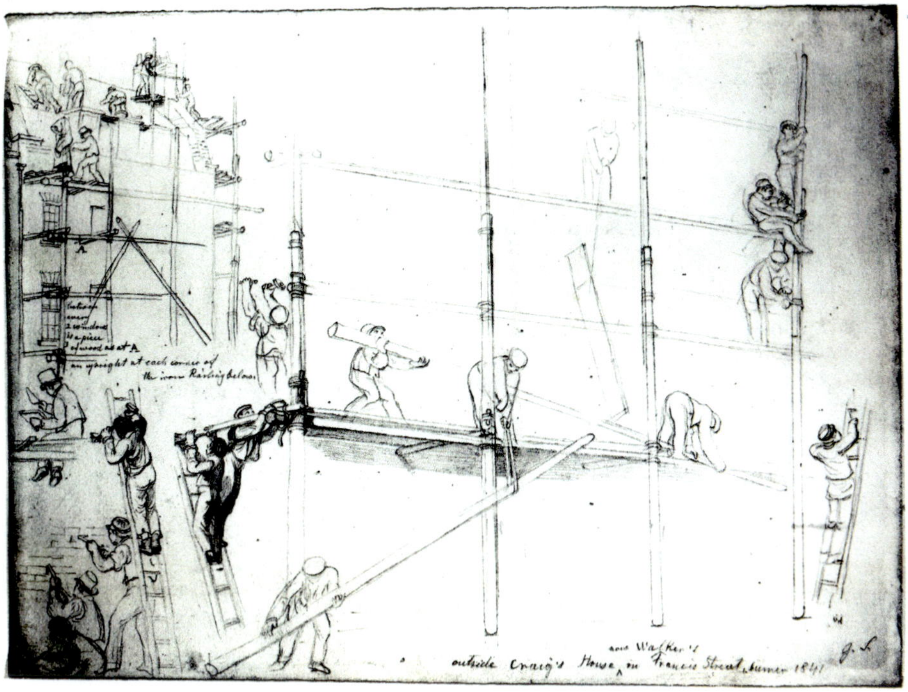

Figure 68: George Scharf, German, 1788-1860, *Outside Craig's house in Francis Street, Summer 1841*, graphite on paper. Private Collection. Bridgeman Images.

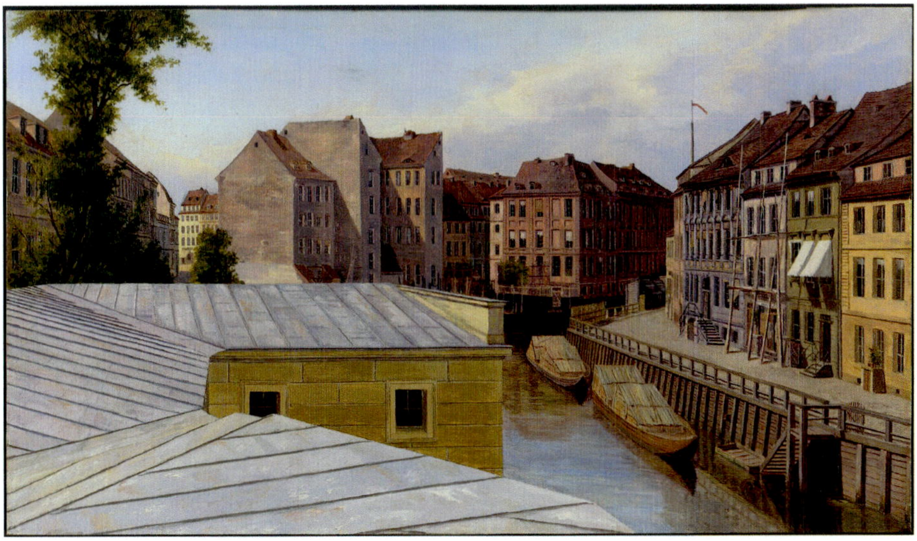

Figure 69: Eduard Gaertner, German, 1801-1877, *The Friedrichsgracht, Berlin*, c.1830, oil on paper laid down on millboard, 25.5 × 44.6 cm. The National Gallery, London.

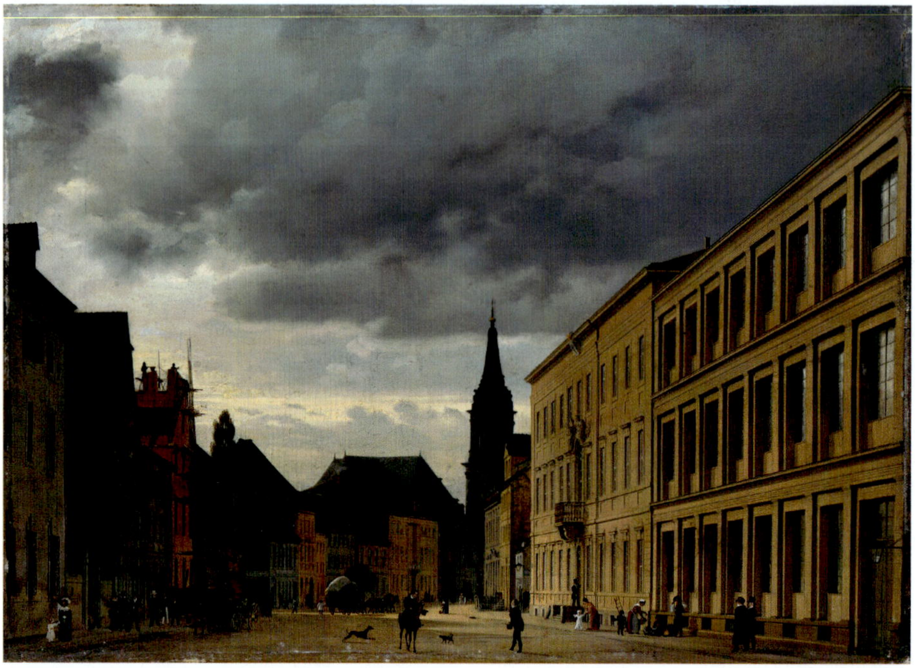

Figure 70: Eduard Gaertner, German, 1801-1877, *Klosterstrasse mit der Parochialkirche*, 1830, oil on canvas, 32 × 44 cm. bpk, Nationalgalerie, Staatliche Museen zu Berlin / Jörg P Anders.

pole above whilst the other two tie on the withies; in the centre of the main image four men are lifting poles, two to each pole and another is hauling up a plank. To the left two workers, both on the same ladder, fix a putlog; one ties it to the ledger whilst the other inserts it into a putlog hole. Above them is another worker raking out mortar, standing on the first lift. Moving to the inserted image at top left; two scaffolders are erecting scaffold around the chimney stack and we see the upper part of the scaffold with cross braces. Four other workers are at the top left but none of these are scaffolders, three of these are roofers working off of boards laid across the roofs of the dormer windows. At the bottom left are five other workers, three of these on the ground are raking out mortar, the one on the ladder appears to be chopping out a damaged brick to replace it, and the one sitting on boards is pointing. Immediately below the man on the top lift who has draped his coat over the standard, below the second lift down, is the letter 'A'. Scharf has annotated the drawing: 'Between every two windows is a piece of wood as at A'; these are putlogs. He continues '[A]n upright at each corner of the iron railing below', meaning the standards but he does not show the railings. Clearly seen on some of the withies are the wooden wedges driven in to tighten them. This drawing shows that work started in earnest whilst the scaffold was still being erected.

Figure 71: Eduard Gaertner, German, 1801-1877, Detail, *Klosterstrasse mit der Parochialkirche*, detail of Figure 70, 1830, oil on canvas, 32 × 44 cm. bpk, Nationalgalerie, Staatliche Museen zu Berlin / Jörg P Anders.

Eduard Gaertner, (1801-1877) another German artist was also a prolific painter of urban scenes, townscapes, panoramas and landscapes. He worked mainly in Berlin and his paintings give a sense of compelling accuracy. In his *The Friedrichsgracht, Berlin*, (Fig. 69) set on the wharf of the canal, the third house in from the right is scaffolded. This is certainly for some form of maintenance, repair or redecoration. The three main standards are set at an angle in towards the building forming a buttressed scaffold with two ledgers attached, together with putlogs, which appear to be set into the building. At the far right end is a single standard at an

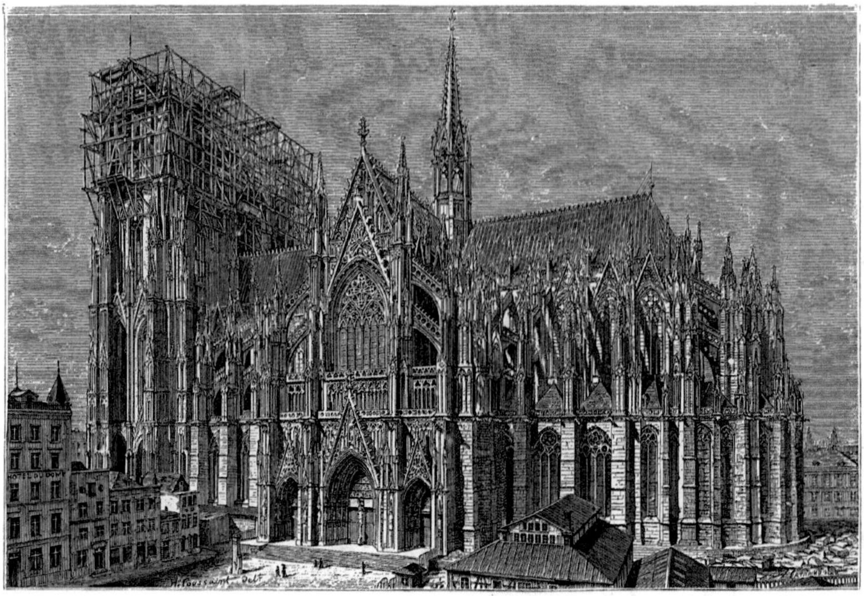

CATHÉDRALE DE COLOGNE EN 1875
Dessin de Toussaint, d'après une photographie de M Schoenscheidt

Figure 72: *Cologne Cathedral in 1875*, wood engraving, drawn by H Toussaint, engraved by Antoine Valerie Bertrand, from a photograph by Johann Heinrich Schonscheidt. Author's collection.

angle leaning into the end of the scaffold, possibly achieving some form of bracing. These standards are long straight poles of Fir or Pine, and the one at the far left must be some 250 to 300mm in diameter at its base. The composition has an early Sunday morning feel, as no one has yet ventured forth and there are no workers; no ladder even, which will be brought along when they start work. In contrast, in the Klosterstrasse, (Figs. 70 and 71) we see a busy scene of street life evolving, and workers are busy on the scaffold which is left of centre. Again the main standards are slightly inclined towards the building, which appears to be a large town house.

Additionally the scaffold is buttressed for added stability, with braces from the pavement up to the second lift. All of the work is at high level, with two men on the fourth lift, which returns along the flank of the building, who appear to be working at eaves level; one man is working off the very highest lift, the sixth at the front at the apex of the gable, and two others are working on the chimney stacks. There is no ladder visible, so the scaffold may return along the far, blind side of the building, where a ladder may be positioned.

Cologne cathedral had an exceptionally long building history. The present Gothic building was begun in 1248 and it was not completed until 1880, a period of 632 years.[52] Within this period, however, some three centuries from 1520 to 1842, building

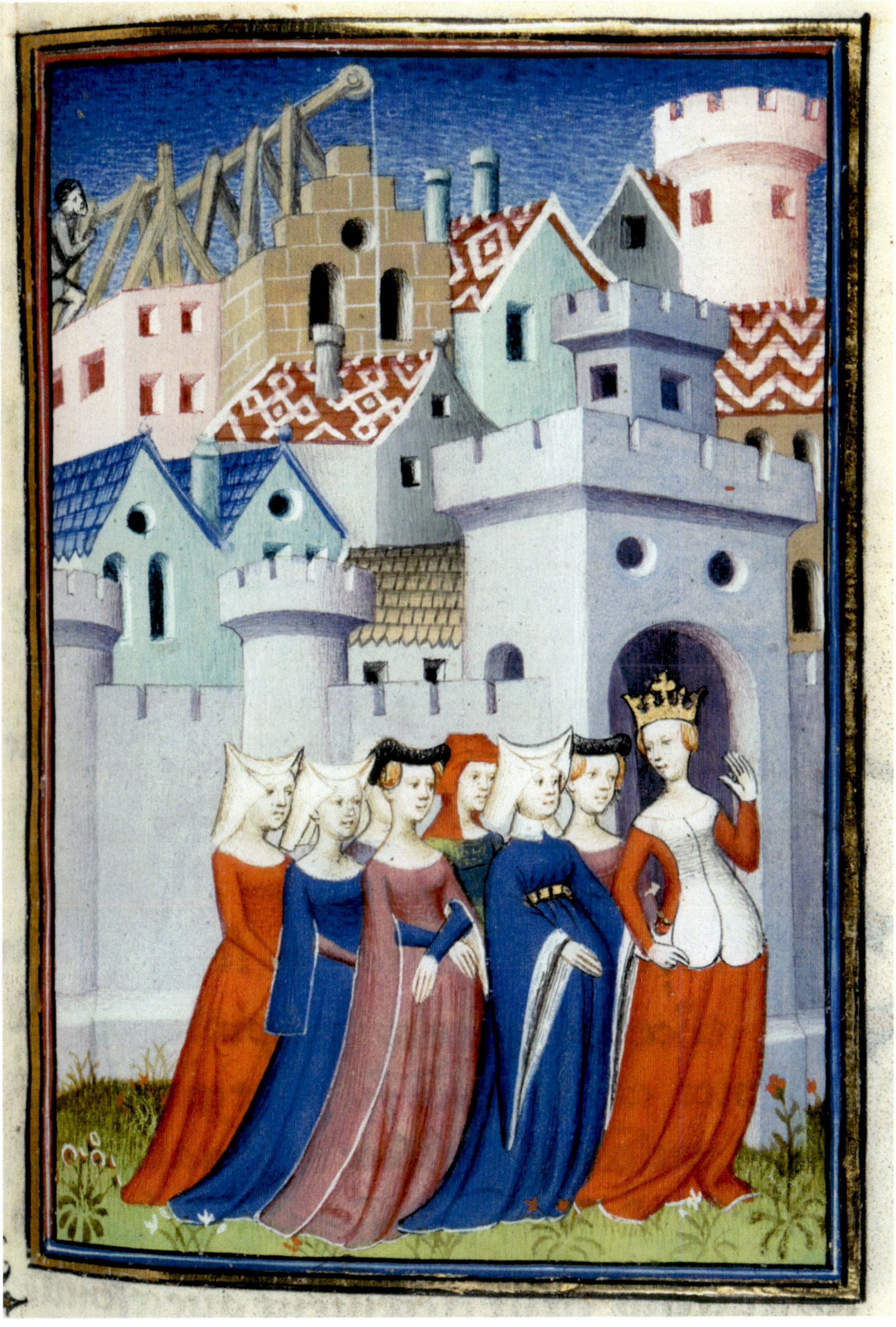

Figure 73: 1410-1414, Master of the Cité de Dames, French, *Rectitude leading Christine and other worthy women into the city*, from Christine de Pizan, Collected Works ('The Book of the Queen'), Harley ms 4431, f. 323r. © The British Library Board, London.

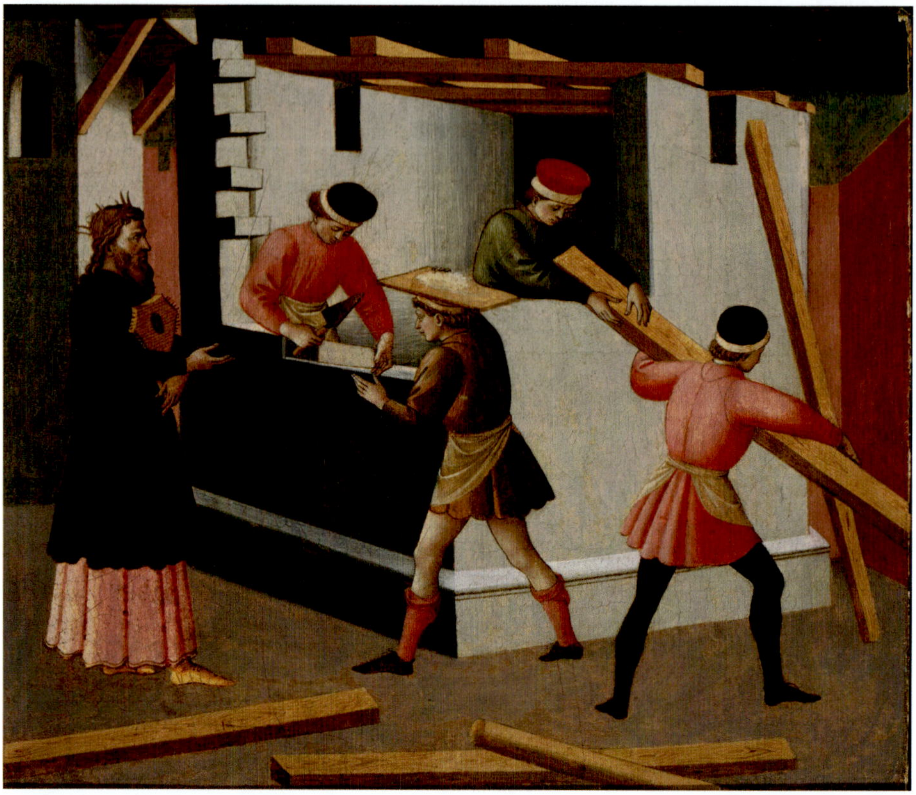

Figure 74: *c.*1445, Pesellino (Francesco di Stefano), 1422-1457, Italian, *The Construction of the Temple of Jerusalem*, tempera on panel, 53.5 × 60.3 cm. (Harvard Art Museum/Fogg Museum, Friends of the Fogg Art Museum Fund, 1916.495. Imaging Department © President and Fellows of Harvard College.)

operations were entirely dormant. We see the cathedral here, in 1875, (Fig. 72) with the tops of the west towers clothed in a robust scaffold, a timber scaffold of stout squared section members rather than one made of wooden poles. It is bracketed out from approximately three quarters of the way up the tower. The towers themselves are near completion and the twin spires are about to be commenced. The great Medieval crane that stood atop the unfinished south tower for more than 400 years has gone; see page 145 and Figures 96, 97 and 98.

Timber Framing and Roof Structure

Here we consider the carpenter's involvement in the actual construction of the building.

In Figure 73 from Christine de Pizan's *La Cité de Dames* we see Rectitude leading Christine and other worthy women into the city. The new city is not entirely complete, and in the left of the background a carpenter is working on one of two roofs. On the roof behind, the ridge board has been left uncut and purposely extended over the gable with a gin wheel rigged at the end to facilitate hauling up timbers.

A similarly exposed and incomplete roof is seen in the fantastically imagined *Rebuilding of the city of Troy* by Jean Colombe. (Fig. 142) Here a crane lifts timbers to the roof which is otherwise unattended by workers. An exposed and unfinished roof structure is also shown in George Scharf's drawing of New Covent Garden Market under construction. (Fig. 67) Here the joists to the roof along the length of the collonade are visible and the roof over the arched entrance is taking shape.

King David contemplates the construction of the Temple of Jerusalem in Francesco Pesellino's painting of *c.*1445, (Fig. 74) as he does in Figures 20 and 21. Here he is identified by the psaltery he holds.[53] It was his son Solomon though who actually oversaw the construction of the Temple (see page 57 above); here carpenters are handling large timbers for either a floor or the roof, and a mason and labourer are also at work.

In the *Bedford Hours* is a miniature of the building of Noah's Ark. Although a depiction of the construction of the superstructure of a ship, it can nevertheless be taken for a timber-framed land building; a simple box frame house or barn. (Fig. 75) Four workers are engaged on the roof, two nailing on boards, one of whom has a box of nails at the side, which has been purpose made so that it does not slide off; another worker carries boards along the far side of the already boarded blind slope; whilst the fourth sits astride the wall plate at the junction of the end truss and drives a peg into the joint of purlin, end rafter and collar.[54] Inside the building, a man crouching works at the joint of a sill beam and stud or post with a wooden handled tool and a mallet; another standing works at the junction of a transverse or cross beam and a post; whilst a third climbs a ladder with a length of timber. Working on the outside of the building, a man on the right drills a hole for a peg using an auger, on the left another is using a mallet but we cannot see what he is doing. In the foreground four other carpenters perform various tasks, from the left: A carpenter uses a large 'trying' plane[55] to produce a straight edge to a large board, which is held upright by being wedged across two timbers with slots cut in each. These boards, stacks of which can be seen in front and behind the carpenter, are intended either for internal floors or for infilling the framing. However such wide boards would be prohibitively expensive to produce from what would have been exceptionally large trees with considerable girth. The narrower planks seen on the left and rear walls are more realistic. Next a carpenter bores or drills a hole in a timber using an auger. Following that a large timber is being trimmed with an axe, and this is held fast in a similar manner to the large board. Finally on the right, a length of timber is

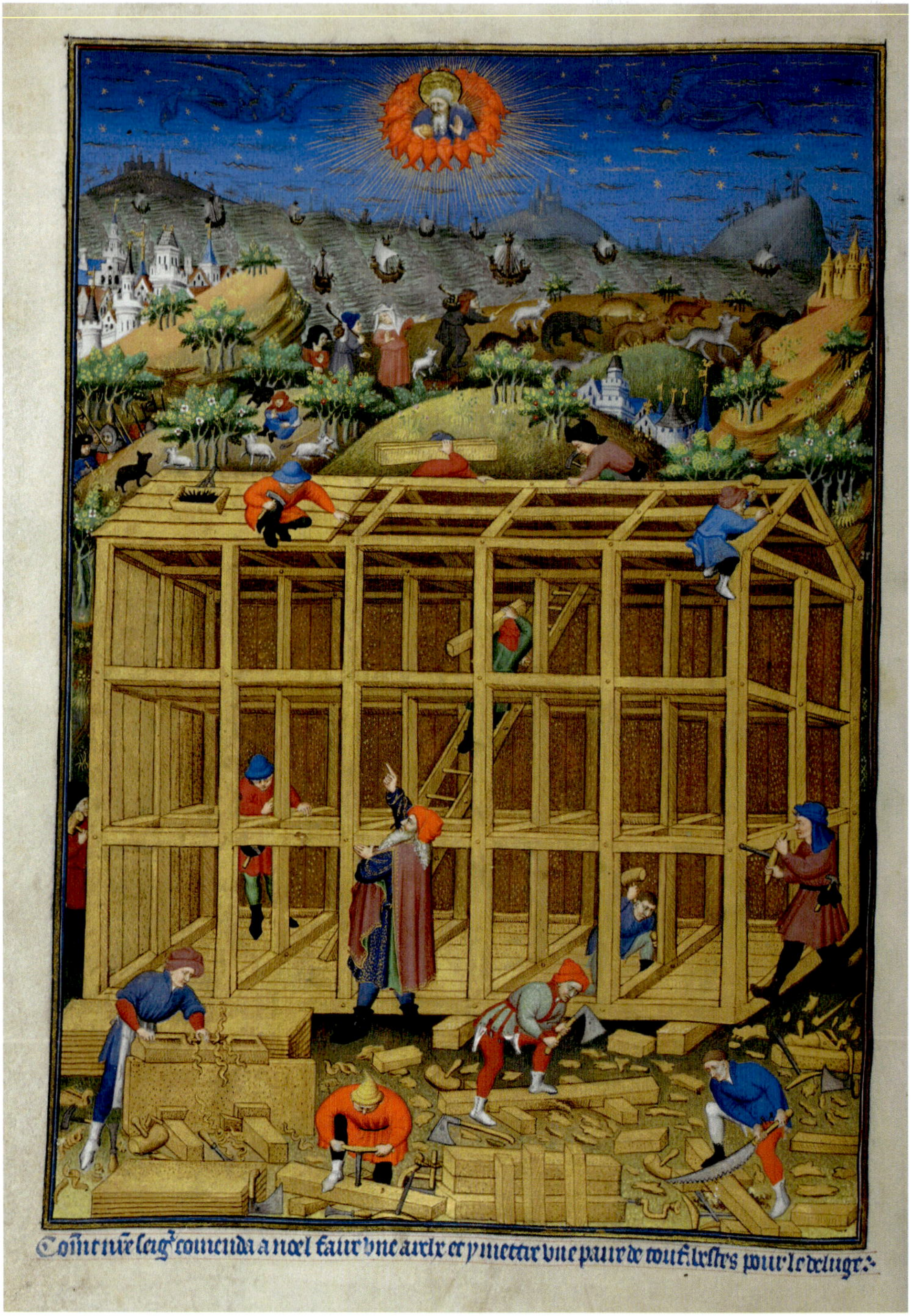

Figure 75: Bedford Master, *Construction of Noah's Ark*, c.1410-1430, from a book of hours of the use of Paris (The Bedford Hours), Add MS 18850, f. 15v. © The British Library Board, London.

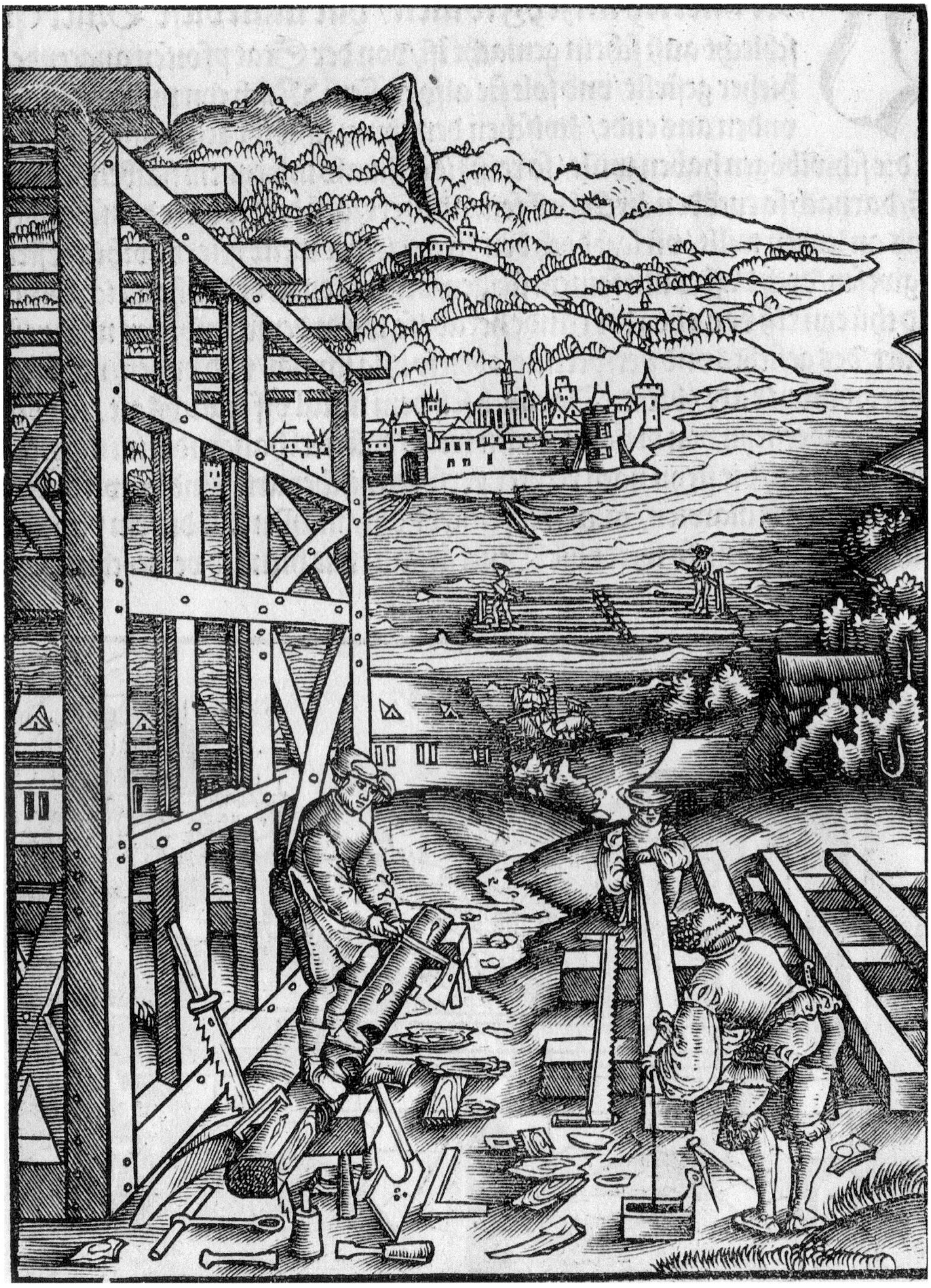

Figure 76: Hieronymus Rodier, Germany, d. 1539, *Perspective: Carpenters Building a House*, woodcut published 1531. Württembergische Landesbibliothek, Stuttgart, Germany.

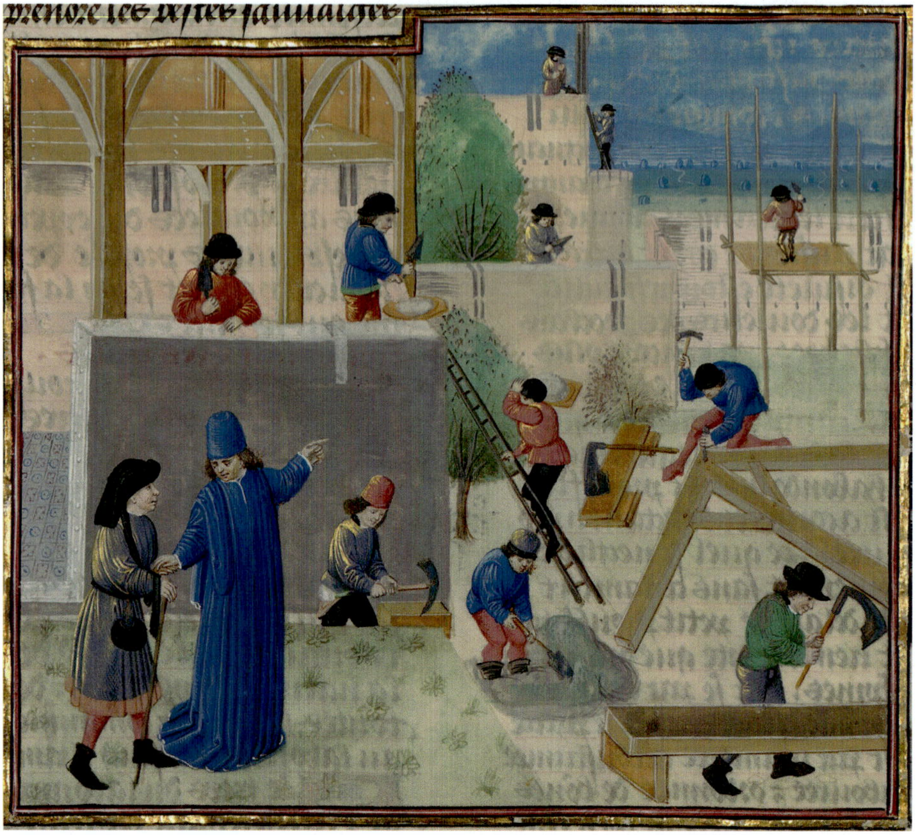

Figure 77: *c.*1480, The Master of Margaret of York, Flemish, active 1470-1480, *The Construction of a Villa*, from Pierre de Crescens, *Rustican*. Arsenal MS. 5064, f. 2. Paris, Bibliothèque Nationale de France.

cut using a crosscut saw, resembling a modern pruning saw. Lying on the ground amongst this activity and pieces of timber and debris are several tools: Augers, a brace, mallets, axes, chisels, a claw hammer, a bow saw and at the extreme left and right of the image two small smoothing planes with scroll shaped handles. In the midst of all this work stands Noah explicitly addressing, it would seem, the carpenter perched precariously on the roof; he may be taken to assume the role of the supervising master carpenter.

A similar image of a century later, and worth considering alongside the Bedford Master's *Noah's Ark*, is Hieronymus Rodier's woodcut *Carpenters Building a House*, published in 1531. (Fig. 76) This is within a book on perspective and shows a similar building but viewed along one side in perspective.[56] Here a carpenter stands on a log, which he is the process of squaring; he cuts notches with an axe in the rough side at centres that he will have marked out and all to the same depth. Once this is complete he will cut the remaining excess to form a straight side, this is repeated by

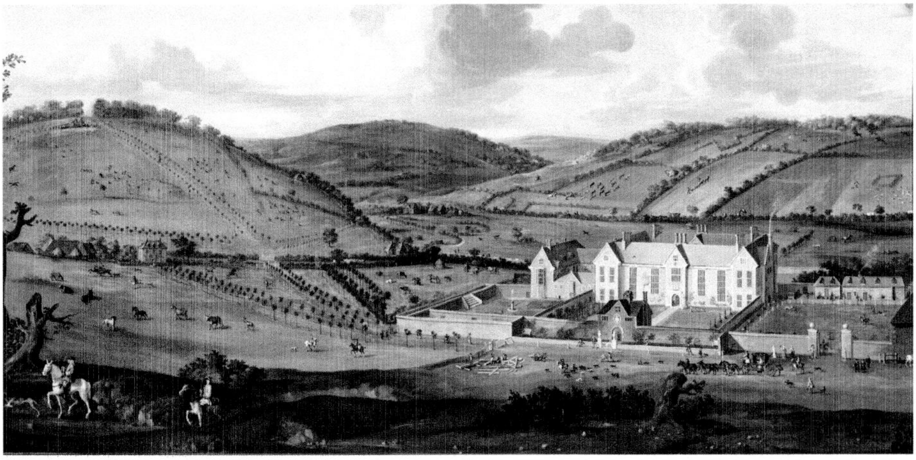

Figure 78: English School, early eighteenth century, *Prospect of Littlecote House from the South*, oil on canvas. (1.269) © Royal Armouries.

turning the log 90° to prepare the next face until he has prepared a square section timber. The log is held off the ground on stools with curved ends and fixed with dogs driven into the stool and log.[57] The two men to the right appear to be marking a timber with a chalk line; a long string coated with chalk, here contained in a box filled with chalk dust. It is stretched taut and pulled upwards at the centre, and when released it strikes a straight line on the surface by leaving a chalk line.[58] Adjacent to them lying across the supporting timbers is a two-man saw. Again various tools are seen in the foreground; leaning against the foremost stool is a socket or paring axe; a crosscut saw leans against the building; on the ground, an auger, a try-square, chisels, a mallet and by the feet of the man on the right a pair of dividers.

In the French manuscript of the third quarter of the fifteenth century, *Livre des profits champêtres et ruraux ou Rustican*, another version of the treatise on agriculture and rural tasks written by Petrus de Crescentiis,[59] seen in Figure 36, here illuminated by the Flemish artist known as the Master of Margaret of York.[60] We see here work on the construction of a rural manor house or villa. (Fig. 77) As previously described, the book is divided into several parts, of which Book 1 describes the best location and arrangement of a manor, villa or farm, hence this illustration. Here carpenters are working on preparing timbers and setting out, and assembling a roof truss; a carpenter is driving in a peg at the apex of the truss. It was the norm for such work to trusses and frames of timber framed building to be carried out on the ground. The frame would then be hauled upright, or in the case of trusses, craned into position. In the foreground another carpenter is preparing a timber using an adze, and an axe rests in another piece of timber beside the other carpenter. Various masons are also busy at work on various parts of the building. The client, the author of the book,

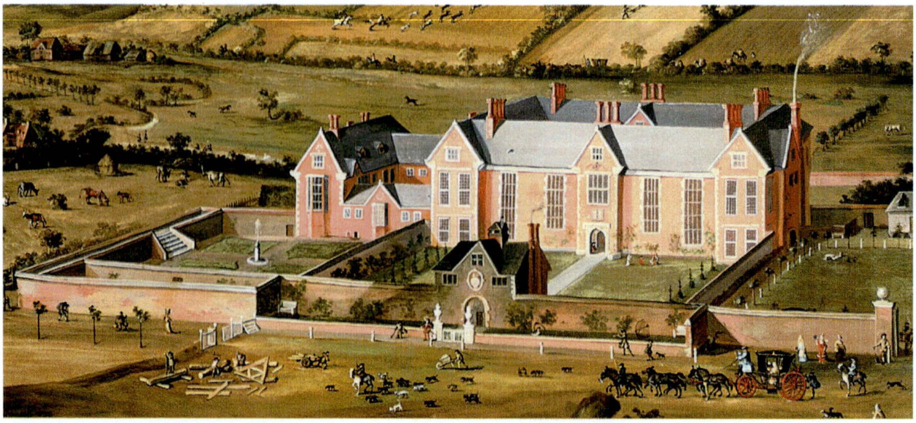

Figure 79: English School, early eighteenth century, details, *Prospect of Littlecote House from the South*, oil on canvas. (1.269) © Royal Armouries.

in the long flowing robe and the master carpenter or clerk of works are seen in the left foreground. Another image in the manuscript, on Folio 282, shows the 'author explaining the properties of places conducive to human settlement'. Here we see the unfinished timber framed part of the settlement, but with no one working on it.[61]

In the previous manuscript of the same date, entitled *LE, LIURE de Rustican des prouffiz ruraulx, compile par Maistre Pierre Croissens, Bourgoiz de Boulongne.* by the Bruges Master of 1482,[62] seen in Figure 36 and discussed for its scaffold on page 78 above, we see two carpenters sawing a timber using a two handed saw. This is a framed pit saw but here the timber is set up on a trestle at one end and a pair of steps at the other. This is precarious for the man on top; where this type of cutting was to be done in bulk a pit or trench would be dug in sound ground and the timber laid lengthways along the top of the pit. The top man would then be at ground level with his partner in the pit. Towards the foreground another carpenter works a length of timber with an axe. As well as the building operations described on page 78, workers are digging out a ditch and others are erecting fencing. In the foreground stand the Lord of the Manor, or again, a representation of the author, with his architect or clerk of works who holds a large level, the precursor of the modern spirit level.

A pit saw is seen in the English School painting *Prospect of Littlecote from the South*, 1701-1730, on the Berkshire Wiltshire border. (Fig. 78) Here carpenters are setting out roof trusses with the pit saw to the left of these activities; the second man is out of sight in the pit below the large timber, with his partner on top. (Fig. 79) These trusses and other timbers are destined for an estate building, but given the extent of the prospect seen, it must be some distance away. So why do this work here in front of the house, which is a strange place for such work? Some of the

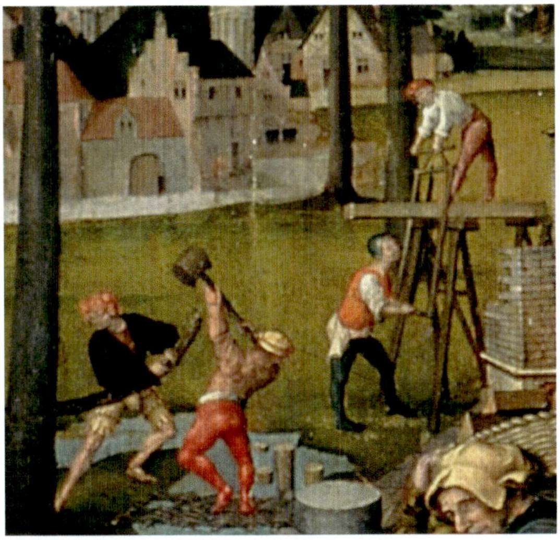

Figure 80: Detail of Figure 161, 1500-1533, Northern French, *Construction of the Abbey of St Bertin at St Omer*, oil on wood panel, lower panel of the right hand leaf of the doors to the treasury of the Abbey © Musee des Beaux-Arts de Dijon. Photo Hugo Martens.

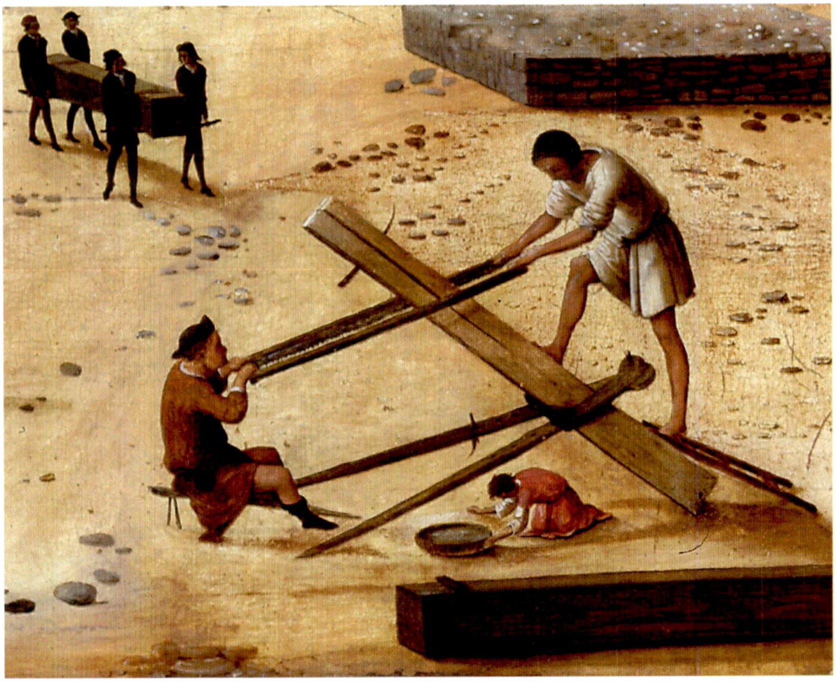

Figure 81: Detail of Figure 165, 1515-1520, Piero de Cosimo, Italian, 1462 -1521, *The Building of a Palace*, 83 × 197 cm, oil on panel, The Ringling Museum of Art, Sarasota, Florida, USA: Bequest of John Ringling, 1936, Collection of The John and Mabel Ringling Museum of Art, the State Art Museum of Florida, Florida State University.

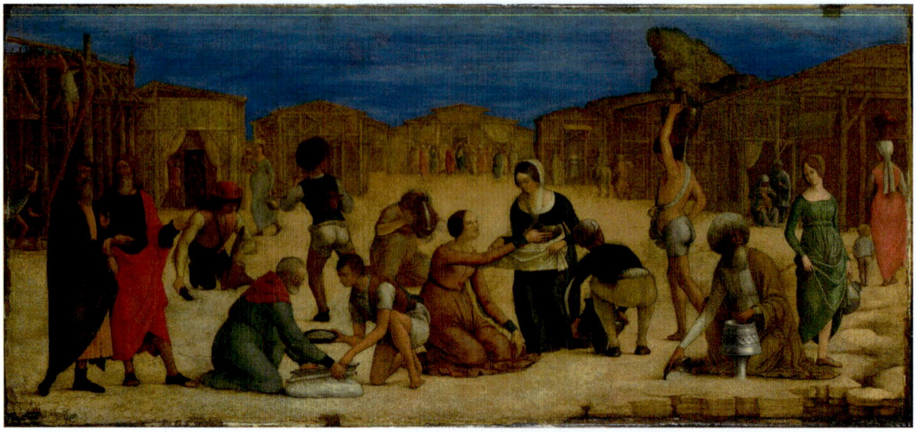

Figure 82: Ercole d' Roberti, Italian, active 1479-d.1496, *The Israelites gathering Manna*, one of two panels from a Predella, 28.9 × 63.5 cm, tempera on canvas, transferred from wood. (NG 1217, London, The National Gallery)

prepared timbers are being hauled away on a hand-cart by two men. To the right of the detail (Fig. 79) above the horse drawn carriage, within the internal angle of the high wall surrounding the house, is a whetstone for sharpening tools.

Figure 80 shows another two-man saw being used to ripsaw a large timber on trestles.[63] Here also is a man felling what looks like an Oak tree, a 'timber' tree suitable for converting into elements, possibly for the construction of the roof. A similar situation with a two-man framed saw is seen in Figure 81. Here the timber is upended and lashed through an 'A' frame trestle resembling a giant pair of dividers, so that one end of the timber rests on the ground with the other end in the air. One man stands with his forward foot on the sloping timber and his other resting on a small wooden step-ladder to stop him slipping backwards; his partner crouches or sits on a stool below the timber. A child is seen below this operation collecting sawdust. The large squared baulk of timber is being quartered into four; it has already been sawn through one face to about half way, and once they reach the same extent along the length the whole process will be reversed and they will cut through from the other end to complete the operation. The spike or wedge inserted into the kerf is to keep it apart and stop the saw binding.[64] Another baulk of timber lies on the ground awaiting the same process whilst a further one is being carried to where the carpenters are working. Two other carpenters are seen in Piero de Cosimo's painting, to the left above the two oxen, preparing large timbers using the adze. For a fuller commentry on this painting see page 220 below and Figure 165.

The Italian Renaissance master Ercole de' Roberti in his painting *The Israelites Gathering Manna*, depicts in the background their sturdy but temporary houses.[65] (Fig. 82) Standing to the left, behind Moses and Aaron, one of these structures is

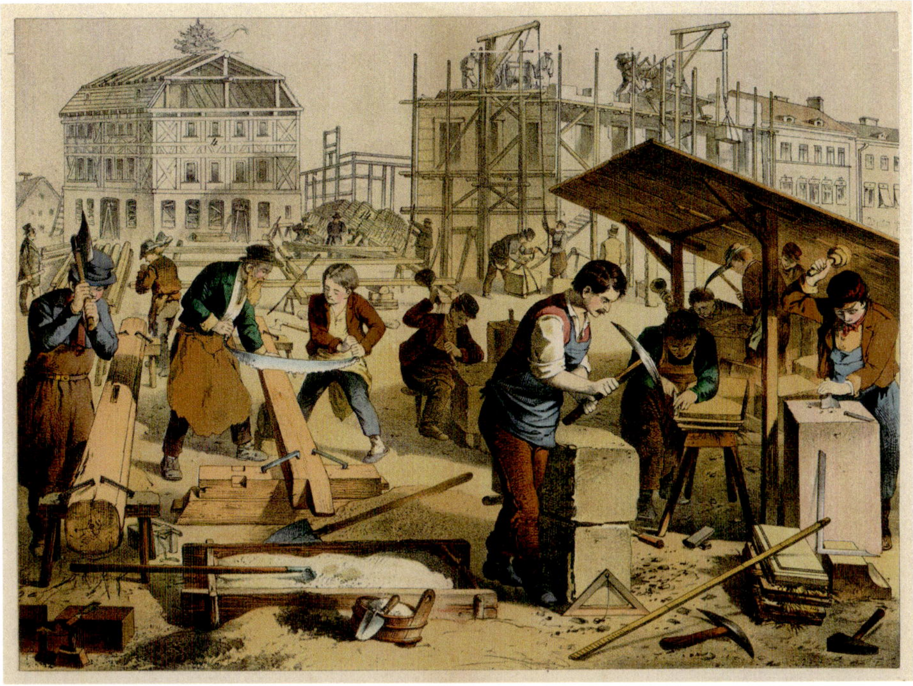

Figure 83: 1876, Illustration from F. Streich and K. de Gerstenberg, *Méthode d'enseignement par l'aspect les métiers et leurs outils*, chomolithograph, published 1876, Paris, editions Bonhoure. Mary Evans Picture Library.

in the course of construction. These are robust timber frames with wattle-like outer walls and internal divisions, similar to 'clays'; some are also covered in fabric. These are temporary structures for the Israelites who were condemned to wander for forty years in the desert (where did they find all this timber in the desert?) before gaining the Promised Land.[66] Two carpenters are working on the building to the extreme left, one up a ladder using a mallet to knock a timber into place; the other at the base of the ladder, but it is not clear what he is holding nor what he is doing with it.

F. Streich and Dr K de Gerstenberg's *Méthode d'enseignement par l'aspect – les métiers et leurs outils*, published in Paris in 1876, is a book that instructs by its illustrations.[67] A coloured lithograph (Fig. 83) shows a large building site, a residential development of mixed half-timbered and masonry houses. This progresses along the street in the middle ground from right to left, and beyond this is a partially finished part timber-framed building. This has been completed to roof level and has been 'topped out', signified by the tree on the ridge, or it may be Christmas?[68] Immediately in front of the house timbers are being set-out for a further frame, a partially erected scaffold is seen beyond, to the left is a stack of timbers and the

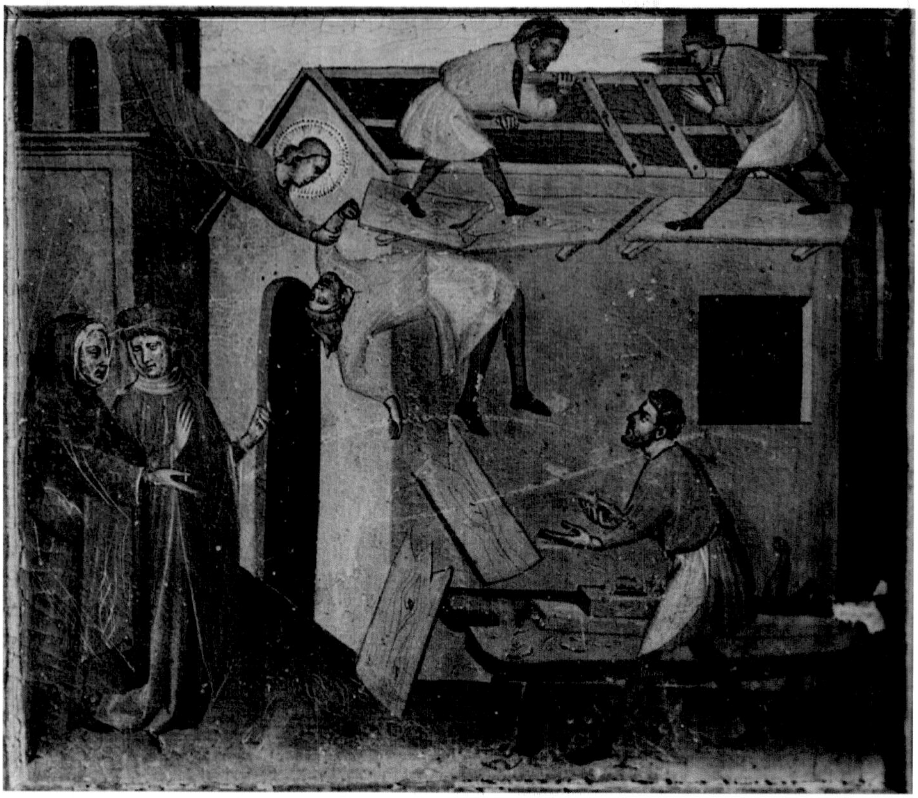

Figure 84: *c.*1402, Lorenzo di Niccolo, Italian, active 1391-1412, *St Fina saves the life of a carpenter who fell from the roof of a church he was building*, oil or tempera on wood panel, lower right-hand panel of the Santa Fina Altarpiece, Museo d'Arte Sacra, San Gimignano, Italy. Archive photograph, Fratelli Alinari, founded 1852. Fondazione Zeri, Universita di Bologna, Italy.

supervising architect has arrived carrying his rolled up drawings. In the foreground carpenters are at work including a master and his apprentice cutting a timber with a two handed crosscut saw. The use of 'dogs' for securing the timbers in place whilst being worked on is evident. To the right are masons, some under cover working away at blocks and mouldings with their tools clearly delineated; in the foreground is a large timber box where mortar has been mixed, with a hoe in the box and a shovel to the side. In front of this, waiting to be taken to the mason fixers or bricklayers on one of the buildings, is a hod of mortar with a mason's trowel. This is a book that educates through its illustrations; this image should be compared, for the carpentry, with Figures 75 and 76. What it teaches us on the cusp of the twentieth century, is that nothing substantial has changed; all is still human powered and the tools virtually identical to those of the preceding centuries.

Accidents and Misadventures

Working on a building site was unsafe and working at height on a scaffold particularly so; indeed, it still is. There were accidents and they must have been numerous. One of the most notable recorded was William of Sens at Canterbury in 1180. William, a French master mason, had worked at the Abbey of St Denis, begun in 1140. He was persuaded to come to England to supervise the rebuilding of the choir of Canterbury Cathedral, which was destroyed by fire in 1174. Gervase was a monk of the cathedral priory and famously records: 'In the year of grace 1174, by the just but occult judgement of God the church of Christ at Canterbury was consumed by fire.' The fire was contained within the choir, the most important part of the church, with the main body being left intact. Gervase goes on to record how the brethren under their Archbishop Richard went about selecting a builder to reconstruct the choir, and later continues: 'And having completed on both sides the triforia and upper windows, he was, at the beginning of the fifth year, in the act of preparing with machines for the turning of the great vault,[69] when suddenly the beams broke under his feet and he fell to the ground, stones and timber accompanying his fall from the height of the capitals of the upper vault, that is to say fifty feet.'[70] The French master mason never recovered and returned to France, where he died soon after.[71]

Lorenzo di Niccolo, a Florentine painter active 1391-1412, depicts a carpenter falling from a scaffold in his *Santa Fina Altarpiece*. This takes place at a church where he is working on the roof. (Fig. 84) The scaffold is boarded and supported on putlogs only; there is no visible access. The carpenter stands on one board, which is splitting across its width, and grabbing a purlin to try and stop himself falling. But this does not work and we see him, in the same scene, having fallen with the board in two pieces below him. A carpenter on the ground working at a bench reaches out in an attempt to help break his fall. However the miraculous intervention of St Fina rescues him mid-fall. St Fina is venerated in the town of San Gimignano in Tuscany where she was born in 1238.[72] A possible explanation of this iconography is that such an accident did take place in San Gimignano during the construction of a church; the carpenter survived the fall and subsequently offered thanks to St Fina. Thus resulting in the local adoption of a miracle to explain the unlikely survival. On the carpenters bench is an adze and a block plane.

In Antonio Vivarini's *Saint Peter Martyr Healing the Leg of a Young Man,* (Fig. 85) a young carpenter has deliberately cut his leg bady with his axe in a fit of remorse at having kicked his mother. His axe lies on the piece of timber he was working on; a try plane and a bow saw hang on the wall behind. Although a carpenter's workshop, it may be one attached to or associated with a building site, such as a mason's lodge which was invariably found on large cathedral building sites. The panel is one of a number of scenes from a dismembered altarpiece.[73] Saint Peter, of Verona (1206-1252, canonized 1253), was himself martyred with an axe.

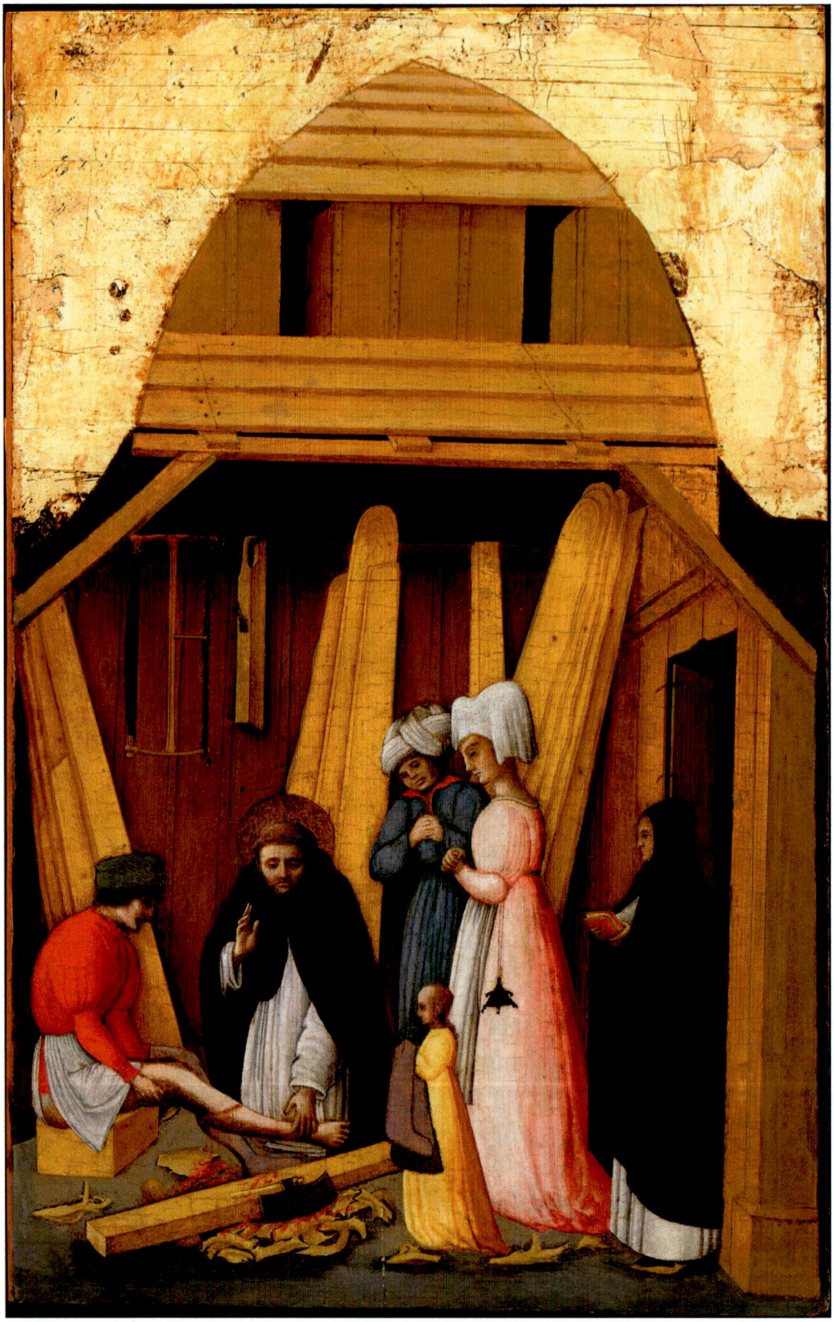

Figure 85: 1450s, Antonio Vivarini, Italian, active by 1441-d.1476/84, *Saint Peter Martyr Healing the Leg of a Young Carpenter*, 53 × 33.3 cm, tempera and gold leaf on wood panel, The Metropolitan Museum of Art, New York, OASC.

PART 2

Cranes, Hoists and other Lifting Equipment

The development of the crane is based on the pulley, which together with the windlass, or winch, were known to Aristotle: '[I]n building construction they can easily raise great weights; for they shift from one pulley to the other, and again from that to capstans and levers; and this is equivalent to making many pulleys.'[74] Archimedes is credited by Plutarch with the invention of both the triple and compound pulley systems utilising the block and tackle to lift heavy objects[75] and for use in military machines.[76] He invented a large and powerful crane-like machine for the defence of Syracuse, his home city, against Roman naval attack, during the siege of Syracuse. Known as the 'Claw of Archimedes' and said to be capable of lifting the Roman ships out of the water, it is described by Plutarch:

> Some of the [Roman] ships were seized with iron hooks, and by a counter-balance were drawn up and then plunged to the bottom. Often was seen the fearful sight of a ship lifted out of the sea by the claws into the air, to be dashed against the fortifications or dropped into the sea by the claws being let go.[77]

Vitruvius goes on to describe different types of crane for the lifting of heavy burdens in his *Ten Books on Architecture*,[78] in Chapter 2 'Cranes and Hoists', Book 10, 'Machines', which includes the following crane.

Roman Cranes

The Romans made significant developments in lifting technology, as can be seen on the second century tomb of the Haterii family, the remains of which are in the Vatican Museum (Fig. 86). One at least of the Haterii family was a builder responsible for many significant buildings and monuments during the Flavian period in Rome. Several of these buildings, including the Colosseum and the Arch of Titus, were portrayed on their mausoleum in Rome. This was originally located on the *Via Labicana*, but is now destroyed. On one of the two surviving stone reliefs is a depiction of a Roman tread-wheel crane so compelling in its accuracy as to enable a full size working replica to be made.[79] This is at present located in a public park in the Castell district of Bonn in Germany.[80] The crane is a clear reference to the Haterii's construction business and is shown in its own right; it is as much to be

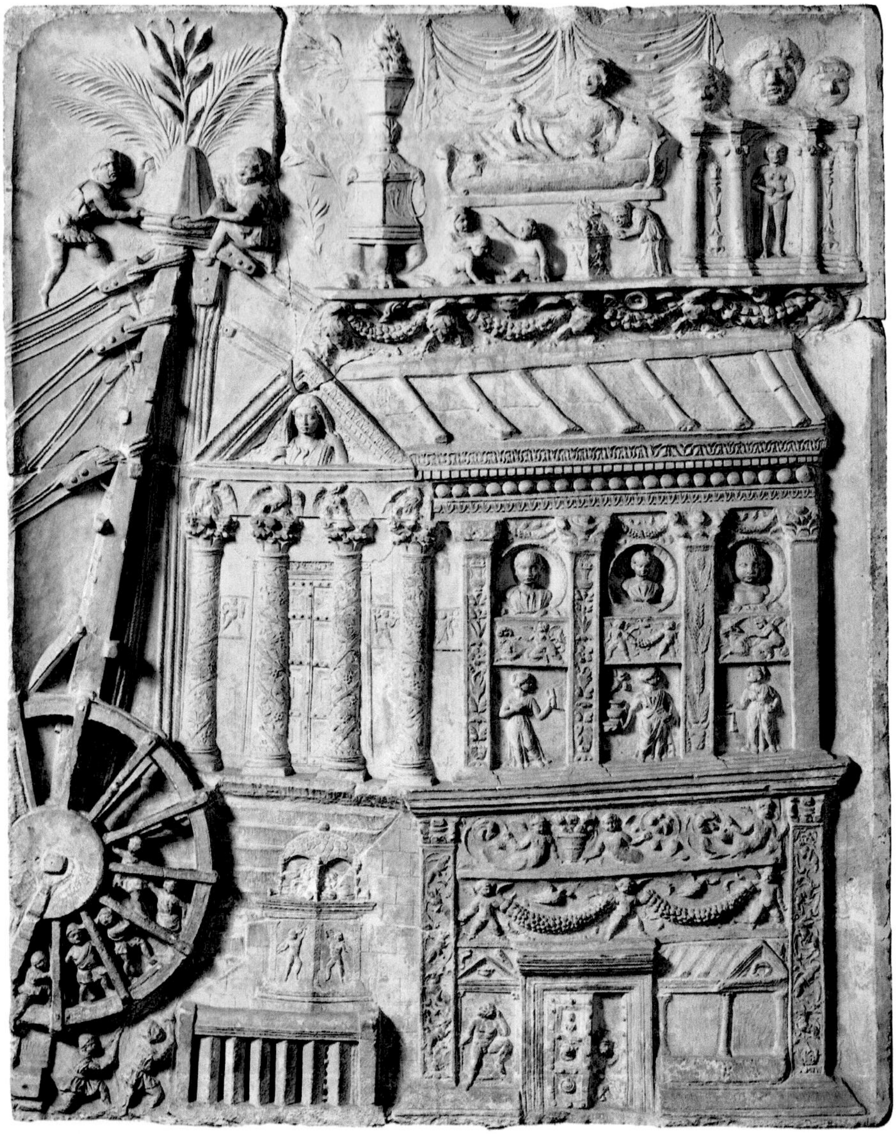

Figure 86: Second Century, Rome, Italy, one of two surviving reliefs from the Mausoleum of the Haterii showing a tread-wheel crane and tombs, originally located on the Via Labicana, now destroyed, carved stone. Vatican Musuems: Greforiano Profano Museum, Cat. 9998.

celebrated as a memorial to the family as the buildings which are also portrayed. It is difficult to ascertain whether the crane is either in the process of being rigged or about to be dismantled.[81] It is unlikely to be engaged in a lifting operation because of the overcrowded tread-wheel; were it rotating men would fall one on top of the

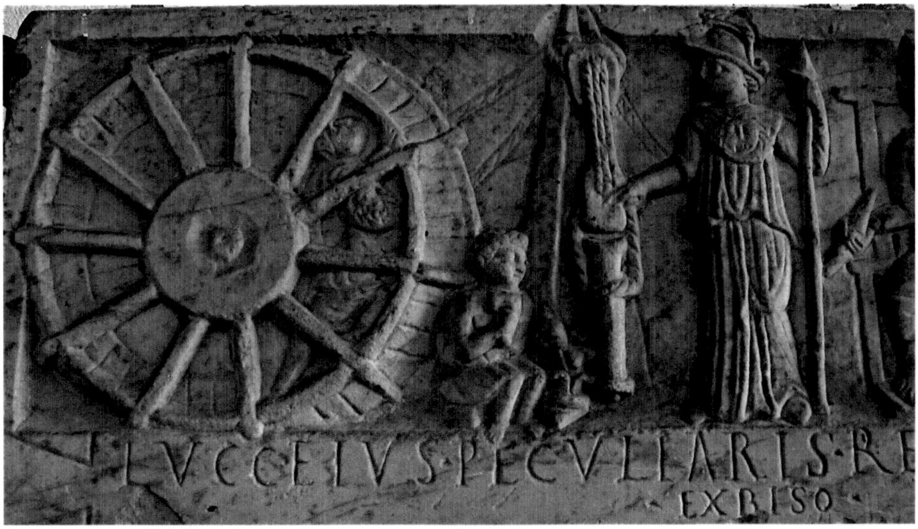

Figure 87: Capua, Italy, Second Century, detail, carved stone relief, part of a sarcophagus, showing a tread-wheel crane lifting a column with a mason carving a capital. Museo Provinciale Campano di Capua, Sarcofagi (Sala III-IV). Photograph: Dan Diffendale.

other and create a catastrophic failure. The two men at the top of the jib appear to be attaching, or detaching, a basket over the head of the jib, possibly to protect it, or the finished work, from an accidental collision. They may also be tying on the foliage and branches visible at the top left to indicate completion of the project as some form of topping out ceremony.[82] It is entirely in keeping that the crane should be shown at the point of celebration on the completion of the building project. That such cranes were dismantled and transported to other sites must be beyond doubt. They were valuable machines and could be readily broken down into component parts. The tread-wheel, for instance, once separated could easily be transported on a cart, or even carefully rolled to the next site. The crew, who were highly trained and skilled workers, would have travelled with it. The relief serves to illustrate that one of the common construction lifting machines of the Medieval and Renaissance periods had its origins in the Roman period, although they in turn had the technology from the Greeks. Vitruvius gives an account of this type of crane. Where he briefly describes the treadwheel, he gives two Greek words: 'Some call this a wheel, some Greeks an *amphiesis* and others a *perithêkion*.' Thus he gives a clear indication of it's Greek antecedents.[83] It is certain that such equipment originated as military machinery for siege and other warfare technology.[84]

Another Roman relief dating to the first half of the second century showing a tread-wheel crane, is from Capua in the province of Caserta. The crane is in the process of lifting a column, and two people are operating the tread-wheel, although probably one would be sufficient for most purposes. (Fig. 87) The relief

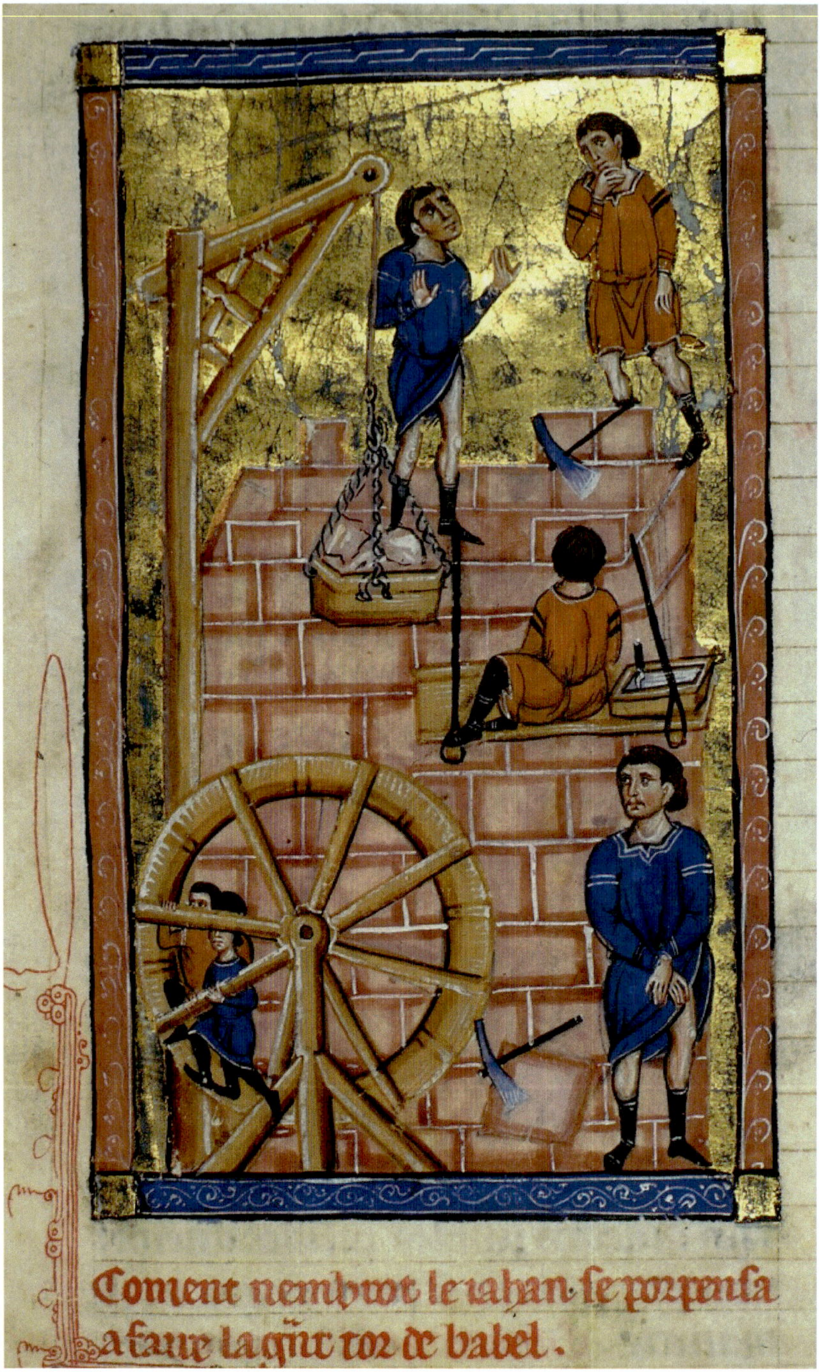

Figure 88: 1260-1270, Holy Land, Acre (before the fall of Acre in 1291) *Construction of the Tower of Babel*, from 'Histoire Ancienne jusqu'à César', ms. 0562, f. 009, Bibliotheque Municipale de Dijon, France.

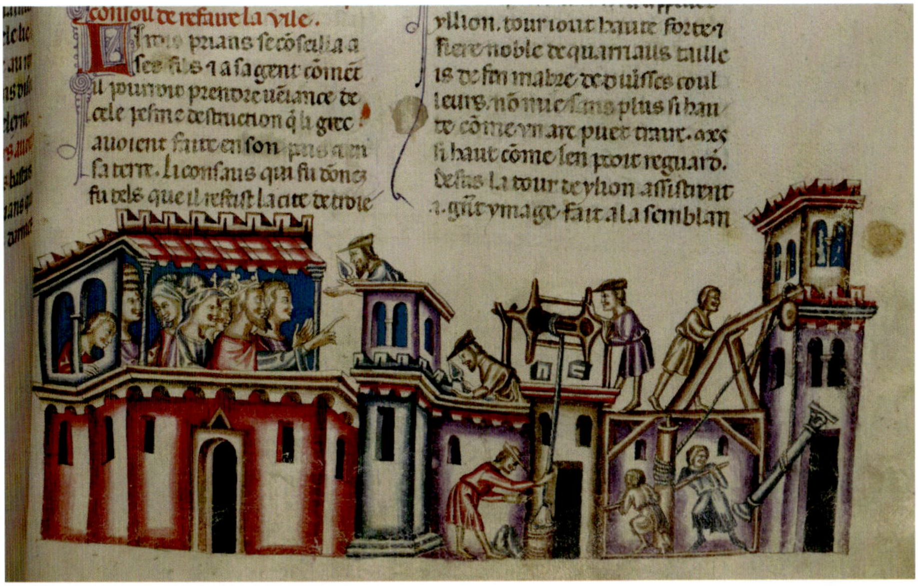

Figure 89: *c.*1330-1340, Italian, Naples, *Rebuilding of Troy*, illumination from *Histoire Ancienne Jusqu'à César*. Royal MS 20 DI, f. 41r. London, The British Library.

commemorates the construction of the stage at the Roman theatre in Capua and is said to be part of a sarcophagus.[85] It bears the inscription *'Lucceius pecularis redemptor prosceni'*.

Medieval and Renaissance Cranes

There are arguments, due partly to the lack of documentary sources including images, that the crane, particularly the tread-wheel crane, fell out of use after the Roman period and was not reinvented until the mid-thirteenth century, coinciding with the date of the earliest Medieval images. The other inference being that stone buildings of the period were constructed of small blocks easily carried by hand and that cranes were not necessary.[86] A brief study of Peter Purton's *The Medieval Military Engineer*, proves that this is manifestly not the case.[87] Also one has only to look at early buildings for further proof. For instance the mausoleum of Theodoric at Ravenna of the mid-sixth century, a two storey structure with a dome fashioned from a single block of Istrian limestone some 10.6m accross,[88] and the Anglo-Saxon church of St John at Escomb in County Durham, England built in the eighth century. Here are massive stone elements incorporated into its construction. Most notably the quoins, the original Anglo-Saxon window heads, jambs, cills and the chancel arch. All of which would have required a crane, windlass or other mechanical device

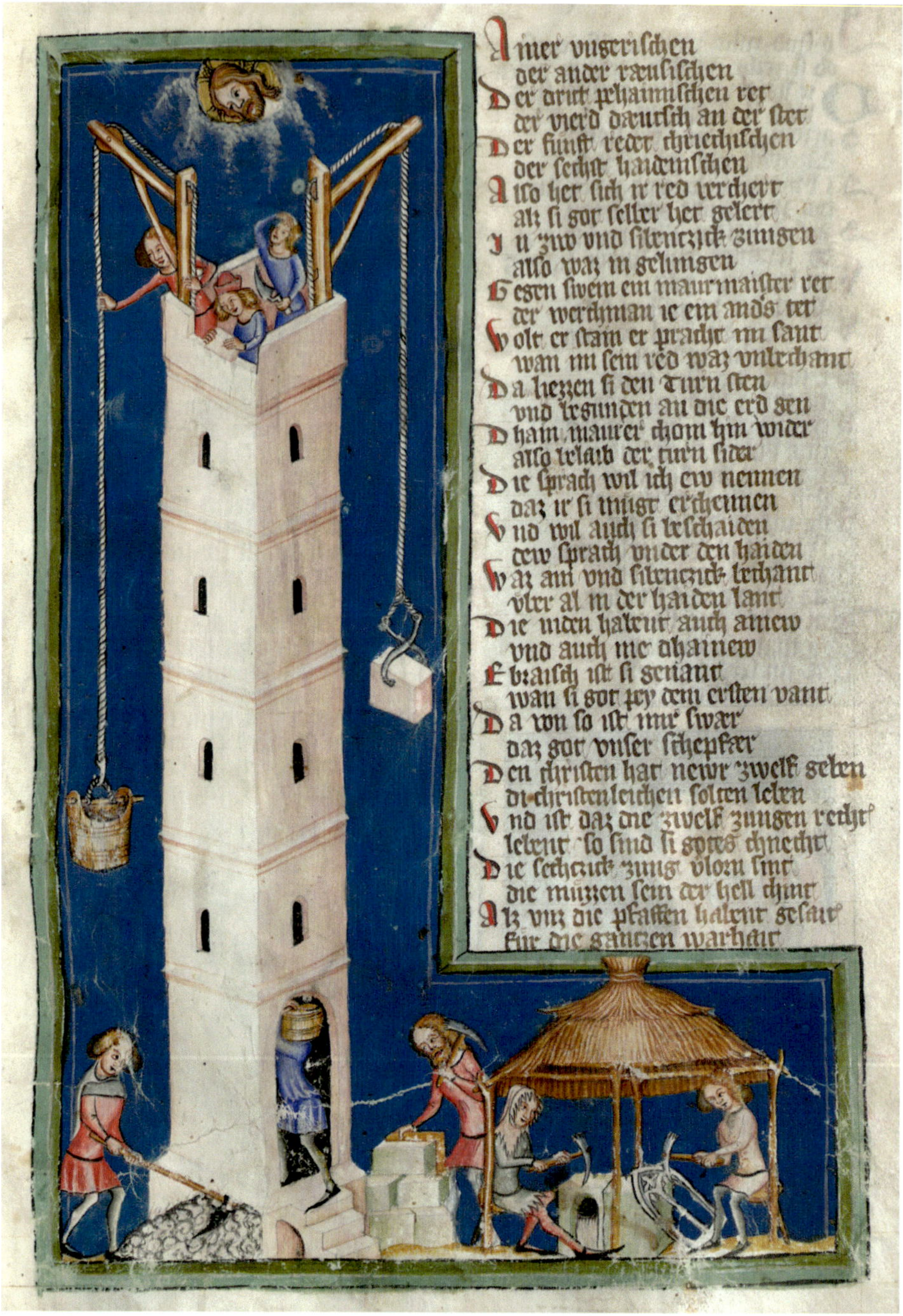

Figure 90: German *c.*1370, *Construction of the Tower of Babel*, from Rudolf von Ems / Jansen Enikel, Jans: *Weltchronik in Versen – Mischhandschrift aus Christ-herre-Chronik*. Bayerische Staatsbibliothek München, BSB Cgm 5, f. 29r.

to lift them into place. St John at Escomb is by no means unusual in incorporating large blocks too heavy for one man to lift, or even two men on a timber ramp.[89] The argument also ignores large and heavy timber elements such as roof trusses, normally assembled on the ground and then hoisted into place. Parts of two early tread-wheels or hoists that predate the earliest Medieval images are extant in England, in Tewkesbury Abbey and Peterborough Cathedral.[90] The former is in the crossing tower begun after 1087 and completed *c.*1150. The evidence at Peterborough is in the north tower of the west end, commenced 1177 and completed 1238.[91] There is no reason to believe that these are not coeval with the early construction phases of their respective sites and are therefore pre-thirteenth century.

So mechanical means of lifting on building projects continued through these 'dark ages'. It is just that there are scant documentary sources and no images, or none discovered yet, until the 1240s, after a gap of some 1,100 years. In a French manuscript, known as the Morgan bible dated *c.*1240, is an image showing a single operative working a tread-wheel.[92] (Fig. 7) This lifts a load via a detached inverted 'L' shaped jib, comprising post, jib and brace, to masons working at the top of a tower. A banksman, just visible to the right, is steadying the load with a guy-rope. The tread-wheel is located on an external platform approximately two thirds of the way up the tower. Here we must assume a scaffold structure, left out for the sake of clarity. Another depiction of slightly later date, 1260-1270 from Acre in the Holy Land by a French hand is *The Construction of the Tower of Babel.* (Fig. 88) This shows a very similar tread-wheel and jib arrangement. The juxtaposition of jib and treadwheel in both of these two early images is awkward. The rope, clearly seen suspended from the jib in Figure 88 with the load, would need to be led back along the jib to the post, and down that to the winch, which is the axle of the treadwheel, all via a series of pulleys, as is shown in Figure 7. In Figure 88 there are two or three operatives in the wheel, three heads and two pairs of legs, with the latter floating outside the wheel, whereas in the earlier image there is just one, which is perhap more realistic. The Medieval illuminator does not always achieve a high degree of detail and accuracy, nor should we seek scale or perspective in these early images. These developments came later. Nevertheless these artists have carefully observed the process of construction, and would have observed and recorded what was going on and sketched it down before working up their images in the studio or scriptorium. There are bound to be errors and technical accuracy was not their main aim. Again in Figure 88 scaffolding is virtually non-existent for the sake of clarity and to show clearly the main actors in this biblical operation. Figures 19 and 20 show similar jibs. In the depiction of Charlemagne and the construction of Aix-la-Chapelle, or Aachen (Fig. 19), the winch is hidden behind the building work. While in the depiction of King David and the Temple of Jerusalem (Fig. 20) there is a hand-operated windlass but there is some confusion with the ropes; to the right a rope rises to the pulley at the front of the jib and thence down to the load with another

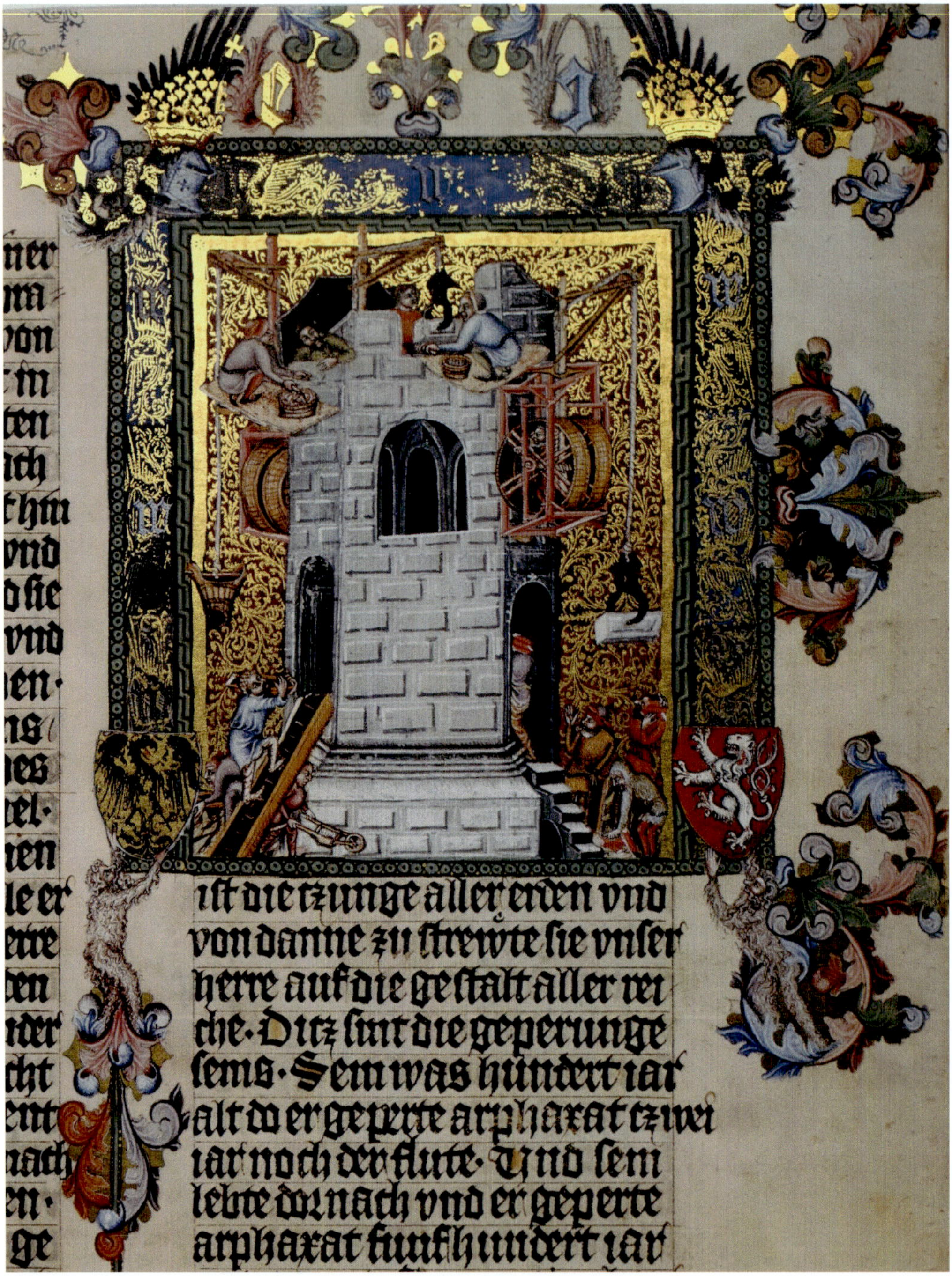

Figure 91: 1390-1400, German, *Building the Tower of Babel*, Wenceslas Bible, Commissioned by King Wenceslas IV of Bohemia, 1361-1419. Cod. 2759, f. 10v, Osterreichische Nationalbibliothek, Vienna.

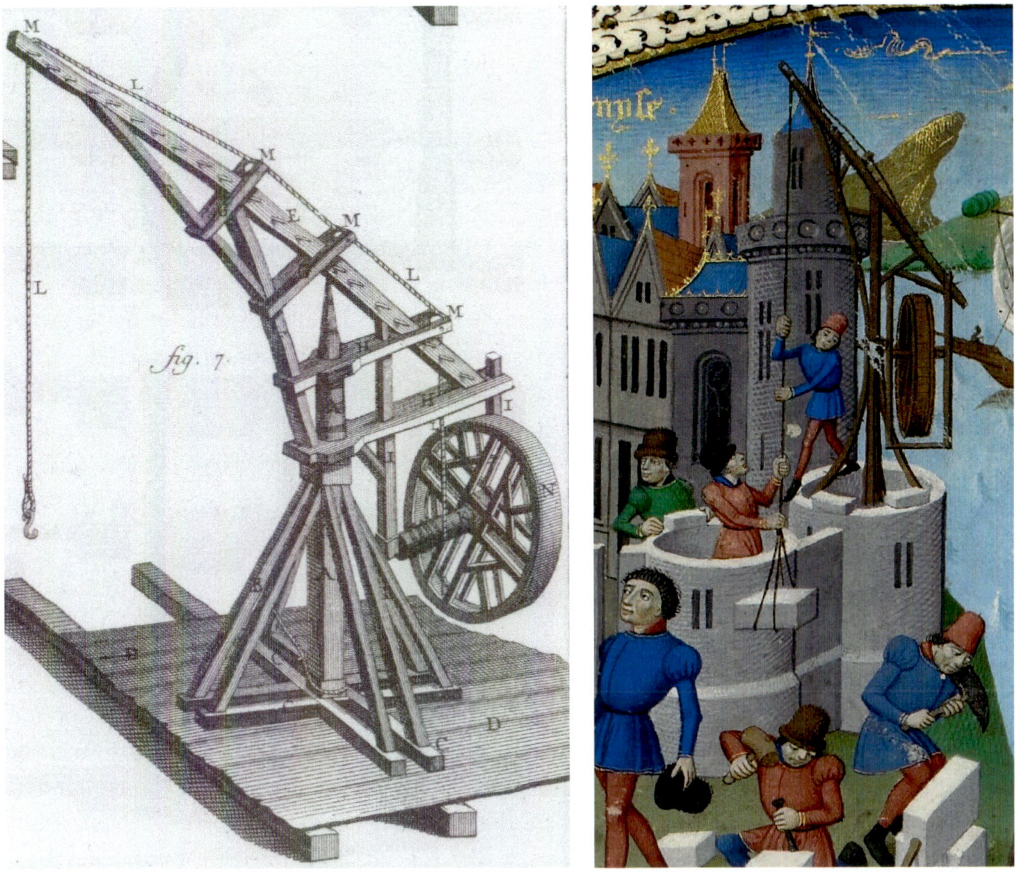

Figure 92: Left: The Perronet crane of the mid eighteenth century, from Diderot, Denis and le Rond d'Alembert, Jean-Baptiste, editors, 1751-72, *Encyclopédie, ou Dictionnaire raisonné des sciences, des arts et des métiers, etc.*, Paris, fig. 7, plate XLVII. Right: Detail of Figure 139 showing the same type of crane in use in the late fifteenth century.

rising over the rear pulley and then dropping behind the operator. This second rope has no function and the artist has drawn it incorrectly. What should happen is that this second rope should pass through the back of the rear pulley and run along the jib to pass around the front pulley and down to the load; there should only be one rope in this operation, the same as in Figure 19.

Other simple pulley and rope arrangements are seen in Figures 16 and 21, respectively from the thieteenth and fifteenth centuries. The earlier image shows a hybrid double pulley on a cross beam supported off a forked pole or mast, with two men hauling on the rope one above the other. There must have been scaffold here, which would have supported this pole, but again it has been left out for clarity. In the later image (Fig. 21) a single pulley and rope is suspended off the scaffold. Such a system is almost universal, right up to the present day on small and medium construction sites in the United Kingdom and is known as a 'wheel and rope' or 'gin-wheel and rope'. They are only suitable for light and medium lifting and are often used for erecting scaffold.

In the Italian manuscript of the first half of the fourteenth century, *Histoire Ancienne Jusqu'à César*, looked at above on pages 48 & 49 and Figures 11-14,[93] and here in the image *Rebuilding of Troy*, (Fig. 89) we are confronted with a very strange hoist and a fairly standard windlass. The hoist or crane has a very odd vertical capstan type winch around the drum of which the rope is being wound, this passes through a pulley on the end of a simple fixed jib. The load is a column complete with base and capital; this is a heavy burden and the two operating the winch are clearly struggling to handle this strange hoist; there are clearly pulleys missing. The other device shown lifting a bucket of mortar is a simple hand winch similar to that seen in Figure 20, although it is without the complication with the double ropes.

Another unlikely crane is shown in Figure 9. There we see what appears to be two hand operated winches incorporated into a single timber superstructure with a single heavy beam like jib with a pulley at each end oversailing the sides of the tower; each is rigged to its own windlass which are out of sight lower down inside the tower. The ropes can be seen running back into the tower, presumably with its own operator. Thus this hoist lifts each side independently; one side has a basket containing bricks, stone or mortar which is thoroughly plausible; on the other side however is a massive piece of stone the weight of which surely exceeds the capacity of the machine. What is thoroughly confusing however is the single hand operated wheel in the centre being turned by an operative, which appears to serve no useful purpose. What did the artist actually see?

In another version of the German *Weltchronik in Versen*, or World Chronicle in verses by Rudolf von Ems there is an illumination of a slender tall tower like a lighthouse or minaret, which is in fact another image of the Tower of Babel. (Fig. 90) There are two crane jibs at the top, one of which is perhaps lifting a bucket of mortar. The other is lifting an ashlar block of stone by means of a pair of iron dogs or nippers.[94] Such a device was normally used for rough blocks, so there is a risk here of the block slipping and falling if it is too heavy, which it certainly appears to be, although both the block and the bucket are vastly out of scale with the tower itself. There are masons working on the tower, as well as in a masons lodge where one is carving a Gothic window. Such a window would not have been constructed like this as a whole piece, and where also will it go, since all the windows have round arches?

Another German manuscript of slightly later date (1390-1400) is the *Wenceslas Bible*. An illumination in this bible shows a pair of double treadwheel cranes attached to the exterior of the tower in suspended scaffold structures; there may be another on the blind side of the tower where only the jib is visible. (Fig. 91) One of these cranes, the one on the right, was reproduced in 2006, using the illumination as a primary

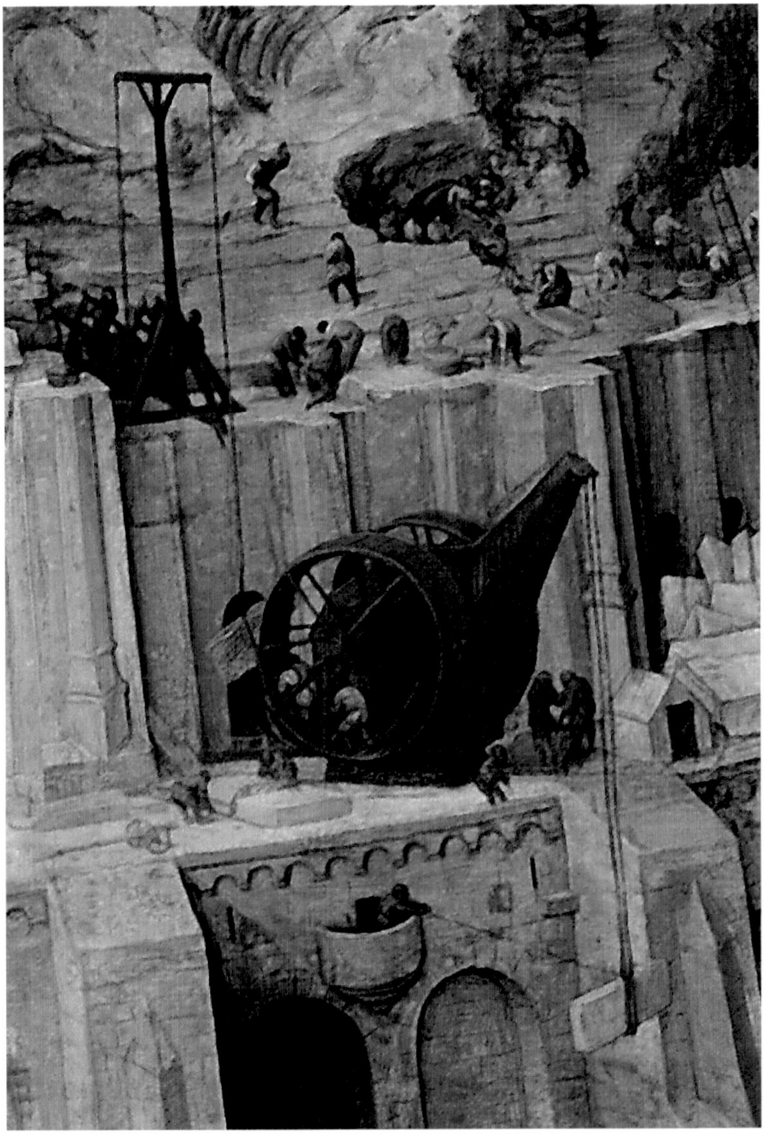

Figure 93: 1563, detail of Figure 166: Pieter Bruegel the Elder, Flemish, 1525-1569, *The Tower of Babel*, Large dockside type crane and simple 'T' hoist, oil on Oak panel. Kunsthistorisches Museum, Vienna: KHM-Museumsverband.

source and utilising Medieval carpentry techniques. First exhibited at Prague Castle it was later moved to Tocnik Castle in Central Bohemia where it was used for repairs to the roof structure.[95] Another similar reproduction crane has been in service, and continues to be in use today at Guédelon in the Yonne district of France.[96]

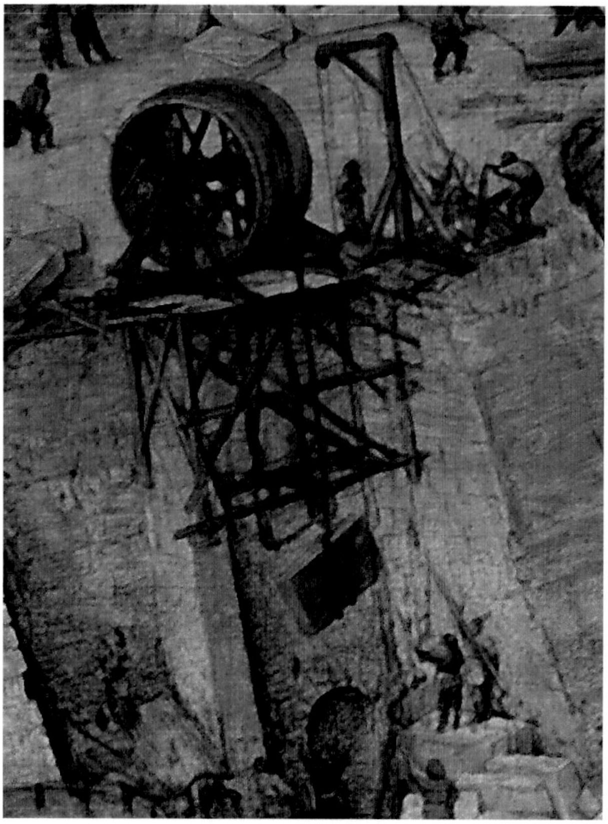

Figure 94: 1563, detail of Figure 166: Pieter Bruegel the Elder, Flemish, 1525-1569, *The Tower of Babel*, treadwheel with simple hoist alongside, oil on Oak panel. Kunsthistorisches Museum, Vienna: KHM-Museumsverband.

Treadwheels were sometimes set up to act on their own without a jib. A case in point is in Pieter Bruegel's *The Tower of Babel* as seen in Figures 166 and 94 (detail) below. (This painting is also discussed at length in the following chapter on masons and stonework.) We have also noted two above, those in Tewkesbury Abbey and Peterborough Cathedral. Several others are also extant in Europe, mainly in cathedrals and large churches.[97] The closest we come to seeing one in art, apart from Bruegel, is Jan van Eyck's *St Barbara*, (Fig 153) although there the treadwheel is attached to a remote jib on top of the tower. If one studies the three windows in the tower we see a treadwheel within. This was used to lift heavy items to the work areas at the top of the tower. As the tower rose in height the jib would rise with it, but the treadwheel could remain in the same position. In the image of *The Building of the Tower of Babel* in the *Rothschild Hours* that we have looked at for its scaffolding and formwork, (Fig. 37) there is a crane that falls into the category

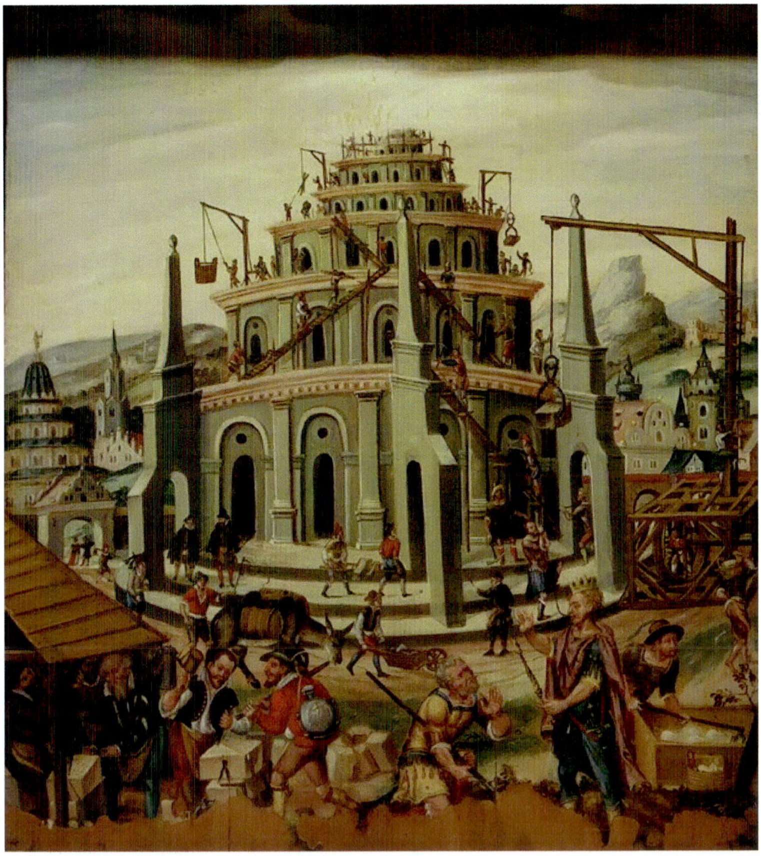

Figure 95: 1587, German, unknown master, *The Tower of Babel*, oil or tempera on panel. Z 2249, Germanisches Nationalmuseum, Nuremberg.

of slewing cranes discussed below. Instead of a treadwheel it has a hand operated winch operated by a large spoked wheel which two men are turning. Another man climbs the jib to ensure the rope is running free.

In most of the images of cranes so far, the jib would have had limited ability to turn or rotate. An important development in the history of cranes was the slewing crane where the jib and treadwheel were mounted on a central rotating post enabling the entire assembly to rotate through 180°. This type of crane was illustrated in Diderot's Encyclopedia of 1751-1772.[98] (Fig. 92) It is also known as a Perronet self-contained-crane after Jean-Rudolph Perronet, engineer and architect, who in the 1750s, designed one for the construction of a stone bridge over the Loire River in Orleans, France.[99] This crane has also been replicated by the same workshop that built the Tocknik crane. Diderot and Perronet notwithstanding, this machine

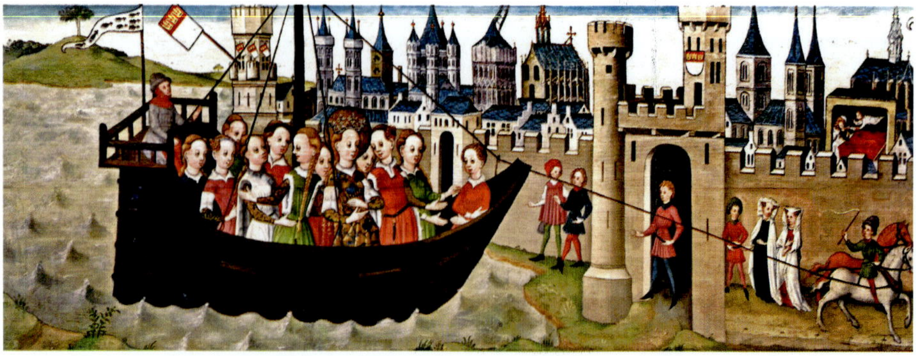

Figure 96: Cologne, *c.*1455-1460, Legend of St Ursula, *Arrival in Cologne and St Ursula's dream*, inside lid of a relic casket, tempera on softwood, 54 × 89 × 248 cm. Collection of Ferdinand Franz Wallraf. WRM 0715. Wallraf-Richartz-Museum & Fondation Corboud, Cologne. Photo: The Print Collector/Alamy Stock Photo.

has much earlier origins and many depictions can be found in fifteenth century illuminated manuscripts. In the image of the *Construction of Venice, Sycambria, Cathage and Rome* of 1475-1500 seen in Figure 139 we see just such a crane in the top left section which depicts Venice. This type of crane is also visible in the following Figures: 24, late fifteenth century; 35, *c.*1478, with a double treadwheel; 130, 1455-1485; 136, *c.*1465 (although the jib is out of the picture); 142, 1490s and 147, late fifteenth century. All are unmistakably the same type as the Perronet crane of at least 250 years later, and its invention cannot therefore be accredited to him as some have asserted. The crane depicted in Gerard Horenbout's illumination *Elijah begging fire from Heaven* (Fig. 40) shows the jib is alongside the treadwheel. It may be a slewing crane and Horenbout has simply decided to cramp the composition so he can get it onto the page, because the juxtaposition does not work. How is the rope rigged from the winch, which is presumably hidden on the right hand side of the treadwheel, to the jib? Some element is missing here, in what is otherwise a well observed building site. Two figures appear to be operating the treadwheel and another has climbed up the jib. Two ropes are rigged over pulleys either side of the jib, one is lifting the load of stone just above the second storey windows; the other is slack and is attached to the large bucket on the first main lift of scaffold. How does the gearing system work that enables two rope lifts to work independently of each other?

In Bruegel's painting, Figures 166, 93 and 94, there are inconsistences regarding the cranes, as well as the sizes of blocks of stone that are discussed on page 222. The double treadwheel crane, (Fig. 93) on the second tier, is of a type usually found on docksides. There is another crane actually on the waterfront on the extreme right of the painting. Such large machines would be unwieldy on a building site, particularly at height but we must remember that the building Bruegel depicts is far from typical, a combination of building site and mountainside. The crane is also

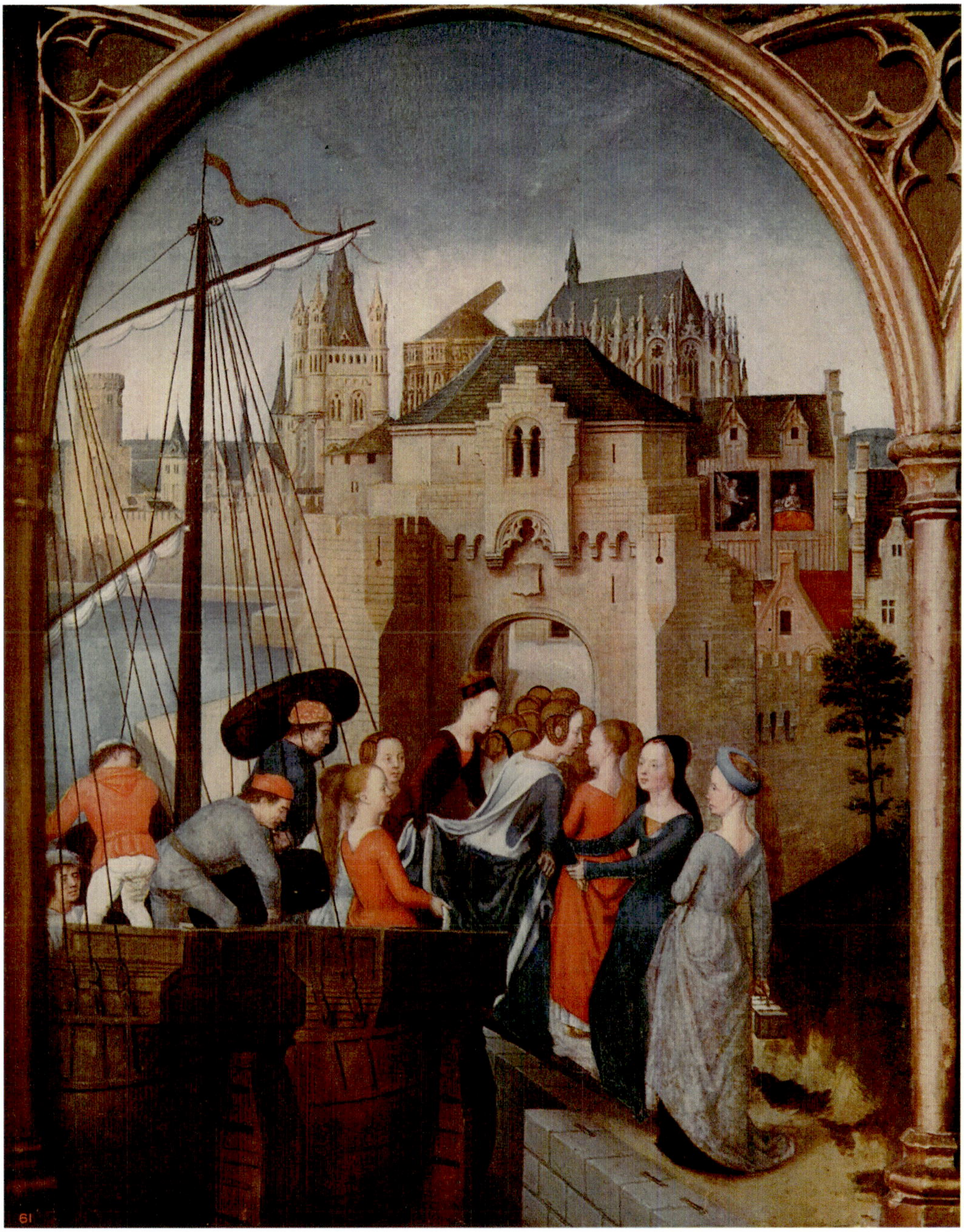

Figure 97: 1489, Hans Memling, Flemish, *The Arrival of St Ursula at Cologne*, Reliquary of St Ursula. Hans Memling Museum, Bruges. Photo: Peter Horree/Alamy Stock Photo.

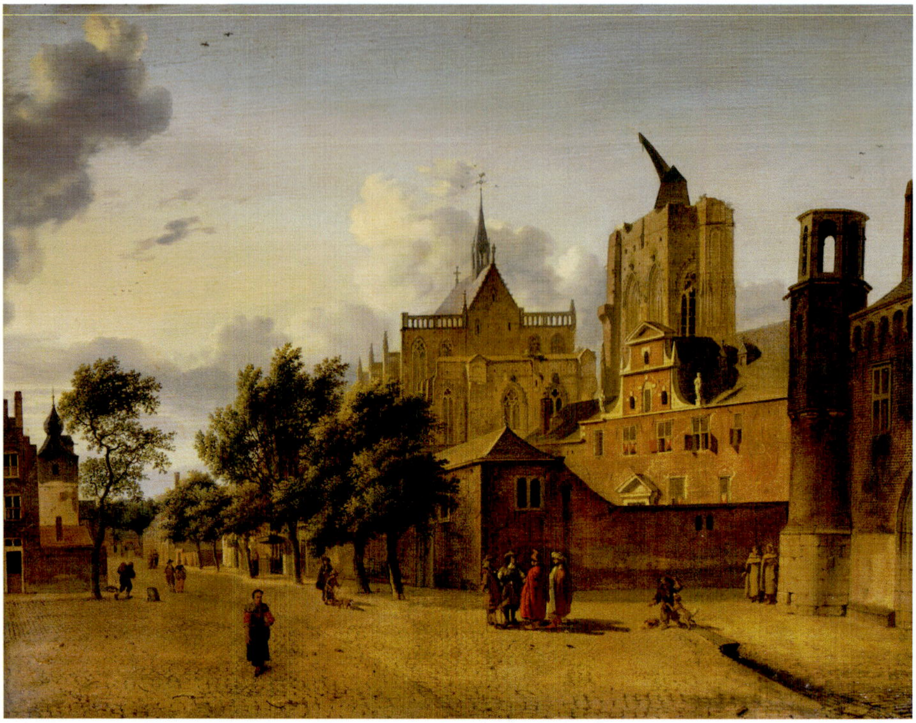

Figure 98: 1684, Jan van der Heyden, Netherlands, 1637-1712, *A Street Scene in Cologne*, 31.6 × 0.6 cm, oil on oak panel, The Wallace Collection, London.

shown at a peculiar angle where the load is uncomfortably close to the buttress below. The encasing of the jib and winch with a semi permanent housing together with its extra weight would not normally have been necessary. But Bruegel's fantasy has surely taken a good 50 years or more of continuous building to get to where we see it. Dockside cranes were larger, heavier and much more permanent and had no place on late Medieval building sites.[100] The Cologne crane below (page 145), was an exception, and was in place for some 518 years, at sometime during this period it was protected with outer casing.

Bruegel's treadwheel windlass (Fig. 94) is even more problematic. Such a windlass can only operate through a trap or opening in the platform on which it is standing. This one operates over the side of the scaffold platform it is standing on and there is no overhang. When the load comes level with the scaffold there is going to be a problem with the burden colliding with the structure. For such a system to work, a jib would be required that overhung the scaffold and would therefore keep the load clear; as is the case with the hand operated winch and jib immediately to the right. Although this does not work either as it is uncomfortably close to the treadwheel arrangement and one will foul the other. The rope is also wound around the drum

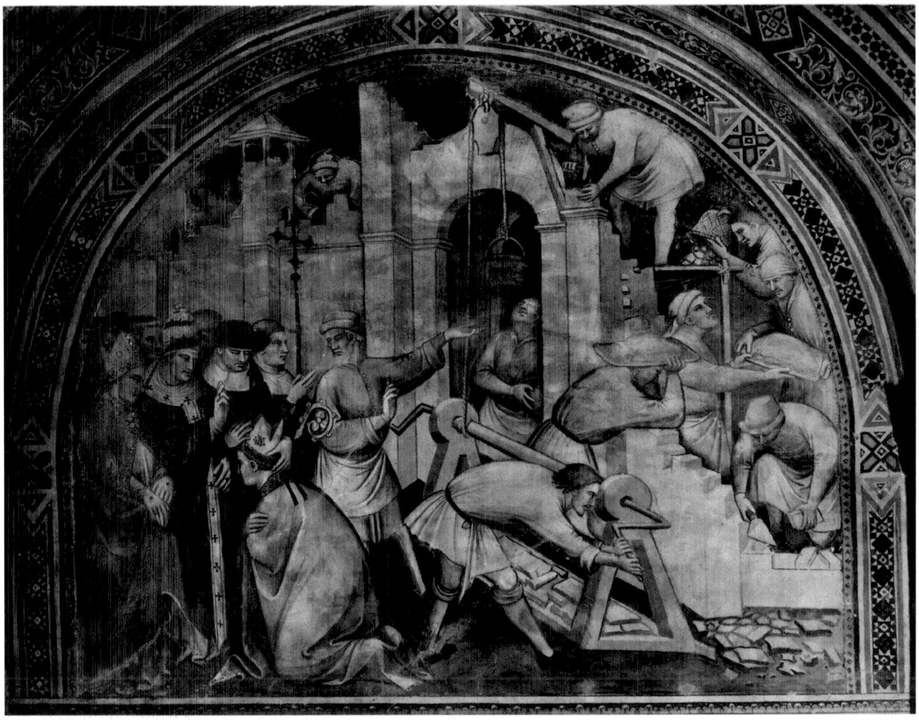

Figure 99: 1407, Spinello Aretino (Spinello di Luca Spinelli), Italian, 1345/52-1410, *The History of Pope Alexandia III: Construction of the Town of Alexandria*, Fesco, Palazzo Pubblico, Siena, Italy. Archive photograph, Fondazione Zeri, Universita di Bologna, Italy.

of the treadwheel, which is also not correct. On the archaeological evidence of surviving treadwheels, and depictions in manuscripts illustrated here, the axle of the treadwheel would extend either side, or in the case of a double treadwheel be in the centre, to form the windlass around which the rope coiled. The rope would then pass through a pulley, usually above the treadwheel, and then pass through a trap-door in the floor of the platform on which the whole assembly stood. This is how the treadwheel in the Bell Harry Tower of Canterbury Cathedral was set up and worked until the 1970s, although it would have been adapted and repaired over time.[101] The basic 'T' shaped mast and jib crane, that we first saw in the thirteenth century image of the *Tower of Babel* in Figure 16 is also here in Bruegel's painting above the dockside crane. (Fig. 93) This is operated by a large hand windlass, with three men hauling up the large block of stone. In many respects this is a more realistic crane than the two large teadwheel machines. These machines have been keenly observed and sketched but have been superimposed here and there without due regard to what would actually work in reality.[102]

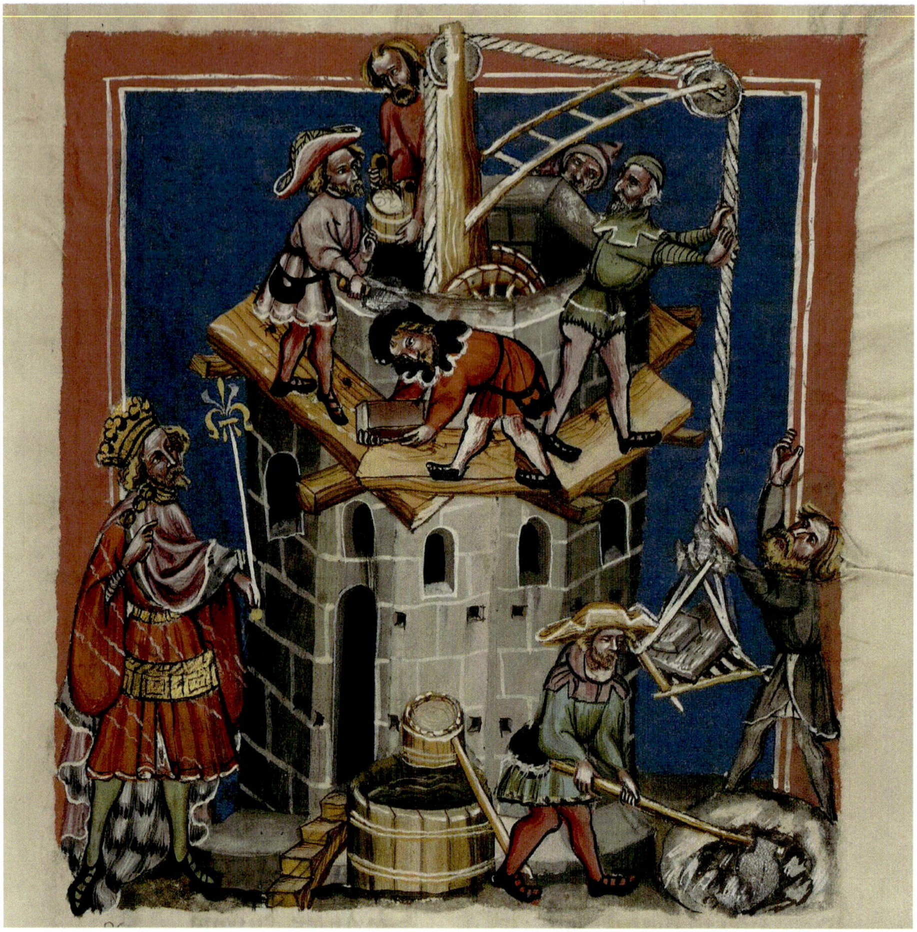

Figure 100: 1400-1410, unknown German Master, Regensburg, Bavaria, *The Construction of the Tower of Babel*, from the Weltchronik, manuscript illumination. 88.MP.70.13, MS. 33, f. 13. J Paul Getty Museum, Los Angeles. Digital image courtesy of the Getty's Open Content Program.

In a wall chest, that once hung in the guild house of the Nuremberg stonemasons and stonecutters guild, is a panel painting of the tower of Babel by an unknown German Master dated 1587.[103] (Fig. 95) This depicts, as one would expect, many stonemasons performing various tasks, some of which may be genuine portraits. The building is accessed by a series of ramps and there is only scaffold at the very top. There are four cranes, including a large treadwheel crane on the ground with a tall post and long jib, the slewing action of which is seemingly limited by the tall pinnacle of the adjacent arched buttress. The treadwheel is housed in an open timber structure or cage, with one man operating it. The arrangement of wheel and windlass is as described above with the latter on the extended axel. Three others

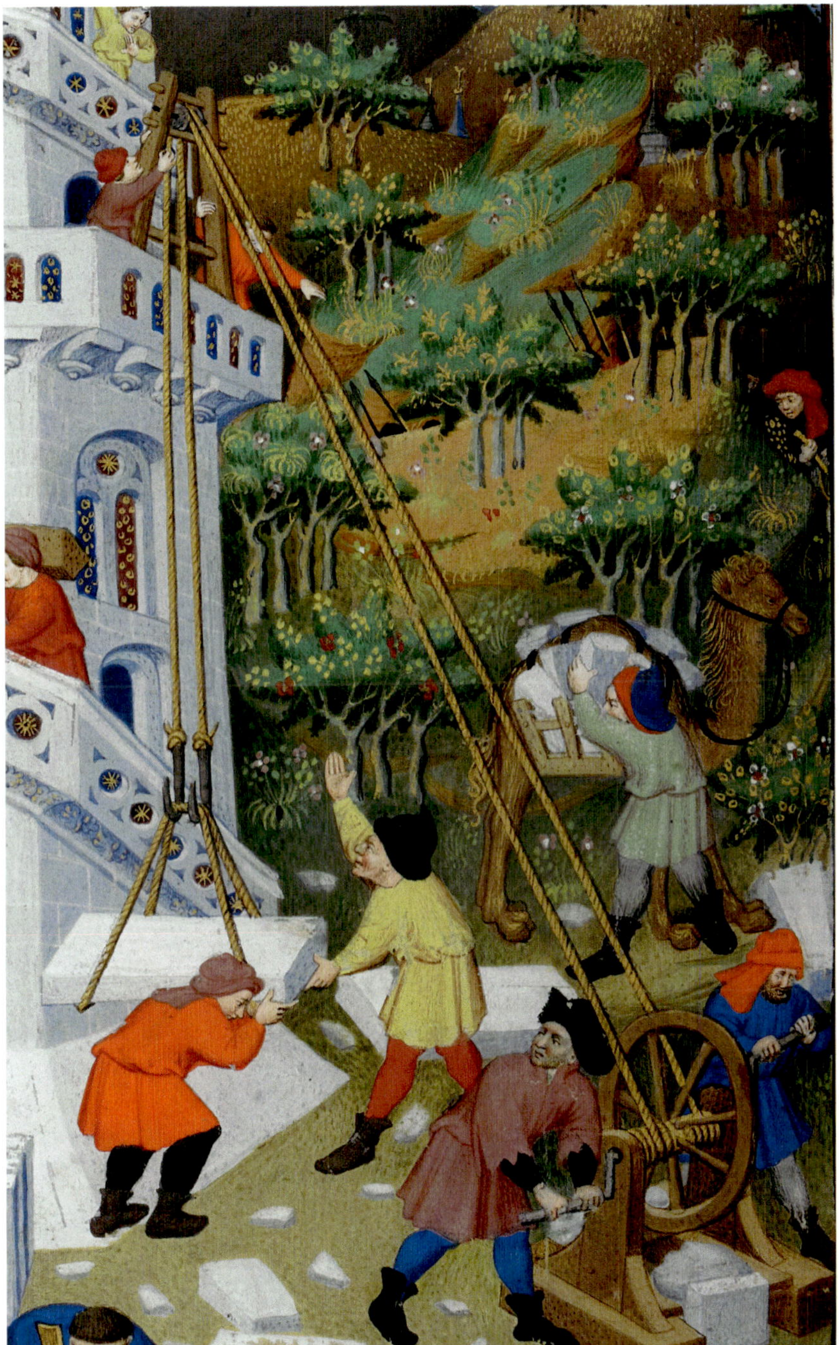

Figure 101: Bedford Master, *The Building of the Tower of Babel*, detail of Figure 144, from a Book of Hours of the use of Paris (The Bedford Hours), © The British Library Board, London, Add. MS 18850, f. 17v.

attend this lifting operation; one climbs the rungs set into the post to ensure free running of the rope, whilst two others, one on the ground and one on the first tier of the building, steady the load. The other three cranes are operated by hand windlasses with inverted 'L' shaped jibs.

Cologne cathedral had a building history stretching from 1248 to 1880, some 632 years. From *c.*1350 to 1868 it had a crane on top of the partially completed south tower, which became a symbol for the city. Work on the cathedral ceased in 1473 when the tower was complete to belfry level and did not recommence until 1842. The crane was dismantled in March 1868.[104] The crane has featured in art at least from the middle of the fifteenth century, (Figs. 96, 97 and 98) initially in depictions of the St Ursula legend, including one by Hans Memling. (Fig. 97) Many eighteenth and nineteenth century images also survive, of drawings, prints and early photographs.[105] The crane was a considerable timber structure with a total height of approximately 25m, on a base 13.5m square with a jib length of 13m. At some point in its history the superstructure was covered in slate to protect it.[106] It is not known for certain what the operating system was, as no image is extant, but the most probable system is a treadwheel.

Hand operated winches feature in many northern manuscripts and paintings. On the other hand, on the basis of evidence from illuminations in manuscripts and other paintings, the Italians appear not to have adopted the treadwheel crane in the Medieval and Renaissance periods. They preferred hand operated winches or windlasses with or without jibs. The fourteenth century Italian illumination depiciting the *Rebuilding of Troy*, as seen in Figure 89, shows a simple hand winch lifting a bucket of mortar and a cumbersome capstan windlass attached to a jib lifting a complete column. A late fifteenth century illumination by Bernardino Butinone does show two jib cranes in use, (Figs. 149 and 150 in Chapter 2) one with an unusually large pulley, but their winches are not visible. Another hand-winch is shown in use in Domenico di Bartolo's fresco at the hospital of Santa Maria della Scala, Siena, depicting construction work on the hospital. (Fig. 26) The winch is set over a trap in the scaffold with the winch handle being turned by a worker, although in the archive image shown in Figure 26 it is difficult to see as this area was damaged, but has since been restored.[107] Spinello Aretino's 1410 fresco *History of Pope Alexandia III: Construction of the Town of Alexandria* in the Palazzo Pubblico, Siena, shows a hand winch on the ground attached to a jib higher up on the building. (Fig. 99)

In an early fifteenth century miniature, from Rodolf von Ems's Weltchronik, of the construction of the Tower of Babel, by an unknown German master, (Fig. 100) there is an interesting and well detailed crane. It has a curved ladder like jib set into a heavy post; a thick rope runs from the winch, which is partially out of sight, through pulleys at the top of the post and at the end of the jib. The load of stone is in a pallet just above the ground, and two men steady it, one on the ground

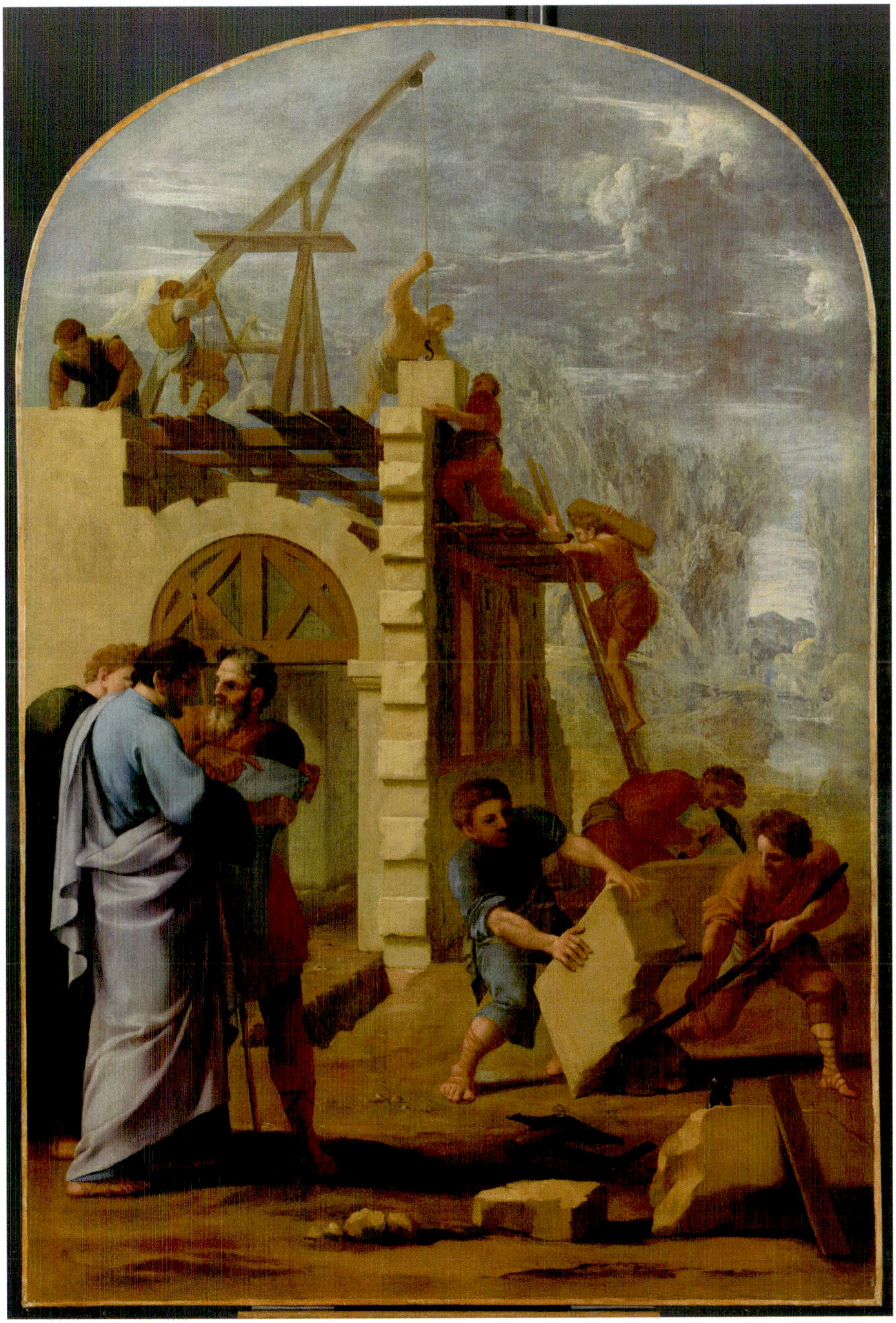

Figure 102: 1645-1648, Eustache le Sueur, French, 1616-1655, *St Bruno orders the Construction of the Monastery*, 193 × 130 cm, oil on canvas, Musee du Louvre, Paris.

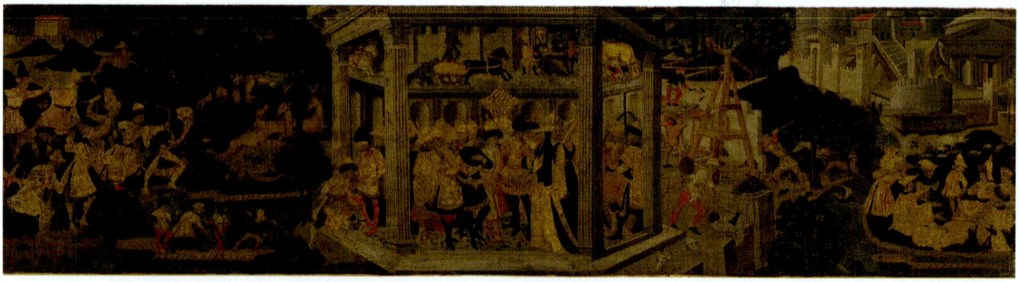

Figure 103: *c.*1450, Apollonio di Giovanni, Italian, 1415/17-1465, *Aeneas at Carthage*, 49.7 × 161.9 cm, tempera on panel, Yale University Art Gallery, New Haven, Connecticut. University purchase from James Jackson Jerves, 1871.35. Open access image.

and the other aloft on the scaffold. The post and rope drop down to within the tower where there are two wheels, on either side of the post. Is this meant to be a treadwheel or is it a hand operated winch? Unfortunately we cannot see. Masons and labourers work on the scaffold, which is composed of enormous thick planks of timber set on squared timber putlogs. This is perhaps a temporary use of such valuable timbers. There must be a ladder on the blind side of the tower, or possibly a staircase is being constructed as the tower rises. Another labourer mixes mortar on the ground where there is a large tub of water and a bucket. Beneath the scaffold are the putlog holes of the earlier lifts of scaffolding.

When Things Don't Quite Work

In Christine de Pizan's *Book of the Queen*, discussed in Chapter 2, is an illumination by the Master of the Cité de Dames (Fig. 132) showing a lifting operation that is a complete muddle, the separate parts do not work together. In combination it is a peculiar contraption of the early fifteenth century master's invention; he has observed but then forgotten how the bits go together. The winch and its frame are realistic enough, although it appears that the operator is winding it anticlockwise; he is therefore letting out the rope rather than winding it in. He is the wrong side and should be the other side operating it clockwise! The winch is attached to a wheel, which resembles a miniture treadwheel, at the top of the mast or post. The mast has rungs inserted, staggered either side of the timber to enable climbing to deal with any snags in the rope. Such snags are inevitably going to happen as the rope is twisted around the mast and in operation the rope could easilly slip off the wheel with disasterous consequences. The base of the mast is behind the wall, inside the building under construction, so is remote from the winch which is outside. Although scale is not present in the image, the mast rises to a height just above the clerestory of the church, so is considerable. All in all this is a clumsy

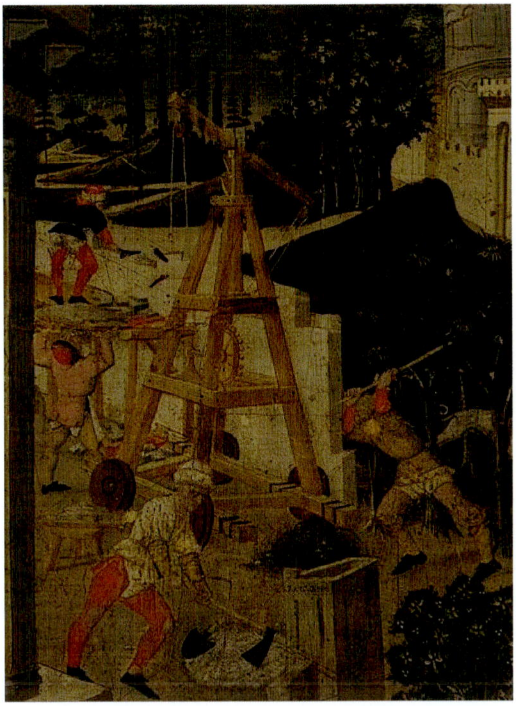

Figure 104: *c.*1450, Apollonio di Giovanni, Italian, 1415/17-1465, *Aeneas at Carthage*, 49.7 × 161.9 cm, tempera on panel, detail of Figure 103, The Construction of Carthage, Yale University Art Gallery, New Haven, Connecticut. University purchase from James Jackson Jerves, 1871.35. Open access image.

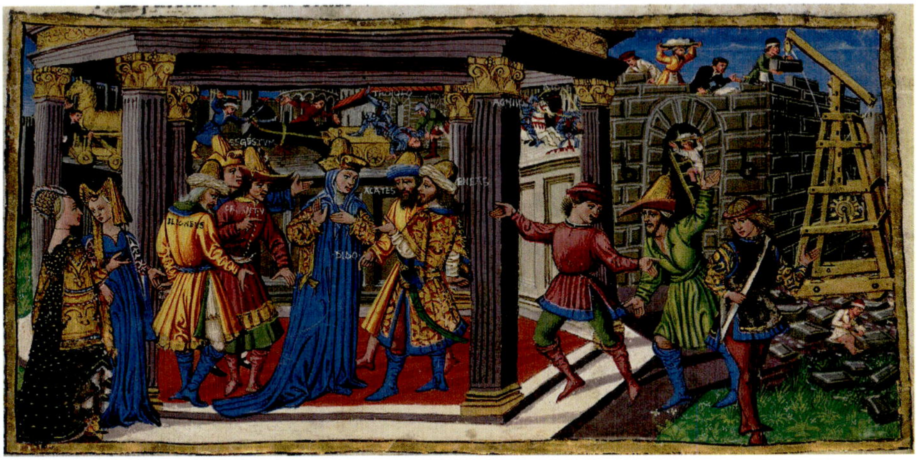

Figure 105: Apollonio di Giovanni, Italian, *Meeting of Dido and Aeneas and the Construction of the Medici Palace*, miniature from *Vergilius Publius Maro, Bucolicon, Georgicon, Aeneis*, 1460-1465. Palazzo Medici Riccardi, Biblioteca Riccardiana, Florence, ms Ricc.492, f. 72r. (©Photo Donato Pineider: 'Su concessione del Ministero dei Beni e le Attivita Cultural')

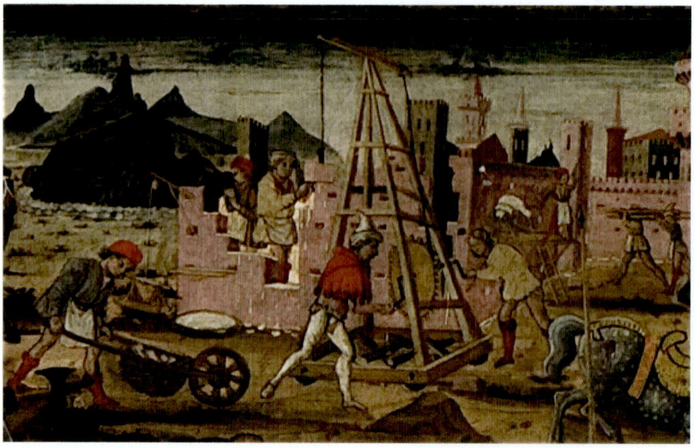

Figure 106: *c.*1460, Master of San Miniato (attributed), detail of Figure 154 *The Victory of Alexander the Great and the Building of the City of Alexandria*, 51 × 160 cm, oil (?) on panel, originally produced for the front of a wedding chest, Ca' d'Oro, cat. D76. Galleria Franchetti, Venice.

and unlikely machine. The illuminator has invented something from fragments he recalls from his observations but does succeed in conveying the strain and difficulty encountered when performing such lifting tasks on the building site.

In the Bedford Master's illumination of *The Building of the Tower of Babel*, (Fig. 101) there are three hand operated hoists; the one in the foreground utilises double ropes on the same winch this enables the heavy load of a slab of stone to be lifted by the two operators. The single wheel serves no purpose and is perhaps another reference to a treadwheel seen elsewhere. If there were a pair of wheels either side of the winch with projecting spokes, they would work for winding instead of the cranked handles. Furthermore the two operators must work in complete unison. If one stops or turns more slowly or quickly, the load becomes immediately unstable and is likely to slide out of the slings. So this is perhaps another rendering of things seen but partly forgotten. These are robust machines however and give us a degree of authenticity, down to the stone counterweighs at the base of the windlass, except for the logistics of what will happen when the large piece of stone reaches the maximum height of the lift. For the jib and pulley system appears not to slew inwards, but it may hinge backwards to allow the load to be hauled inwards. The hoist with the bucket, which utilises only one rope, is easier to comprehend as it is easy for the man at the top to reach over and lift it over the balustrade as it is a lighter load. The third hoist is largely hidden on the blind side of the tower. More convincing is a similar windlass with a single winch seen in the *Ruralia Commodorum*. (Fig. 36) The post, jib arm, and stay are higher than the working level and set further back, with the jib overhanging the side of the building, thus enabling easier handling once the load is at the required height. Eustache le Sueur's painting *St Bruno orders*

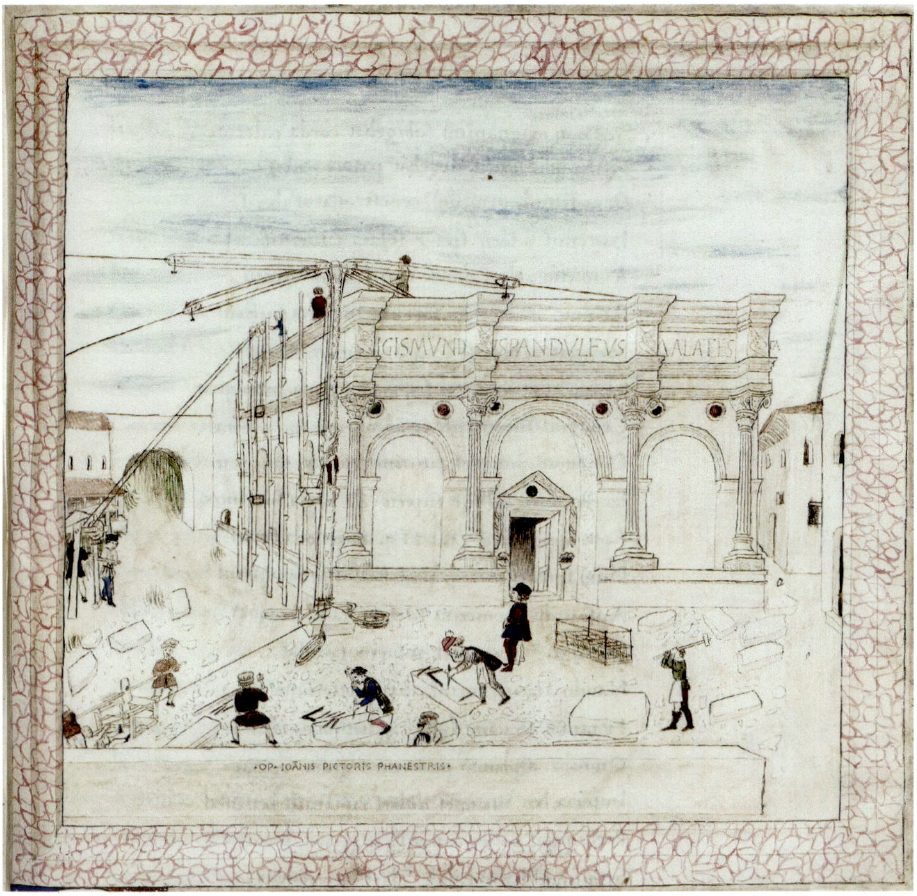

Figure 107: 1457-1468, Giovanni Bettini da Fano, *c.*1450-?, *Il tempio malatestiano di Rimini under construction*, a temple designed by Leon Battista Alberti, miniature from Basini parmensis, *Hesperis*, Book XIII, code: Ms 52 B L, f. 126r, Bibliotheque de L'Arsenal, Paris.

the construction of a Monastery of 1645-1648, shows a crane set in a fixed position over the immediate area of work being undertaken. (Fig. 102) The jib is similar to the slewing crane described above, but here the entire crane assembly will have to be moved along to accommodate the next piece of stone to be fixed as it has no pivoting or slewing capacity. The winch is on the same plane as the jib and is operated by two men. The rope which is just seen behind the first man winding it must pass through a pulley hidden behind his arms and head.

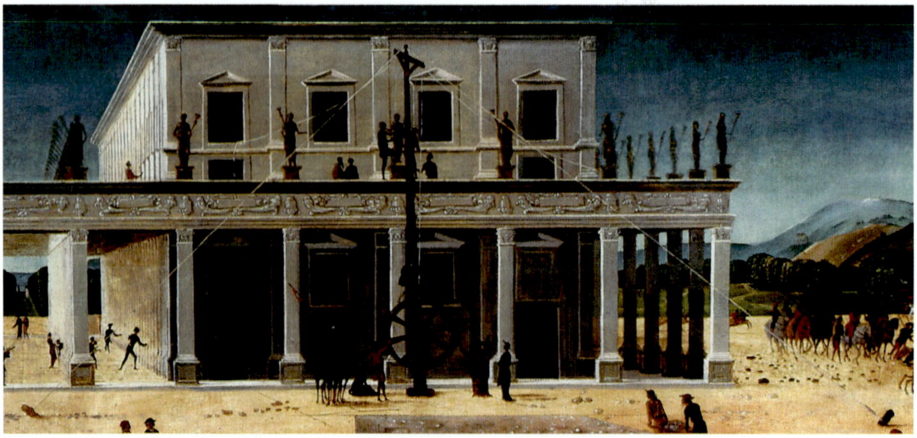

Figure 108: 1515-1520, Piero de Cosimo, Italian, 1462 -1521, detail of Figure 165, *The Building of a Palace*, 83 × 197 cm, oil on panel, The Ringling Museum of Art, Sarasota, Florida, USA: Bequest of John Ringling, 1936, Collection of The John and Mabel Ringling Museum of Art, the State Art Museum of Florida, Florida State University.

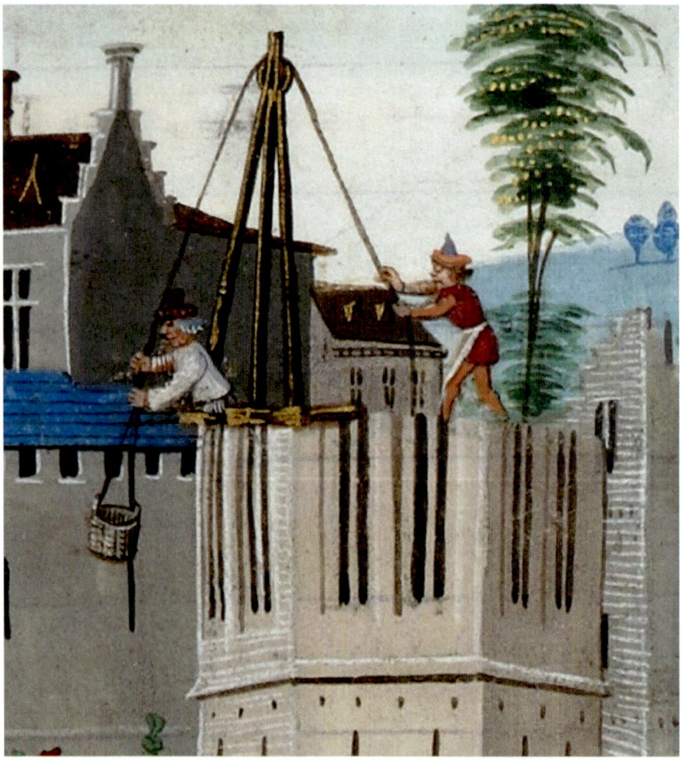

Figure 109: Flemish, *c.*1480, Loyset Liédet, detail of Figure 23, *Construction of Alexandria*, from *History of Alexander the Great*, Paris, Bibliotheque National ms. Fr 2247, f. 76r.

The Mobile Crane

It is safe to posit, on artistic visual evidence, that the Italians invented the mobile crane. In particular these appear in Florentine paintings of the mid fifteenth century. They take the form of pyramidal timber framed structures with a pivoting jib attached to a post with internal hand winches. There are four wheels, one at each corner of the base frame. (Figs. 103, 104, 105 & 106) These cranes would only have been serviceable at ground level, as indeed they are all shown, and would have required a prepared hard standing or roadway. In the panel painting by Apollonio di Giovanni (Fig. 103), originally a panel from a cassone depicting a scene from the *Aeneid*,[108] there is just such a crane to the right of the central hall where Dido receives the Trojans. The crane has wheels at the base of the frame for basic manoeuvring to and fro. The winch, of which the main element seen is a large cogged wheel, is centrally placed within the frame; this controls both the pivot of the jib and the lifting operation, although how the two are separated within the mechanism, and any gearing system, are not seen. No one attends the winch although the crane is in the process of lifting a stone into position on the wall being constructed. (Fig. 104)

Also by Apollonio di Giovanni is a second miniature, here we see a near identical crane, (Fig. 105) although it appears bigger and the wheels are inside the frame rather than externally as in Figure 104. Again, the winch has a cogged wheel, but there is no indication of any gearing mechanism. The rope to the pulley at the end of the jib and attached to the load can be seen all the way down to the winch drum around which it coils. However, again, no worker attends this winch. The central post may have been adjustable up and down, evidenced by the notches seen in the side of the post, into which some form of tenon could be fitted to hold it in place.

In the Master of San Miniato's painting (Figs. 106, 154 & 155) a very similar crane is being used. There the base is in-filled with planks upon which are what appear to be a number of counter weights to steady the operation of the machine. The winch, as with the others, consists of a large cogged wheel with the winch drum beyond, although here two men are operating it, and are lowering a piece of stone into place. The jib is attached directly to the apex of the triangular structure without a central post, which has been overlooked, or else perhaps there wasn't one. (Fig. 106) There must have been rachet mechinisms combined with the cogged wheels of these cranes to act as brakes. There is another simple hoist at the corner of the building consisting of a gin-wheel at the end of a strut supported by a diagonal brace or stay.

These cranes, notwithstanding the possibility of raising the central post, would have had height restrictions and were probably not capable of practical use above a ground storey. As indeed, taking into account imperfections of scale and perspective, they are all shown. To operate higher they would have to be moved to very robust scaffold platforms of level planks firmly fixed to the scaffold superstructure.

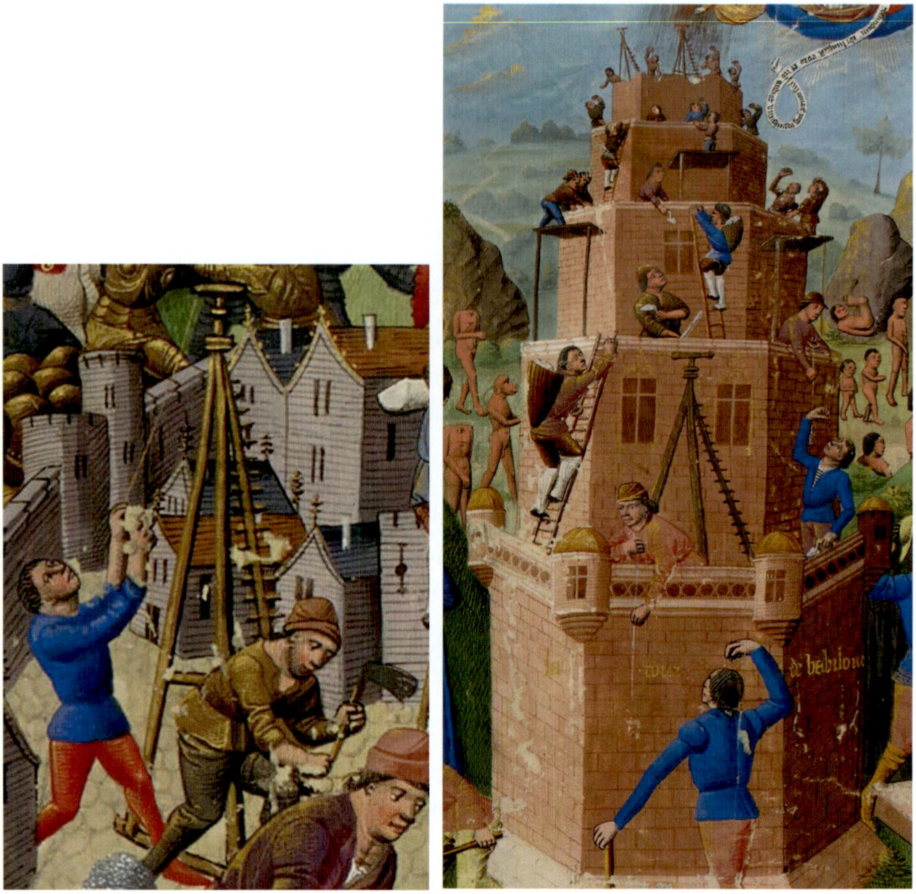

Figure 110: *c.*1480, French, details, a) left, of Figure 140, *Construction of the City of Enoch*, and b) right, of Figure 141, *Construction of the Tower of Babel*, illuminations from 'La Cité de Dieu', ms 0002, folios 065, and 094 respectively. Bibliotheque Municipale de Macon, France.

A better, and perhaps cheaper, method of higher lifting was the *stella* 'star'or *falcone* 'falcon', (Fig. 107) a machine widely used in Renaissance Italy with Roman antecedents originating from ships rigging. They consisted of a long vertical pole, probably a semi mature pine or fir trunk, set into a block of stone or timber. This was braced upright by four radiating guy ropes from the top to the ground, thus the term 'star'. These had block and tackle for straining taught, much like a ships rigging; the ropes, or sheets, for this adjustment can be seen neatly coiled at the base of the pole. There are two manuscripts which illustrate the same hoist and building site: Basini Parmensis, *Hesperis*, Book XIII, illustrated by Giovanni Bettini da Fano; one copy is in the Bodleian Library, Oxford, while the the other is in the Bibliothèque de l'Arsenal in Paris. They both contain illuminations depicting *The*

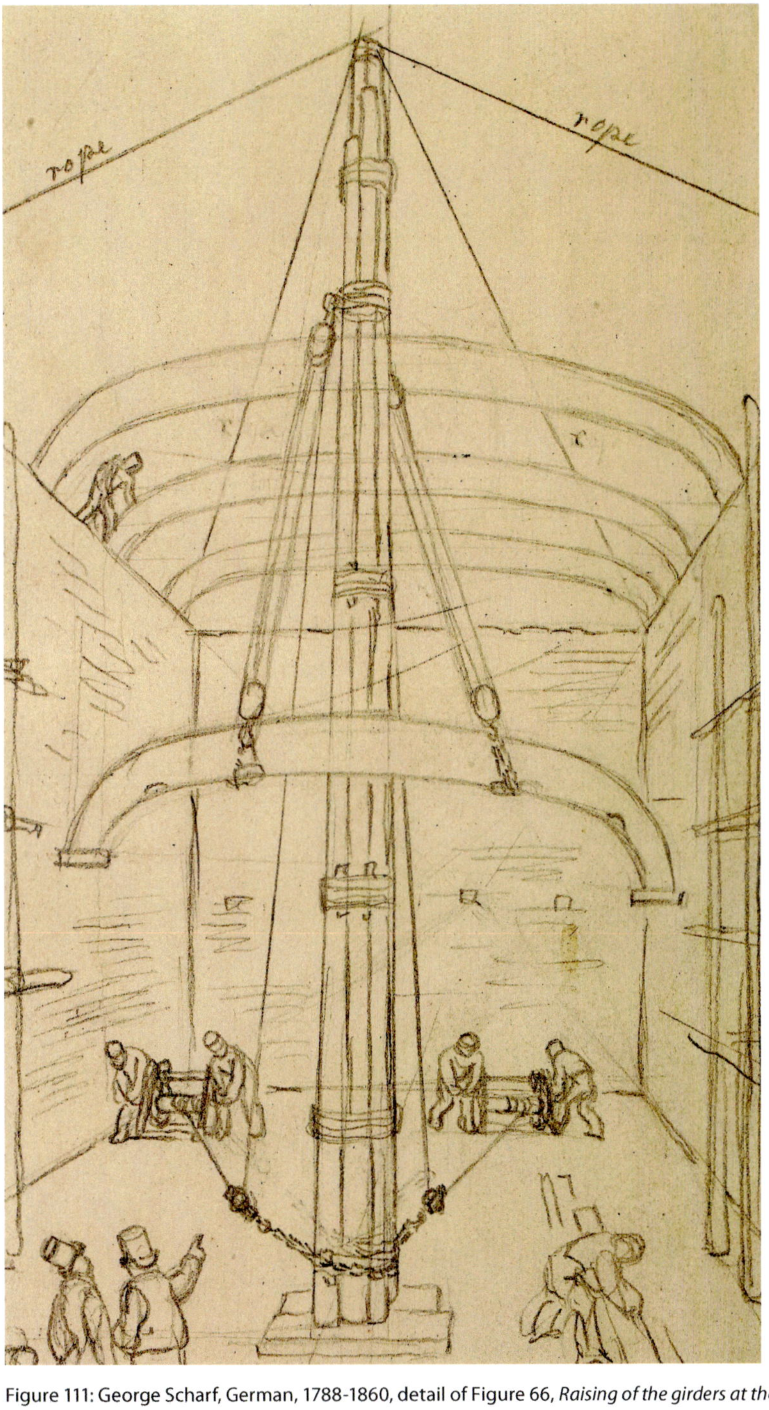

Figure 111: George Scharf, German, 1788-1860, detail of Figure 66, *Raising of the girders at the College of Surgeons*, 1835, graphite on paper. © The Trustees of the British Museum

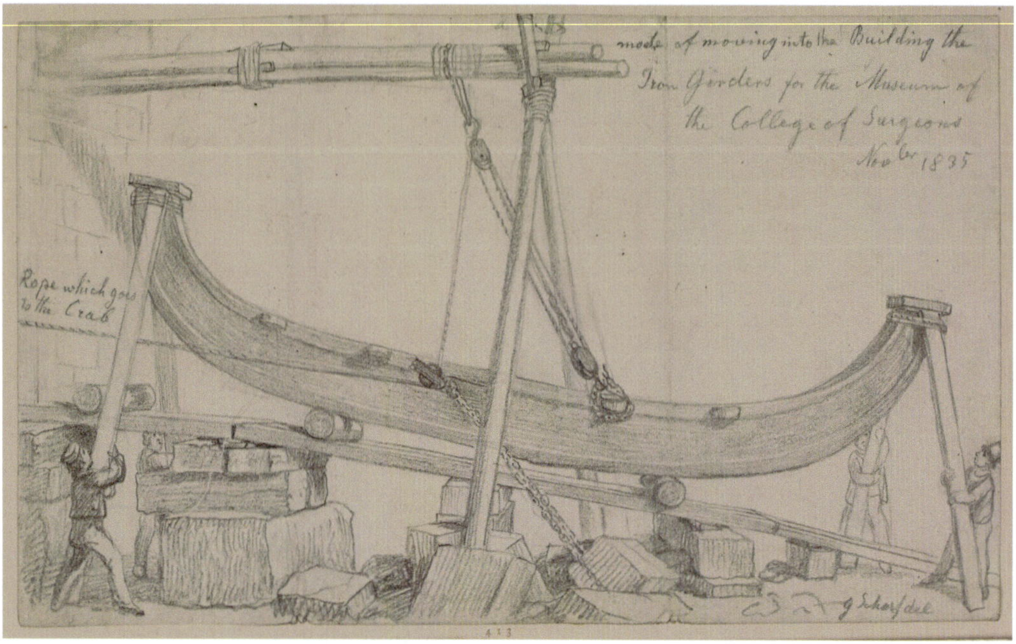

Figure 112: George Scharf, German, 1788-1860, *Mode of moving into the building the iron girders at the College of Surgeons*, 1835, graphite on paper. © The Trustees of the British Museum

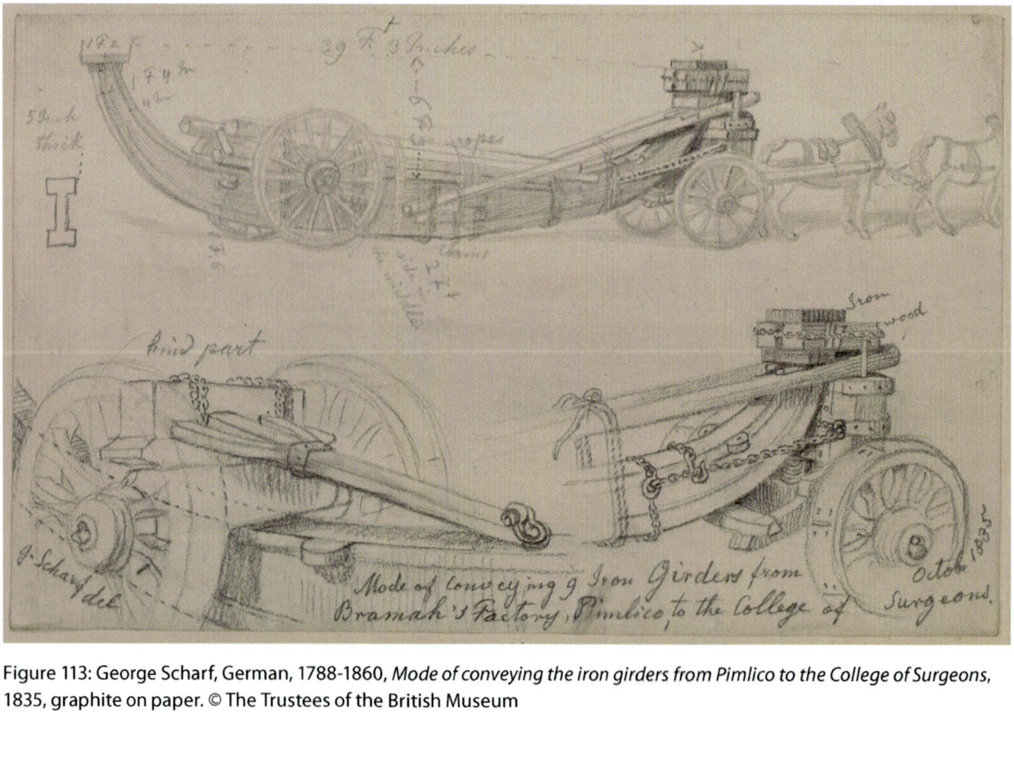

Figure 113: George Scharf, German, 1788-1860, *Mode of conveying the iron girders from Pimlico to the College of Surgeons*, 1835, graphite on paper. © The Trustees of the British Museum

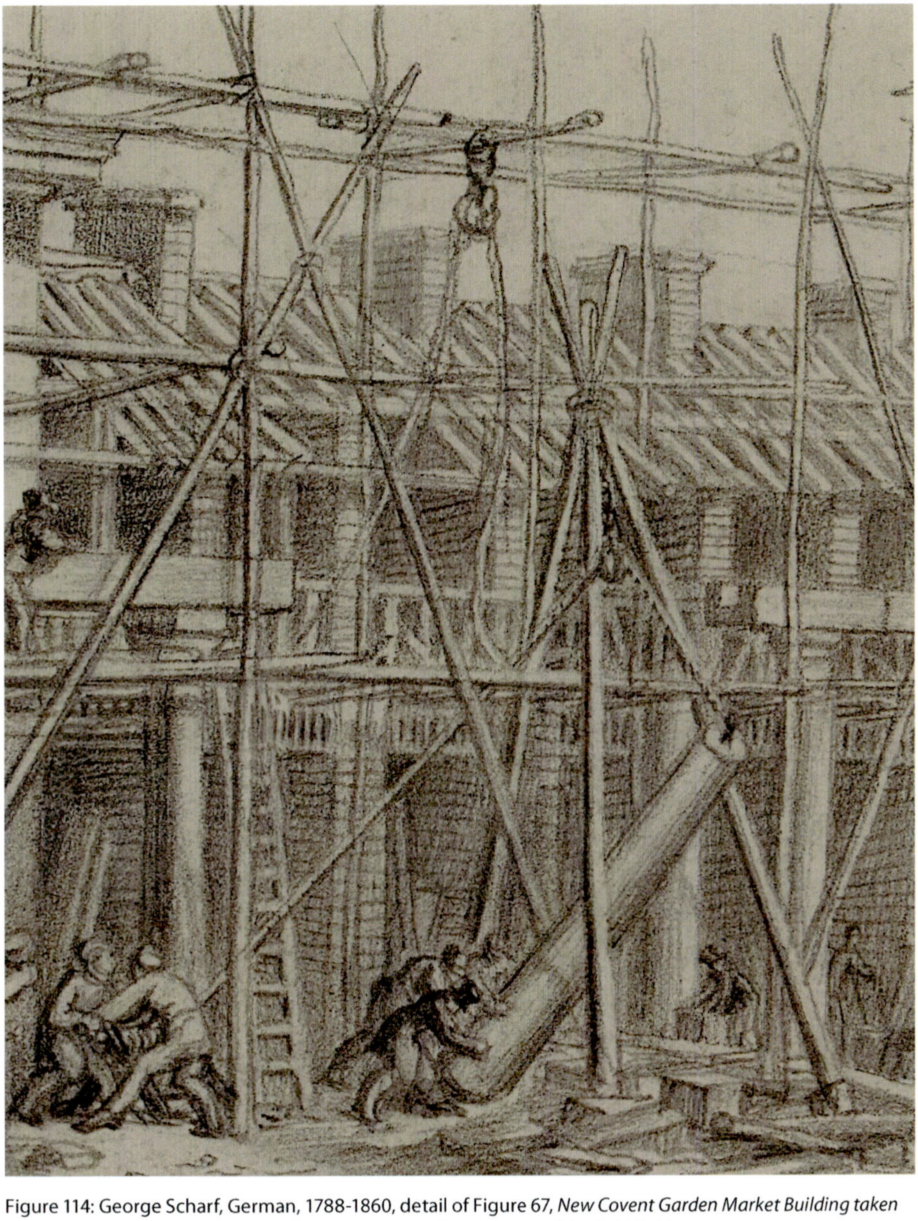

Figure 114: George Scharf, German, 1788-1860, detail of Figure 67, *New Covent Garden Market Building taken from Southampton Row*, 1829, graphite on paper. © The Trustees of the British Museum

Tempio Malatestiano in Rimini under Construction. The one shown here is from the Paris illumination. (Fig. 107) The pole is equipped with rungs at opposite sides to enable climbing and a man is seen approximately half way up the pole, either to gain access to the top of the construction or to check the various rigging. The windlass for hoisting is an upright capstan in the left foreground of the image, turned by two hand spikes with a man at each. A hole can be seen in the shaft just above the spikes for insertion of a further spike for heavier loads; the men have to step over the cable running just above ground to pulleys at the base and top of the pole or mast. Another man sits at the base of the capstan possibly to engage a brake. The load, which is a section of cornice, is just below the feet of the climbing man. The Oxford illumination is a near identical image.[109] Both images show a second hoist, also operated by an upright capstan windlass, but connected directly to pulley arrangements on the scaffold; this is a double rope system and two loads are being hauled up. Various masons and stone cutters are also working on the ground. The Temple of Malatesta in Rimini was never finished but is largely extant as built.

A 'star' forms an important feature of Piero di Cosimo's painting *The Construction of a Palace*. (Figs. 108 & 165) This is placed centrally against the right hand wing of the palace. It is engaged in hoisting what is probably the last of the many statues on the balcony of the building. Close inspection tells us that this is a single large tree trunk with its bark still in place. It is not a prepared post or pole and is probably still green. The guy ropes are clearly seen, the front pair taken down to the ground and lashed around large stakes driven into the ground at an angle. The rear pair are tied around parts of the building, but are worryingly slack. The hoist is operated by a type of small treadwheel upon which a man walks or treads the wheel, whilst holding onto the post, to lift or lower the load, which here is being lowered into its final position. At the top of the post is a crosstree with diagonal stays; this has pulleys at each end to take the lifting rope. This is also clearly seen with one end attached to the statue and the other disappearing behind the treadwheel to the winch drum.

Other similar machines or hoists are seen in Flemish (Figs. 23 & 109) and French (Figs. 140, 141 & 110) manuscript illuminations although these are formed of shearlegs. The Flemish image by Loyset Liedet, *History of Alexandra the Great: Construction of Alexandria*, (Fig. 23) is certainly a hoist as it is shown engaged in a lifting operation. (Fig. 109) It comprises shearlegs with a pulley at the top set on a timber frame. This sits on top of the actual level of work being undertaken, it would be better set just inside the tower so that it does not interfere with on-going work, otherwise it has to be repeatedly taken down and re-erected. It requires one man to haul on the rope, whilst the other leans over to hold the basket away from the building, as it can only be used for light loads.

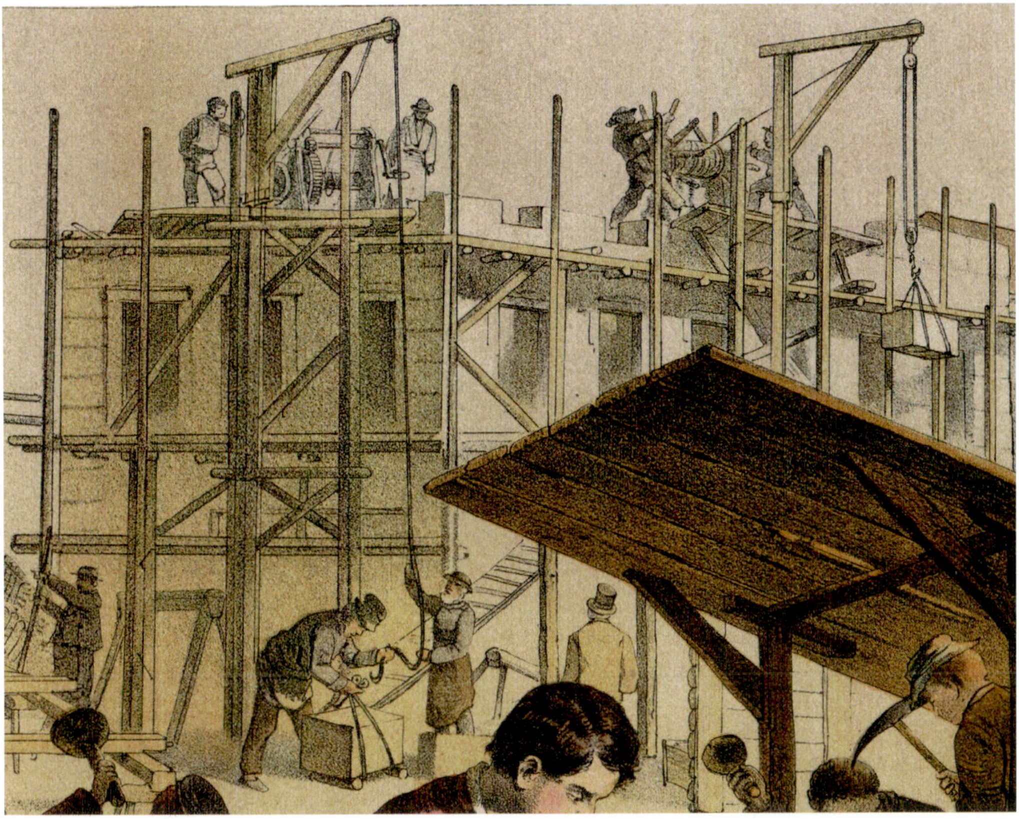

Figure 115: 1876, detail of Figure 83, Illustration from F. Streich and K. de Gerstenberg, *Méthode d'enseignement par l'aspect les métiers et leurs outils*, chomolithograph, published 1876, Paris, editions Bonhoure. Mary Evans Picture Library.

Two illuminations from a copy of the *La Cité de Dieú*, a French manuscript of *c.*1480 have similar shearleg apparatus although these have a form of crosstrees at the top. One of the legs of each have rungs inserted for climbing. The first illumination, *The Construction of the City of Enoch*, (Fig. 110a) has a man hauling on a rope, which runs to the crosstree, but there is nothing at the other end. In his hands he appears to hold an object as well as the rope but this is only a loss of paint from the parchment surface. The shearlegs are so placed as to serve no useful purpose for the building operations going on around him. In the second, *Construction of the Tower of Babel*, (Fig. 110b) shearlegs are set up on the first tier of the building, but it is not rigged up for hoisting. The man leaning over the parapet holds a plumb line, which should not be confused with any rigging mechanism. There are another two sets at the very top of the building, but they also do not appear to be rigged up but are nevertheless too far distant for such detail to be shown. There is a certain

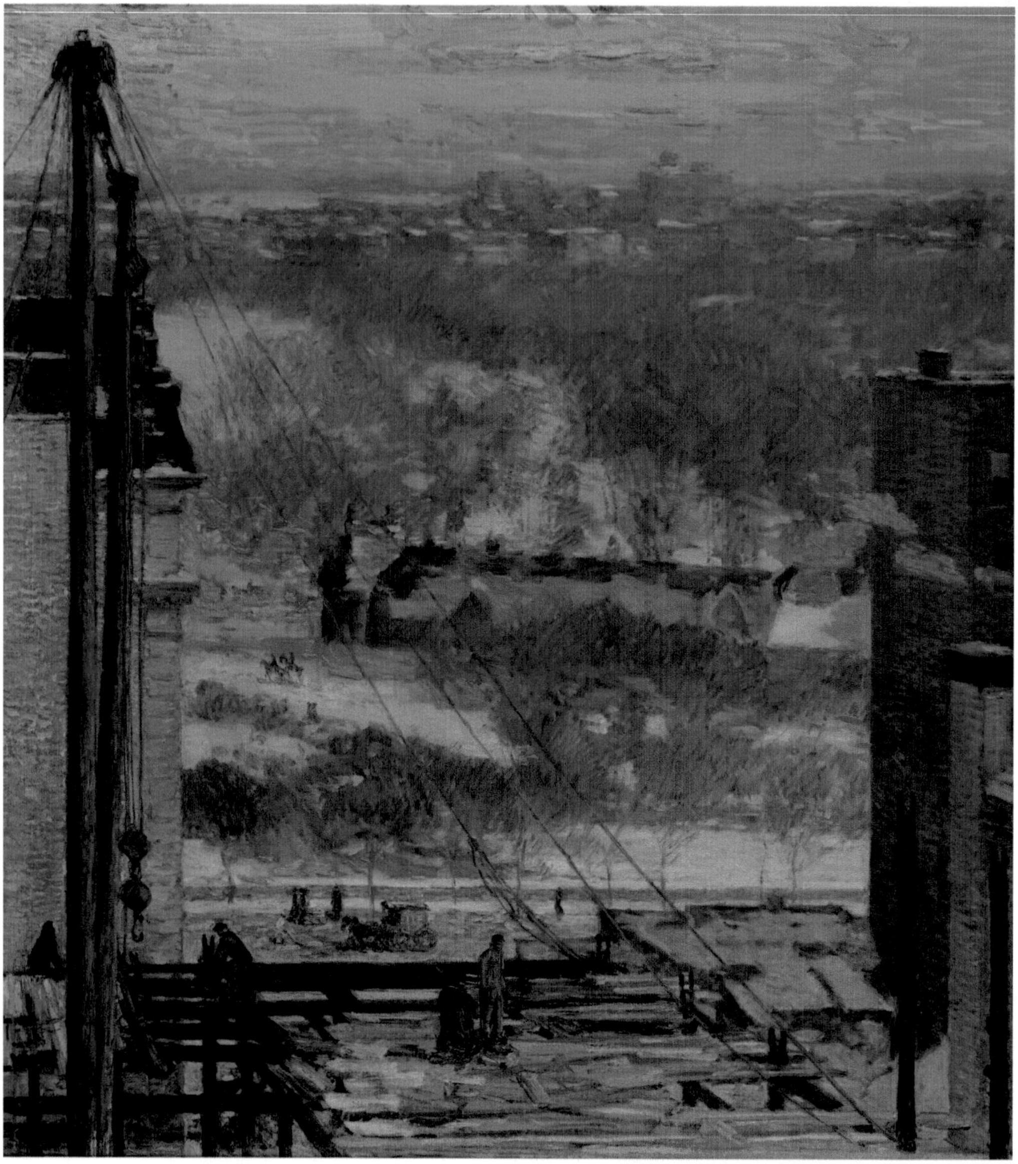

Figure 116: Childe Hassam, American, *The Hovel and the Skyscraper*, 1904, oil on canvas, 88.26 × 78.74 cm. Acc. No. 2003.1.5 Courtesy of the Pennsylvania Academy of the Fine Arts, Philadelphia. The Vivian O and Meyer P Potamkin Collection. Bequest of Vivian O Potamkin.

amount of fear and agitation at the approaching wrath of God, which is about to come down on these workers.

George Scharf's drawings of the roof girders for the College of Surgeons in London, built in the 1830s, show a 'star' being used to hoist these into position; also the method of transporting them from the foundry and into the building. (Figs. 66, 111, 112 & 113) The 'star' is very similar to that employed by Alberti at the Temple of Malatestiano in Rimini which is seen in Figure 107. The post is composed of a number of poles lashed together with wedges driven in to tighten the bindings. There are also the requisite guy ropes holding it upright, these disappear out of the frame of the picture. Around the base of the post is a chain with pulleys at each end, with ropes running to winches each operated by two men; the ropes run to blocks rigged sufficiently high on the post to allow clearance to lower the girder into position. The post is in the centre of the hall, the walls of which have been built to full height. The girder is lifted at a skew or angle to allow clearance of the walls. Once it is above the height of the walls it is turned square on and lowered onto the wall tops. (Figs. 66 & 111) The girders were hauled upsidedown into the building, presumably through a window opening at ground level, on a series of large planks and rollers utilising a system of a horizontal beam, shearlegs, pulleys and block and tackle. This was all held in position and stabilised with large blocks of stone and balks of timber. Additionally the girders are held from tipping over by four men, at each end and either side who handle lengths of timber lashed around the legs of the girders, a precarious operation. (Fig 112) The girders were transported to site from Bramah's factory at Pimlico on specially constructed horse drawn carts. (Fig. 113)

How precarious these operations were can be seen in Scharf's drawing of an accident when hoisting a girder during the building of the British Museum, which is eloquently recorded in Figure 65. Another drawing by Scharf of the construction of the New Covent Garden Market in London shows shearlegs in use for raising a column. (Figs. 67 & 114) The rope attached to the top of the column runs through a pulley system suspended from the shearlegs to a winch on the ground operated by two men. Three men steady the base of the column, but it will have to go behind the scaffold ledger and brace. At present these are behind the column, although they will need to end up in front as already is the case with the other columns. Also the rope attached to the top runs in front of the scaffold whereas it runs behind the ledger on its way from the pulley to the winch. We must assume that the column passes beneath the ledger and around the brace, and also that it passes between the ledger and the front leg of the shearlegs. The column base also must be manoeuvred further back from where it is, and is perhaps the task of the three men at the base. Higher on the scaffold another pulley is rigged; the rope seen slack at present must be attached at the top, before the other is released, to finally lift the column into position. Not a straightforward operation but it has obviously worked for the other columns.

The closest modern equivalent to the 'star' are the type of shearlegs seen in the American artist Childe Hassam's painting *The Hovel and the Skyscraper* of 1904, at the end of the period under study.[110] (Fig. 116)

The treadwheel continued in use on new build sites into the nineteenth century, as well as utilising existing ones such as at Canterbury cathedral. Images of Sir John Soane's site for the new infirmary at Chelsea Hospital in the first decade of the nineteenth century, show several treadwheels (Figs. 52 & 55) on the ground, either waiting to be used or having been taken down after use. There is a whole stack of them, at least five, in R C Cantrell's pen and watercolour sketch of 15 October 1810. (Fig. 52) Similarly a hand winch is shown in another of Cantrell's sketches, (Fig. 56) and in George Schraf's view of the New Gallery of the British Museum in the middle foreground, either finished with or not yet required. (Fig. 64) In front of the winch is what he calls a winding Jack marked X, Schraf notes, at the bottom of the page: 'The winding jack is too large. The body of it is 3 ft long 1 ft wide and 69 in thick.' This information given to him in conversation with the builder, Mr Baker. The hand operated windlass is also much in evidence in the lithograph of 1876 from Streich and de Gerstenberg's, *Méthode d'enseignement par l'aspect les métiers et leurs outils.* (Figs. 83 & 115) The one on the left is a more up to date geared machine, but the other could be found on any Medieval building site of several centuries earlier.

2. Stone Masons and Bricklayers: Materials, Tools and Equipment

Masonry is one of the three major crafts in building, the other two being carpentry and the smith. Masons have figured in art throughout most of history, from ancient Egypt to the present day. Throughout, the term 'mason' includes all crafts associated with stone and brick, including quarrying and brick-making.

Ancient Egypt

The eighteenth dynasty, New Kingdom tomb of Rekhmire at Thebes in Upper Egypt, contains rare depictions of construction work. Rekhmire was vizier, or chief minister, to Thutmosis III who reigned 1479-1425BCE, and the tomb is dated *c.*1450BCE. Here are seen masons at work in three different categories: cutting and smoothing blocks of stone; finishing a statue using a lightweight scaffold, and making bricks.

The masons working on the blocks of stone (Fig 117) are using mallets and chisels, the mason on top of the block (6) is using a punch or point, whilst the other two are using wider tools such as boasters or pitching tools.[111] Of the other masons, two (1 & 3) are using a type of boning-rod to check the level or flatness of the surface. Boning-rods are still in use today by ground workers and paviours to check levels. The other mason kneeling on the ground, appears to be measuring across the diagonals of the block with a cord, or it may be a straight edge to check the surface. The stone working tools were almost certainly hardened copper as the process of converting iron to steel was not known to the Ancient Egyptians.[112] The stone therefore must have been a soft to medium hard limestone or sandstone, both readily available in the Nile valley. Copper or bronze chisels would have worn down too quickly on a hard stone such as granite.

The masons finishing a sculpture (Fig 118) are using some form of abrasive or rounded pieces of rock, and one of them is also painting the back of the statue. It is known that the Ancient Egyptians used dolerite for working hard stones. This was formed into rounded lumps big enough to be held in both hands, or made into hammers and then used for pounding, although in this scene rubbing is more likely. Remains of such tools have been found at quarry sites in Egypt and there are depictions of these 'balls' of stone being used elsewhere in the Rekhmire tomb.[113] The masons are using a lightweight scaffold of wooden poles tied together with cords.

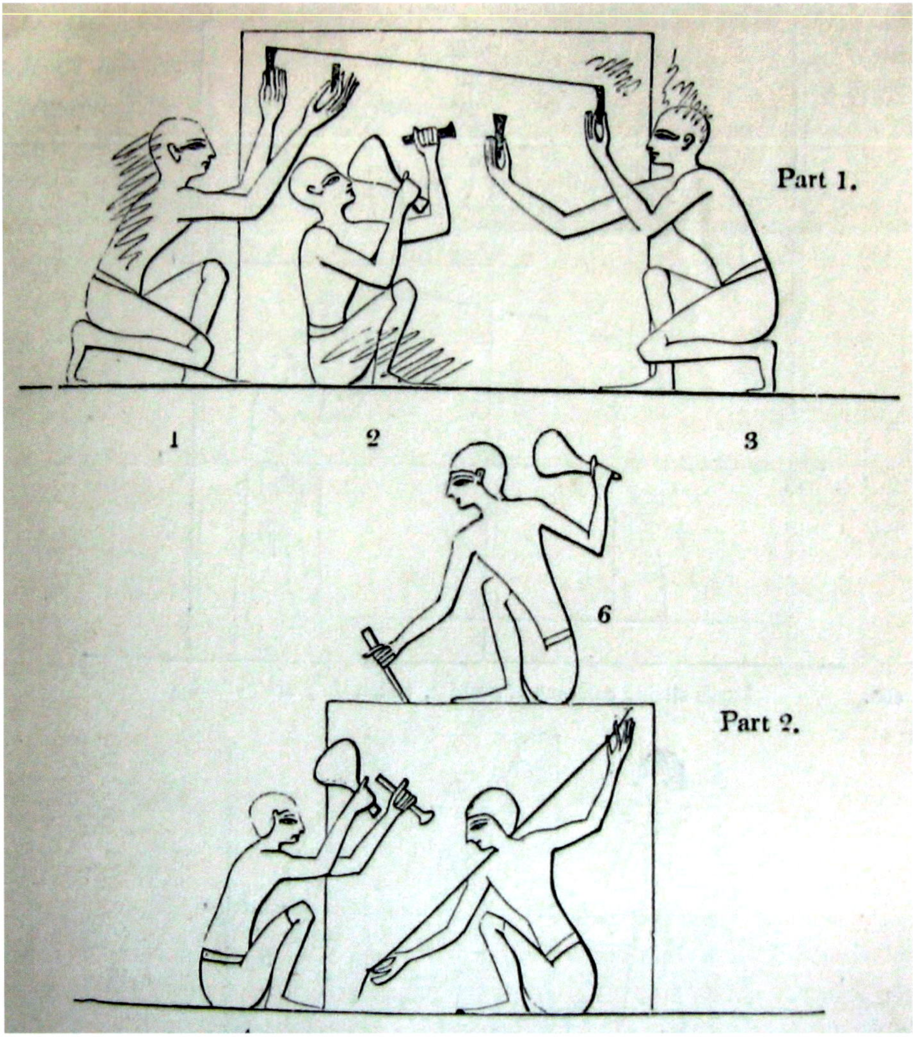

Figure 117: Masons levelling and squaring blocks of stone, Tomb of Rakhmire, Thebes, *c.*1450BCE, woodcut, drawn and copied from the original, from Sir John Gardner Wilkinson, *A Popular Account of the Ancient Egyptians*, 1854, London, John Murray, p.313 (Wikimedia Commons, Public Domain)

Unfired brick was a standard building material from pre-dynastic times throughout Ancient Egyptian history. Bricks were produced from mud and sun dried, brick kilns being unknown until the classical period. These mud-bricks were thoroughly subservient to stone for royal palaces and tombs. They were used mainly for the vernacular houses of the common people and were thus widespread, and sufficiently important to be depicted in Rekhmire's tomb. Brick makers are shown in Figure 119, and the accompanying inscription tells us that the workers are captives of 'His

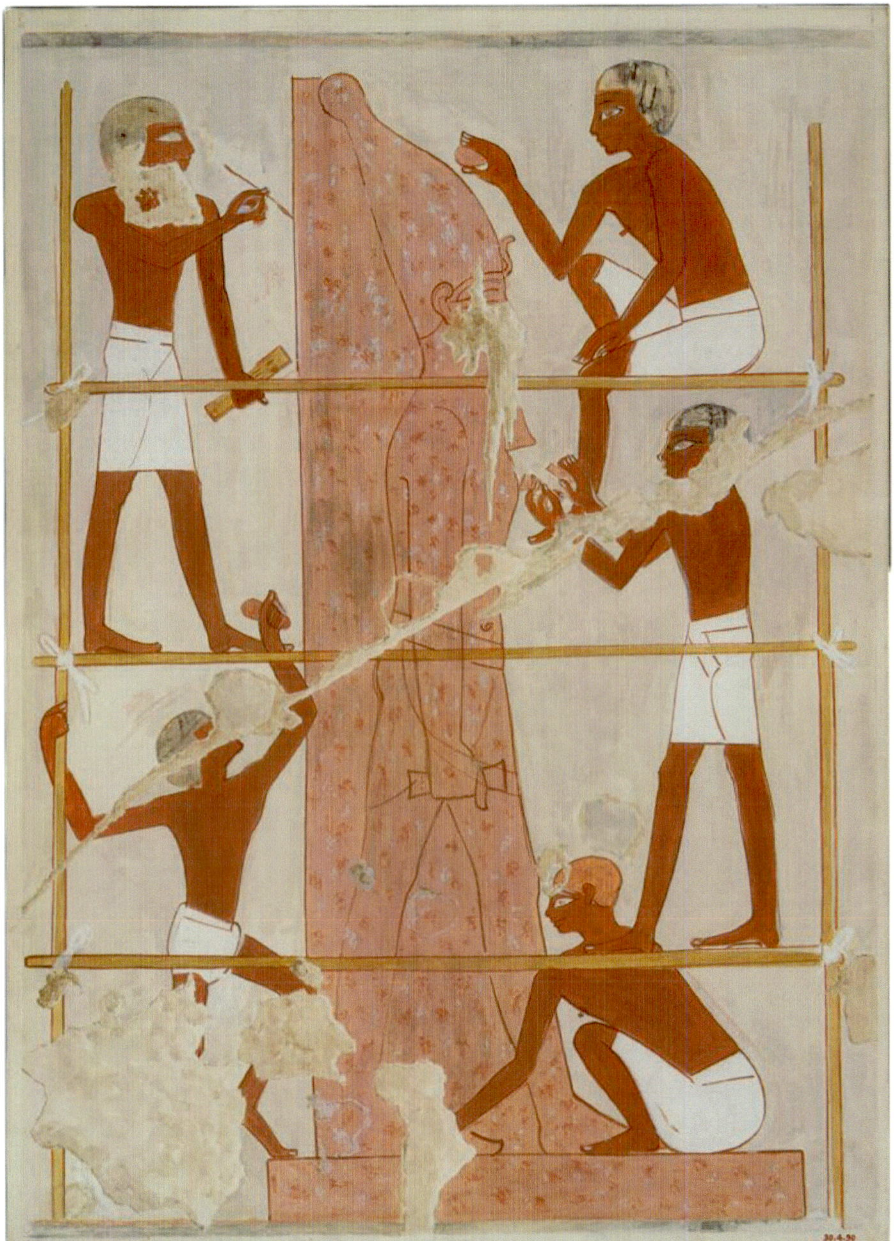

Figure 118: Nina de Garis Davies (1881-1965), *Sculptors at Work, Tomb of Rakhmire*, 1927, facsimile painting from the original at the Tomb of Rakhmire, Thebes. Tempera on paper, 54 × 38.5cm. Metropolitan Museum of Art, New York, Rogers Fund, 1930. (Public Domain) (http://www.metmuseum.org/art/collection/search/544655)

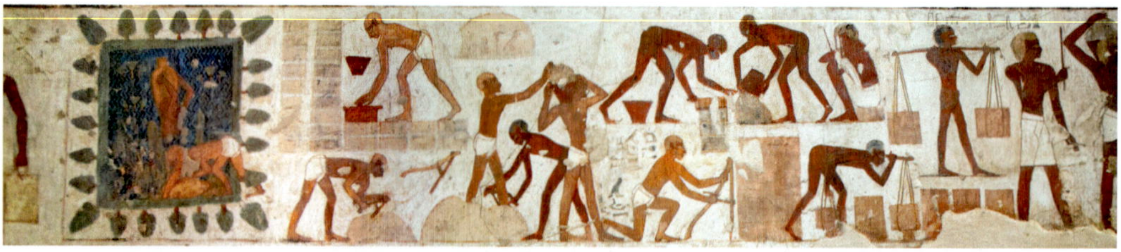

Figure 119: Brick makers, tomb of Rakhmire, Thebes, Sheikh Abd el-Qurna. Photograph, November 2016. (Kairoinfo4u, https://www.flickr.com/photos/manna4u/32513653582) (Creative Commons: https://creativecommons.org/licenses/by-nc-sa/2.0/)

Majesty', prisoners of war from Nubia and Syro-Palestine, bought to build workshops for the temple of Amun at Karnak. The inscription continues with an explanation: From the left water is gathered from a tree-lined pool and mixed with clay on the lower left. The mixture is worked with a hoe and additionally kneaded with the feet. It is then passed up in baskets to the brick makers above who form the bricks with moulds; the bricks are then left to dry in the sun. To the right the hardened bricks are carried to the building site. There are two Egyptian overseers present with rods, one sitting and the other standing to the far right. These workers are not necessary skilled masons but deserve their place as they are akin to masons and labourers such as the mortar mixers and hod carriers of later Roman and Medieval practice.

The Roman Period

This artistic progress now leaps one and a half millennia to the early second century CE and Trajan's Column in Rome. This is a victory column with a spiral narrative celebrating Trajan's victories in the Dacian wars. Dacia is broadly equivalent to modern Romania. The column as a building is a masterpiece of engineering by the Roman architect Apollodorus[114] and the carved narrative on the exterior is no less a masterpiece of Roman art.[115] There are several scenes of construction on the rising bas-relief to the exterior of the column, which are exclusively scenes of military fort building by Roman legionnaires and auxiliaries. Masons are shown building walls and using hammers and chisels. In all the scenes the bricks are of precisely the same size and laid in a simple bond; though intriguingly, the bricks the builders are carrying are bigger. (Fig. 120) It is known that the Roman army established brick-works in Dacia, for the very purpose of building forts, during these campaigns.[116] On the Column of Marcus Aurelius, less than a mile away and one hundred years later, is a similar scene, but in this instance it is the only one showing construction on the column.[117] The Column of Marcus Aurelius is almost an exact copy of Trajan's Column, in building and engineering terms, but not in

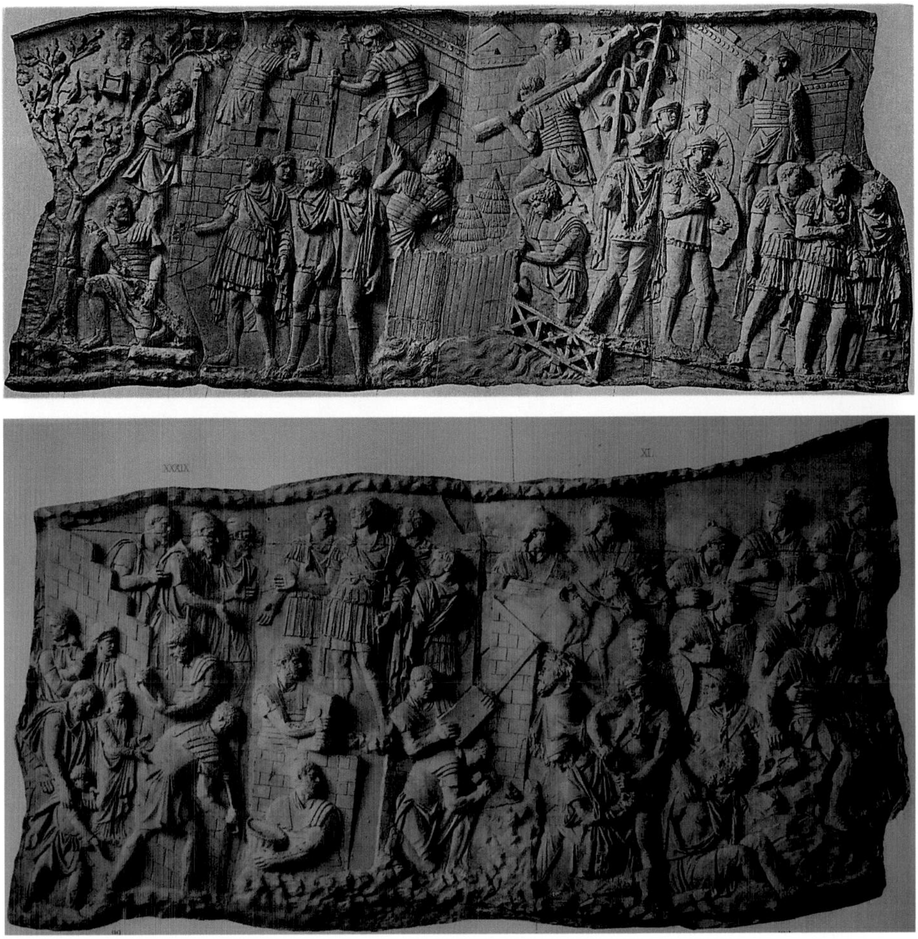

Figure 120: Soldiers building forts, Trajan's Column, Rome, stone bas-reliefs. Photographed by Conrad Cichorius published in *Die Reliefs der Traianssäule*, Vol. 1, *Die Reliefs des Ersten Dakischen Krieges*, 1896, Verlag von Georg Reimer, Berlin, top, plate number 14, bottom, plate number 30. (Public Domain) (https://commons.wikimedia.org/wiki/Category:Trajan%27s_Column_-_Cichorius_Plates)

the art. Wilson Jones calls it 'perhaps the most unashamed instance of copying in the whole history of Roman architecture'.[118] There is debate over whether the reliefs, of both columns, were carved on the ground before their erection, or were instead carved once the heavy engineering of construction had been completed. It is certain that the stairs were cut within their separate drums before erection. It is most probable that the setting-out of the design concept for the exterior art, together with some roughing out by the sculptors, was achieved on the ground, followed by finishing *in situ* on the columns.[119]

Another Roman relief showing a mason at work is from Capua in the province of Caserta dating to the first half of the second century. Against the backdrop of a

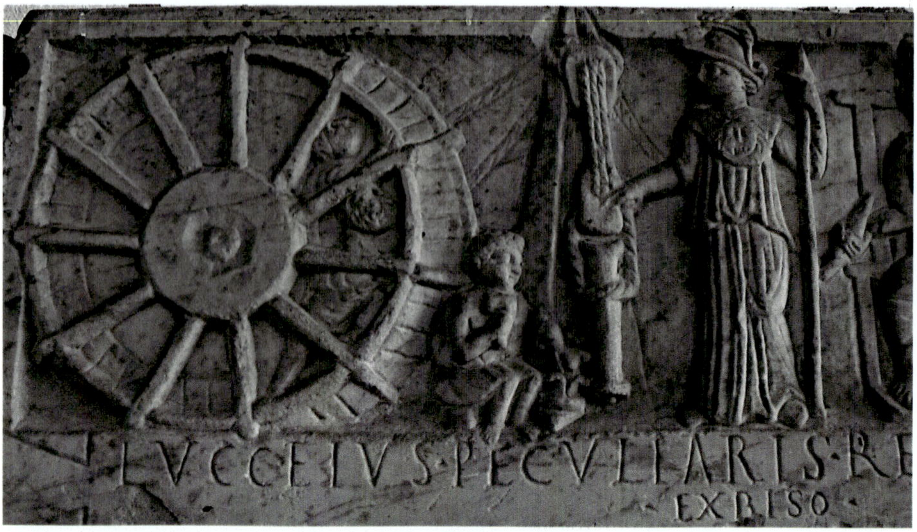

Figure 121: Capua, Italy, Second Century, detail, carved stone relief, part of a sarcophagus, showing a tread-wheel crane lifting a column with a mason carving a capital. Museo Provinciale Campano di Capua, Sarcofagi (Sala III-IV). Photograph: Dan Diffendale.

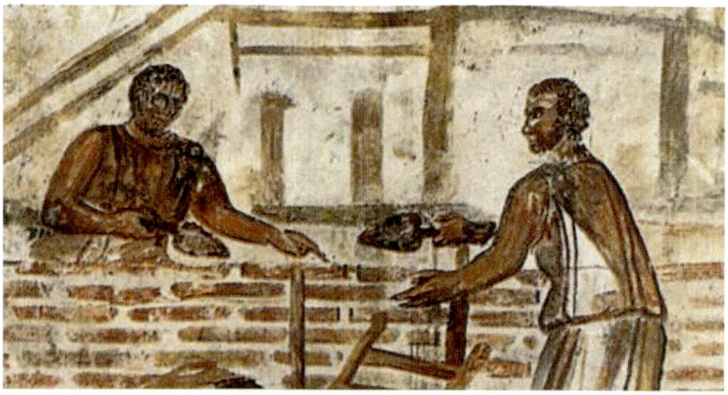

Figure 122: Early Christian Painter (active fourth century in Rome) Tomb of Trebius Justus, Rome, fourth century AD, detail of Figure 3.
https://www.wga.hu/frames-e.html?/html/zearly/1/2mural/5vialati/latina7.html

tread-wheel crane in the process of lifting a column, a mason sits carving a capital, doubtless destined for the top of the column once erected. (Fig. 121) The relief commemorates the construction of the stage at the Roman theatre in Capua and is said to be part of a sarcophagus.[120] It bears the inscription *Lucceius pecularis redemptor prosceni.*

The mason's trowel does not appear in art until the fourth century CE, this is in the Roman wall painting of builders at work (Figs. 3 and 122) from the Hypogeum of Trebius Justus in Rome. The painting shows two brick masons at work on a scaffold

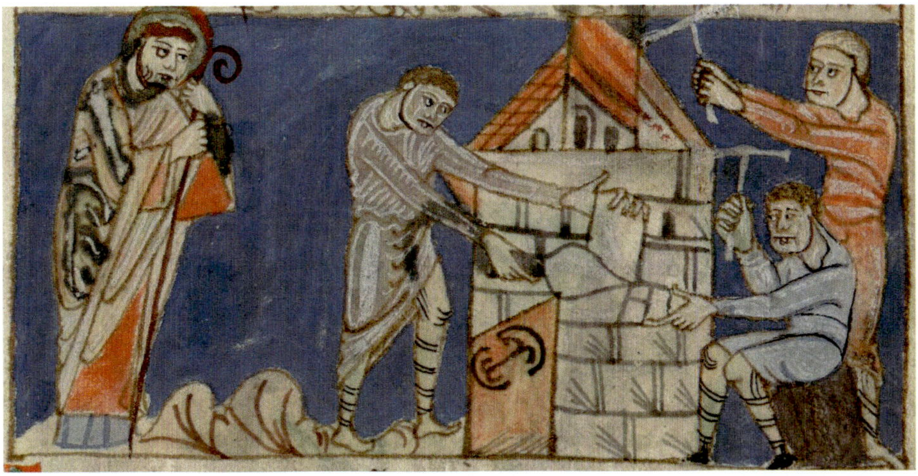

Figure 123: Flemish miniature *Flemish Converts Building a Church under the Supervision of St. Amand*, from the manuscript *The First Life of St. Amand*, produced in the Abbey of Saint Amand in 1066-1107. (Bibliotheque Municipale de Valenciennes, France, ms. 0502, f. 018.)

building a wall, both of whom are using bricklayers trowels which are immediately recognisable. The form of the trowels is almost exactly the same as one encounters today. The precise metallurgy is not known, but the forging and working of iron would have been straightforward for Roman smiths with their long experience and expertise in producing weaponry.

Medieval and Renaissance

In the Medieval period virtually all depictions of building work show at least one mason at work.

The eleventh century Franco Flemish manuscript *The First Life of St. Amand*, produced in the Abbey of Saint Amand in 1066-1107, contains an image *Flemish Converts Building a Church under the Supervision of St. Amand.*[121] (Fig. 123) The miniature is primitive and naïve in that the figures are vastly out of scale with the church, which appears to be either complete or near complete.[122] The mason's trowel is unmistakable, held in the right hand whilst the left places the stone he is laying. The other two masons are using axes; these were used for cutting stone to shape and even for carving. They appear unwieldy but given experience and skill were used to great effect.[123]

The fine mosaics in the late twelth century Arab-Norman-Byzantine cathedral of Monreale in Sicily include a depiction of the building of the Tower of Babel on the south wall of the nave above the arcade. (Fig. 124) This is part of the largest cycle of such mosaics in Italy. Two masons work at the top, one with an axe, while

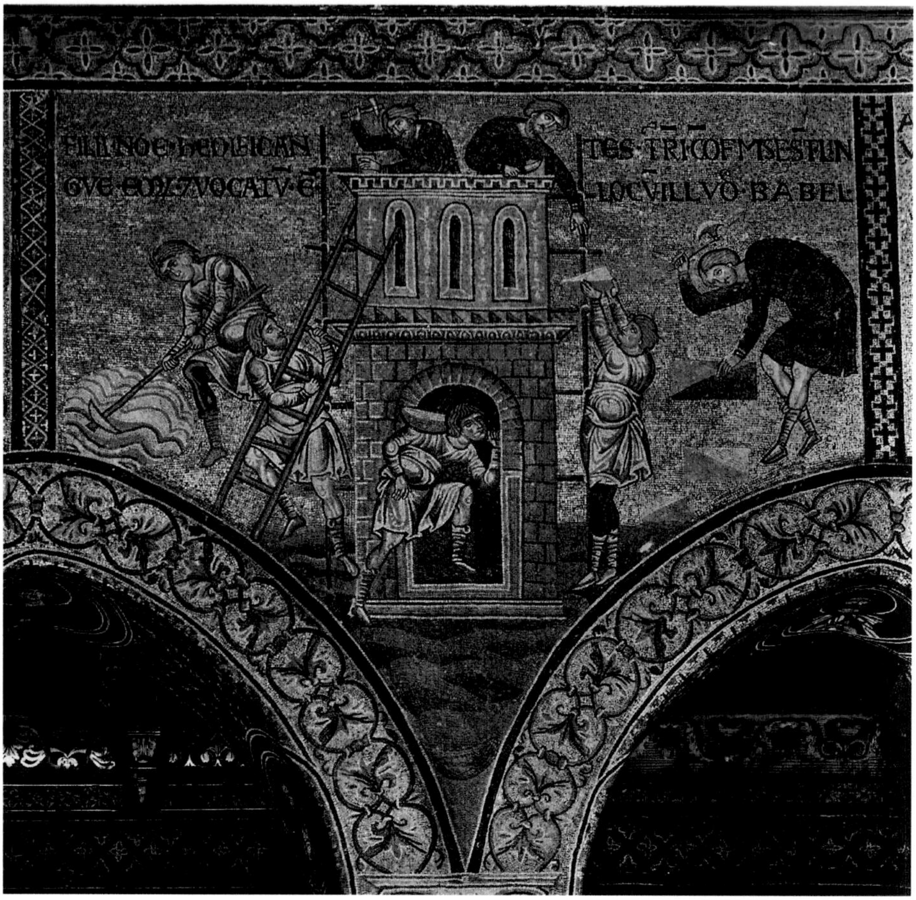

Figure 124: Mosaic artist active 1180s, Monreale Cathedral, building the Tower of Babel. Archive photograph *c.*1910-1940, Fratelli Alinari. Fondazione Zeri, Universita di Bologna, Italy.

the other holds a trowel whilst leaning down to take a piece of stone from a worker below. On the ground mortar is being mixed, a mason works stone with an axe and a labourer enters the building with a hod of mortar on his shoulder.

Similarly in the basilica of San Marco, Venice is another scene of the tower under construction. (Fig. 125) Here a mason is mixing mortar and loading it into a container on a stand, a mason's labourer also loads stone or bricks into a similar container whilst another carries a load up the ramp on his back to the two masons above at the top of the scaffold. One of these masons prepares to use a container of mortar from which the handle of his trowel is seen protruding, while the other at the edge of the scaffold appears to lean on his upturned axe. At the bottom is an overseer or architect with a measuring stick with two other figures, one of whom passes up a bucket; all three are suspended in mid-air.

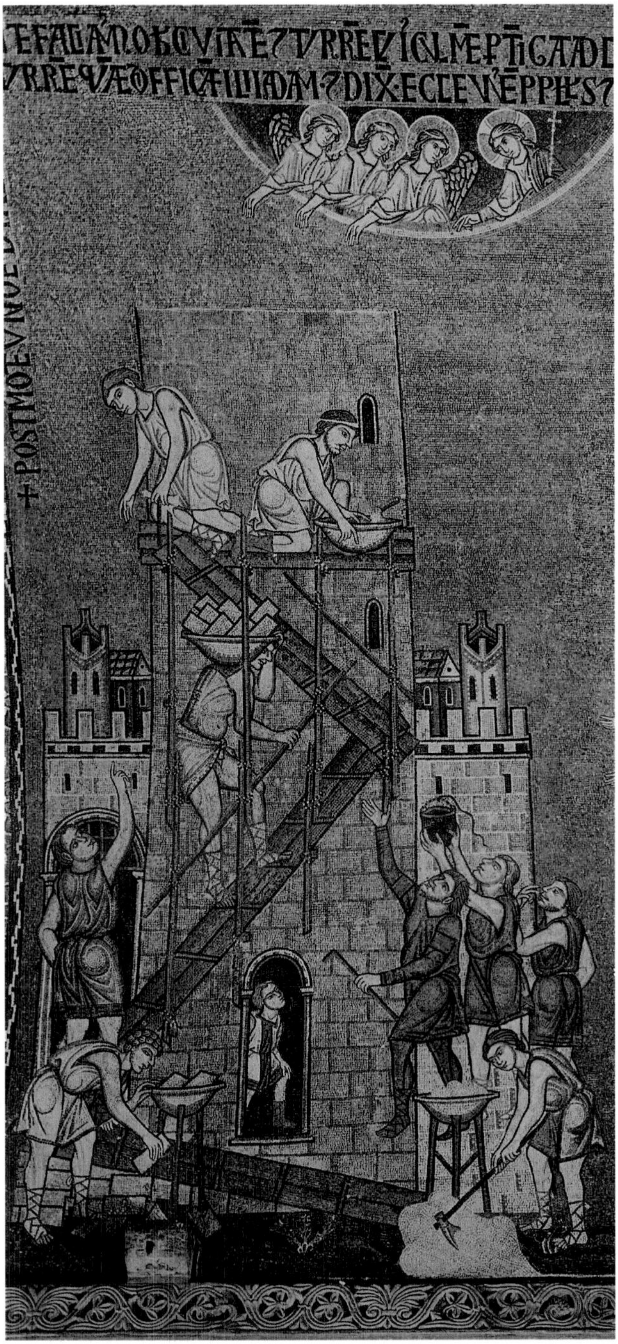

Figure 125: Mosaic artist active in the twelth century, Basilica of San Marco, Venice, *building the Tower of Babel.* Archive photograph *c.*1910-1940, Fratelli Alinari. Fondazione Zeri, Universita di Bologna, Italy.

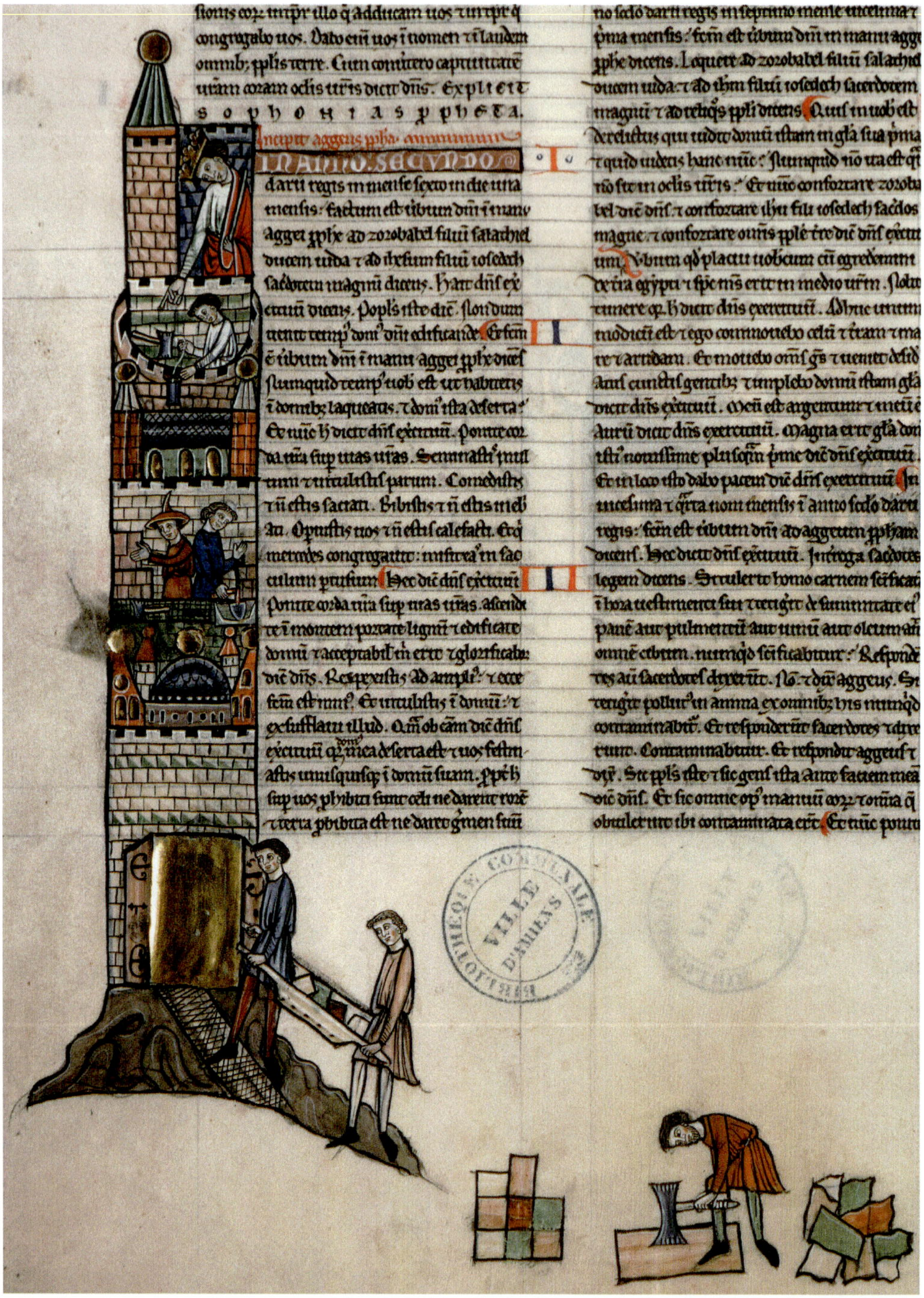

Figure 126: French, thirteenth century Bible, *Cyrus orders the rebuilding of the temple of Jerusalem*: Historiated initial 'I' (Bibliotheque Municipale de Amiens, France, MS. 0021, f. 167v.)

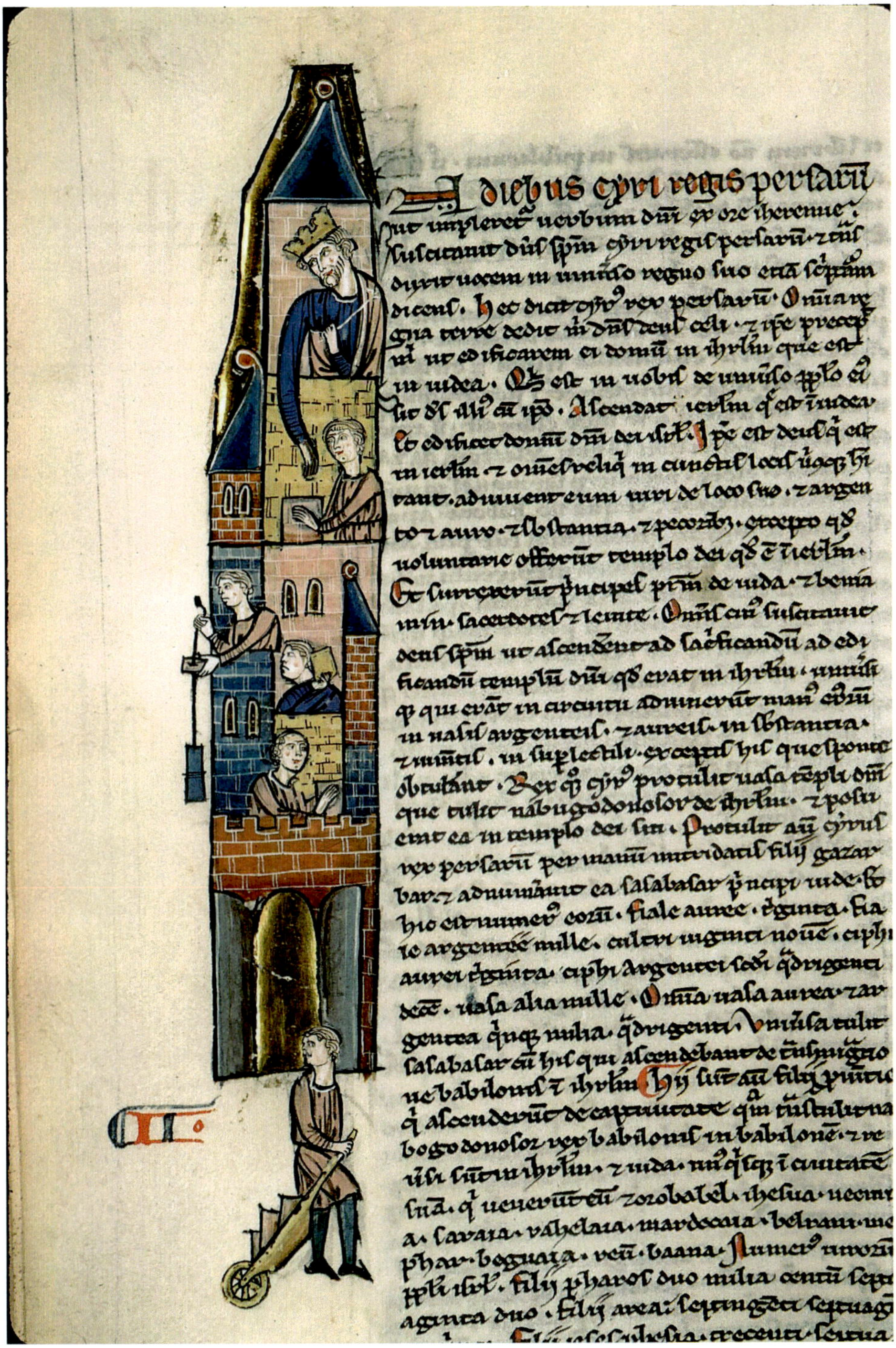

Figure 127: Studio of Blanche, French, 1220-1230 Bible, *Cyrus orders the rebuilding of the temple of Jerusalem*: Historiated initial 'I' MS 1185, f. 127v Bibliotheque St Genevieve, Paris, France.

Illustrations of the edict of Cyrus *'In anno primo Cyris regis'*,[124] when Cyrus King of Persia proclaims the rebuilding of the Temple in *c.*539 BCE, are common in early bibles, as already seen in Chapter 1 and Figure 22. In another early thirteenth century French bible Cyrus directs operations from the top (Fig. 126) in a building, that for all intents and purposes, does not yet exist. But in these early bibles it is the norm for many such illuminations to show the building under construction as being virtually complete. There are various activities: Below Cyrus at the top a mason has set aside his axe and is preparing to use a plumb line. Below him are two masons building a wall, one with a curiously bent trowel. On the ground about to enter the building are two labourers with a barrow of stones, whilst another mason cuts stone with an axe. In a similar image of around the same date we see an early depiction of a wheelbarrow at the base of the construction.[125] (Fig. 127) Other masons are at work, including one grappling with a plumbline.

In Pietro Lorenzetti's Beata Umiltà Altarpiece painted *c.*1340, which is now in the Uffizi, Florence, one of the panels depicts St Humility supervising the building of a church or convent. (Fig. 128) She arrives with a donkey laden with bricks and is in conversation with a mason who leans over a partially completed wall. In his outstretched hand he holds a bricklayer's or mason's trowel. The trowel is interesting as the sheet of metal it has been forged from is bent in half at 90°. The business end, the blade, is the standard 'V' shape whilst the upper half has rounded shoulders or a 'U' shape and would normally form part of the shank to which the handle is attached again at 90°. The mason at the top is checking the verticals of the window opening with a plumb line, whilst a labourer carries a basket of bricks on his back into the building.

The universal chronicle *The Historie Ancienne Jusqu'a Cesar* covers a period of time from the moment of creation to the time of the Roman Emperor Julius Caesar. An early thirteenth century copy in the British Library, London contains a scene of the construction of Rome.[126] (Fig. 129) Thought to be compiled in the crusader Kingdom of Jerusalem in the scriptorium of St-Jean-d'Acre it shows a largely complete and eclectic group of buildings which are decidedly Frankish-Medieval rather than classical in appearance. To give the scene a feeling of construction various masons range over the buildings. Apart from two in the foreground who are actually cutting stone, the rest are only really posing with their tools, two with axes and one with a trowel. Whilst two others grapple with materials and one is seen hurrying into the building in the foreground, almost it seems retreating from the two figures on the left who are of a larger scale, and therefore higher status.

Other chronicles and epics where masons are busy include the *Grandes Chroniques de Saint Denis*, 1300-1349; (Fig. 19) the fifteenth century French *Histoire d'Outremer*; (Fig. 130) the Flemish *History of Alexandra the Great*, also dated to the fifteenth century; (Fig. 23) the French *Grandes Chroniques de France* of the early fiteenth century, (Fig. 131) the French *Book of the Queen*, 1410-1414; (Figs. 132

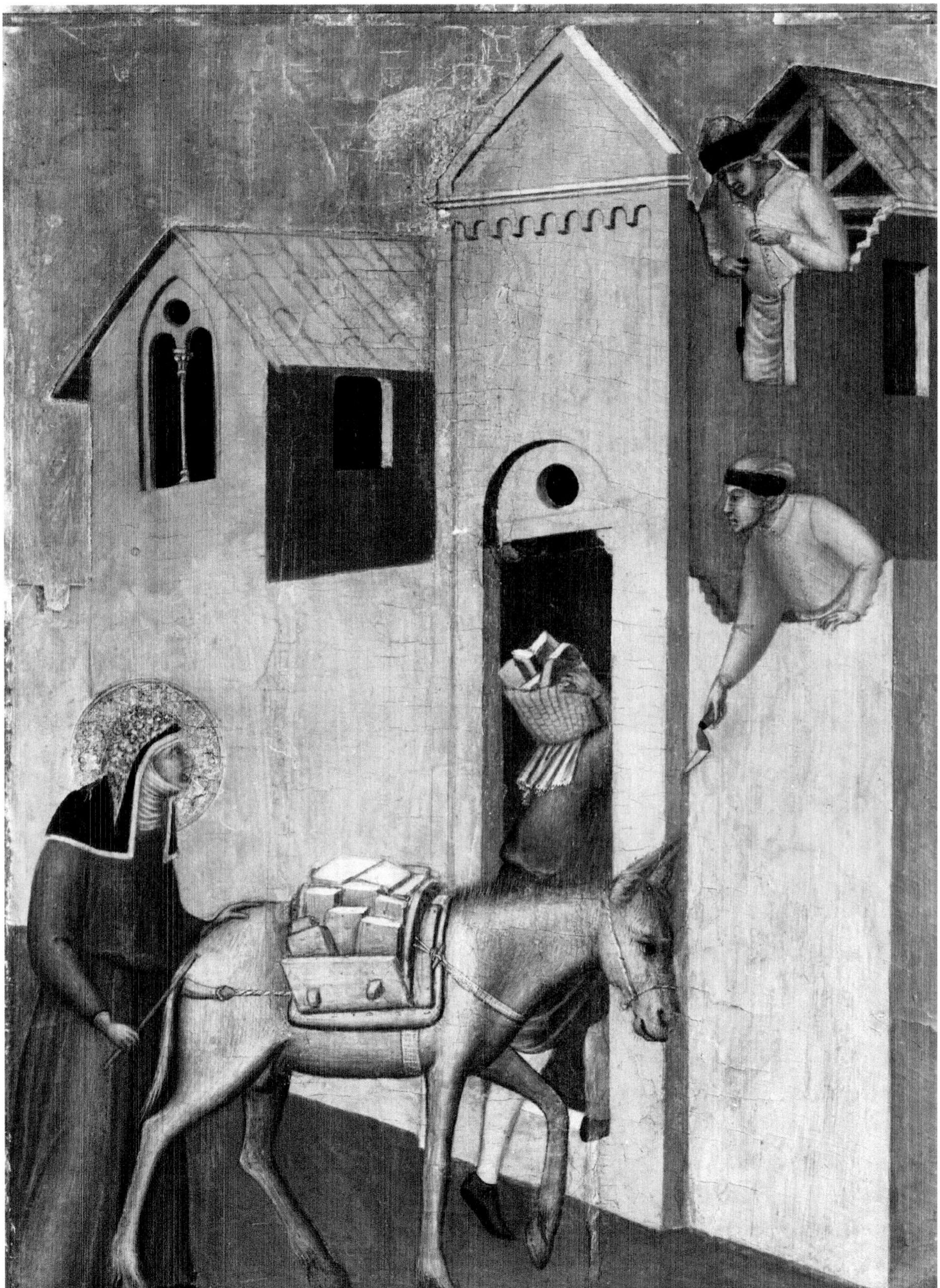

Figure 128: Pietro Lorenzetti, *Saint Humility Transports Bricks to the Convent She is Building*, c.1341, Tempera on wood app. 40 × 30 cm, part of the now incomplete Beata Umiltà Altarpiece, Uffizi, Florence. (Archive photograph by Brogi, 1900/40, Fondazione Zeri, Universita di Bologna, Italy.)

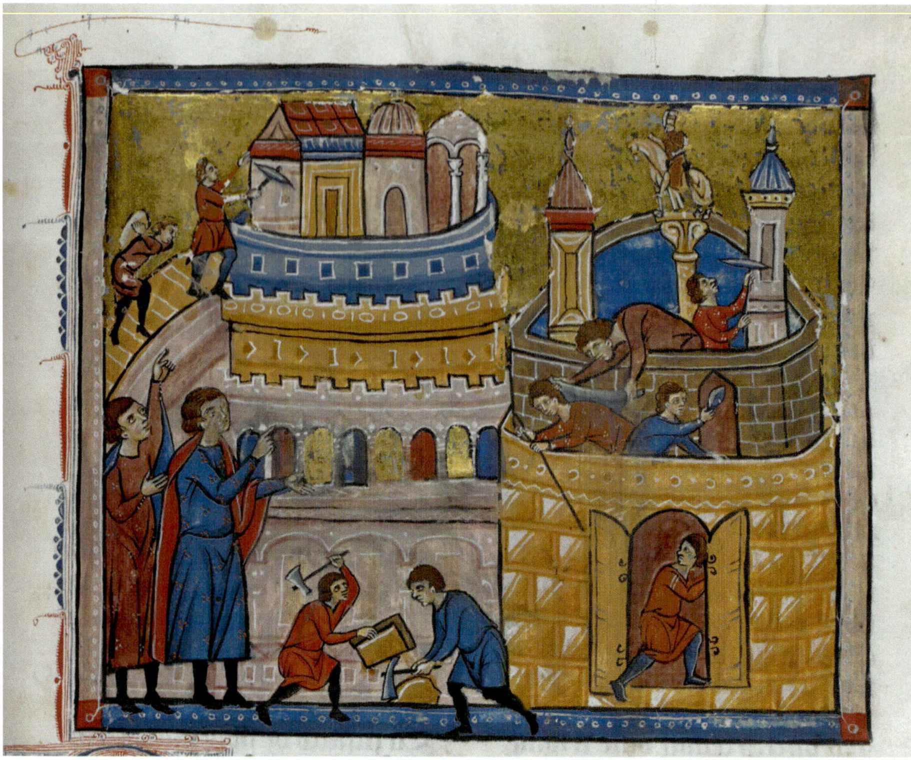

Figure 129: *Construction of Rome*, late thirteenth century, before the fall of Acre in 1291, French, from *Historie Ancienne Jusqu'a Cesar*, © The British Library Board, London, Add. Ms 15268, f. 156r.

& 133) the Burgundian *Roman de Girart de Roussillon*, *c.*1460; (Fig. 134) the Italian *Aeneid*, *c.*1460-1465, illustrated by Apollonio di Giovanni; (Figs. 105 & 135) the French *Antiquites Judaiques* by Josephus, *c.*1465, possibly illustrated by Jean Fouquet; (Fig 136) the Flemish *Romuléon*, 1466-1467; (Figs. 29 & 137) Livy's *Ab Urbe Condita*, French, 1467-1470, illustrated by Jaques de Besançon (attributed); (Fig. 138) the French *Chronique de la Bouquechardiè*, of the third quarter of the fifteenth century, illustrated by the Master of the Echevinage de Rouen; (Fig. 139) St Augustine's *La Cité de Dieu*, Burgundian, *c.*1480; (Figs 140 & 141) and the French *Recueil des Histories de Troie*, 1490, illustrated by Jean Colombe. (Fig. 142)

The following books of hours, breviaries and commentaries on, or summaries of, the Bible also show masons at work: The French *Bible Historiale Complétée*, 1411; (Fig. 143) the French *Bedford Hours*, 1415-1430; (Fig. 144) the Flemish *Grimani Breviary*, 1515-1520; (Fig. 145) the French *Breviary of Charles de Neufchâtel*, 1498; (Fig. 146) a French Book of Hours, late fifteenth century; (Fig. 147) and the Flemish *Spinola Hours*, 1510-1520. (Fig. 40) Finally the architectural treatise: The Italian *Trattato*

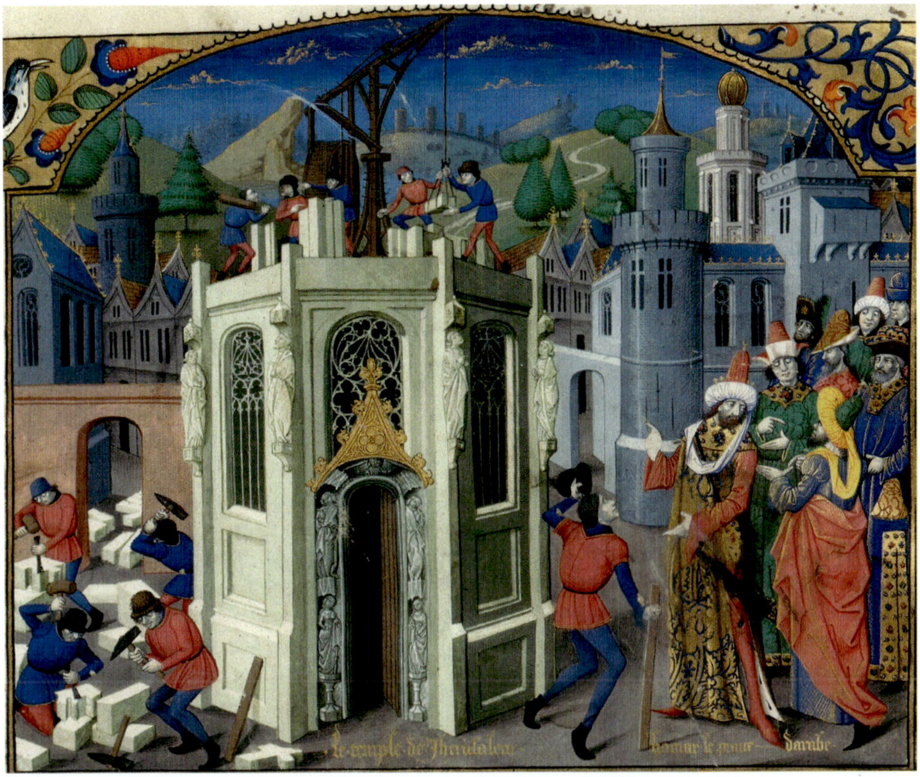

Figure 130: Master of the Rouen Echevinage, active 1455-1485, French, *Construction of the Temple of Jerusalem* from the manuscript *Histoire d'Outremer*, written by the chronicler and Archbishop of Tyre *c.*1130-1186. Bibliotheque Nationale de France, ms 2629 f. 17.

d'Architettura, by Antonio di Pietro Averlino, known as Filarete, *c.*1465, (Fig. 148) and from a later copy of the same treatise, 1488-1489. (Figs. 149 & 150)

In the image of Charlemagne supervising the construction of his chapel, at Aachen or Aix-la-Chapelle, Figure 19, there are three masons and two labourers; a carver works on mouldings using a mallet and chisel, while below him is a mason squaring off a piece of stone using an axe, and beside him are a square and a rule or straight edge. On the scaffold a fixer-mason lays a piece of stone tamping it down with the wooden handle of his axe, although one would expect him to be using a trowel; in his left hand he holds a level or straight edge. A labourer on the ground mixes mortar, whilst another carries it up the ladder.

The Master of the Rouen Echevinage, also know as the Master of the Geneva Latini, was a French illuminator active 1455-1485 who painted the scene in Figure 130, the *Construction of the Temple of Jerusalem*. Here we see masons both on the ground and also constructing the second storey of the building. Those on the ground are

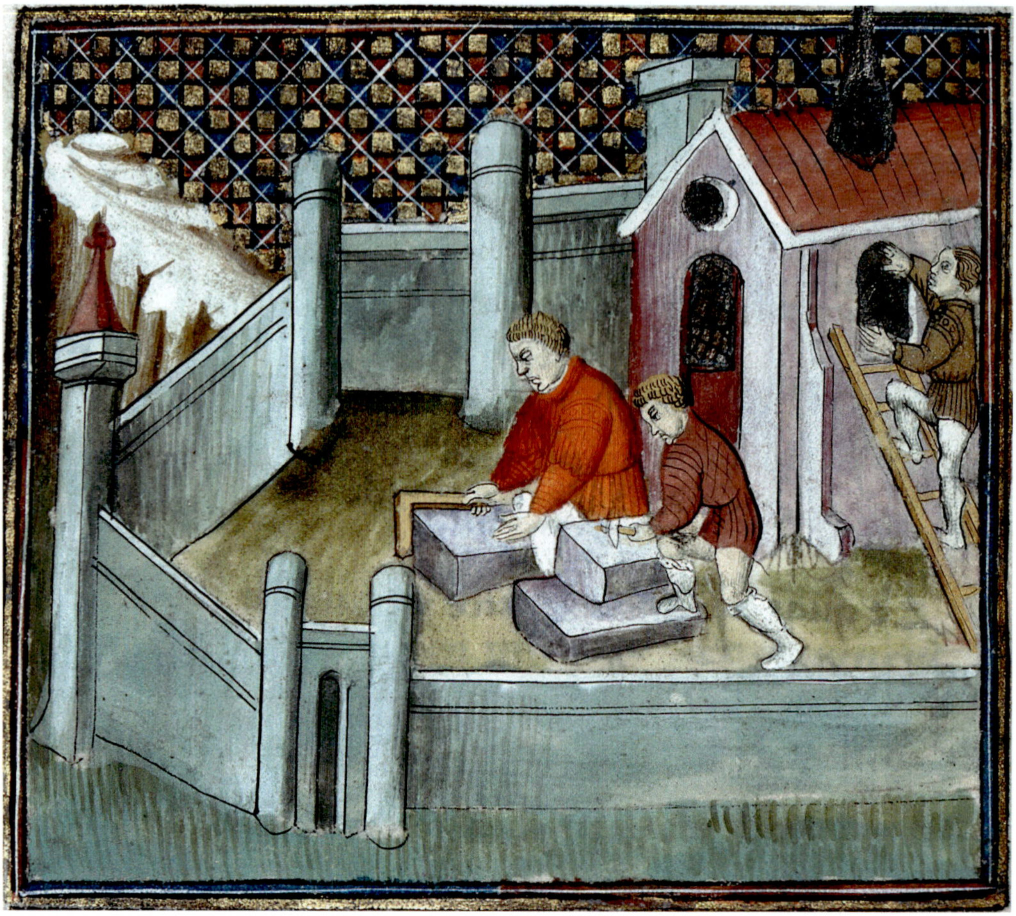

Figure 131: Early fifteenth Century, French, *Construction of the Abbey of St Denis*, illumination from the *Grandes Chroniques de France*, ms. 0863, f. 056v. Bibliotheque Municipale de Besançon, France.

squaring blocks with axes and carving mouldings with chisels and mallets, and beyond an archway can be seen more blocks awaiting work. On the upper storey are masons fixing the blocks that have been lifted by the crane. The architect or master mason pays obeisance to King Solomon, who is accompanied by his lavishly dressed retinue, and who has commanded this work.

In Figure 23 the *Construction of Alexandria*, only mason fixers and bricklayers are seen working, although mortar is being prepared and ready worked stone is waiting in the left foreground. The somewhat naïve illumination of the early fifteenth century, of the construction of the Abbey of St Denis in the *Grandes Chroniques de France* (Fig. 131) shows two masons, one of whom is testing a block using a try-square, and the other is working with an axe. The glazier and his ladder

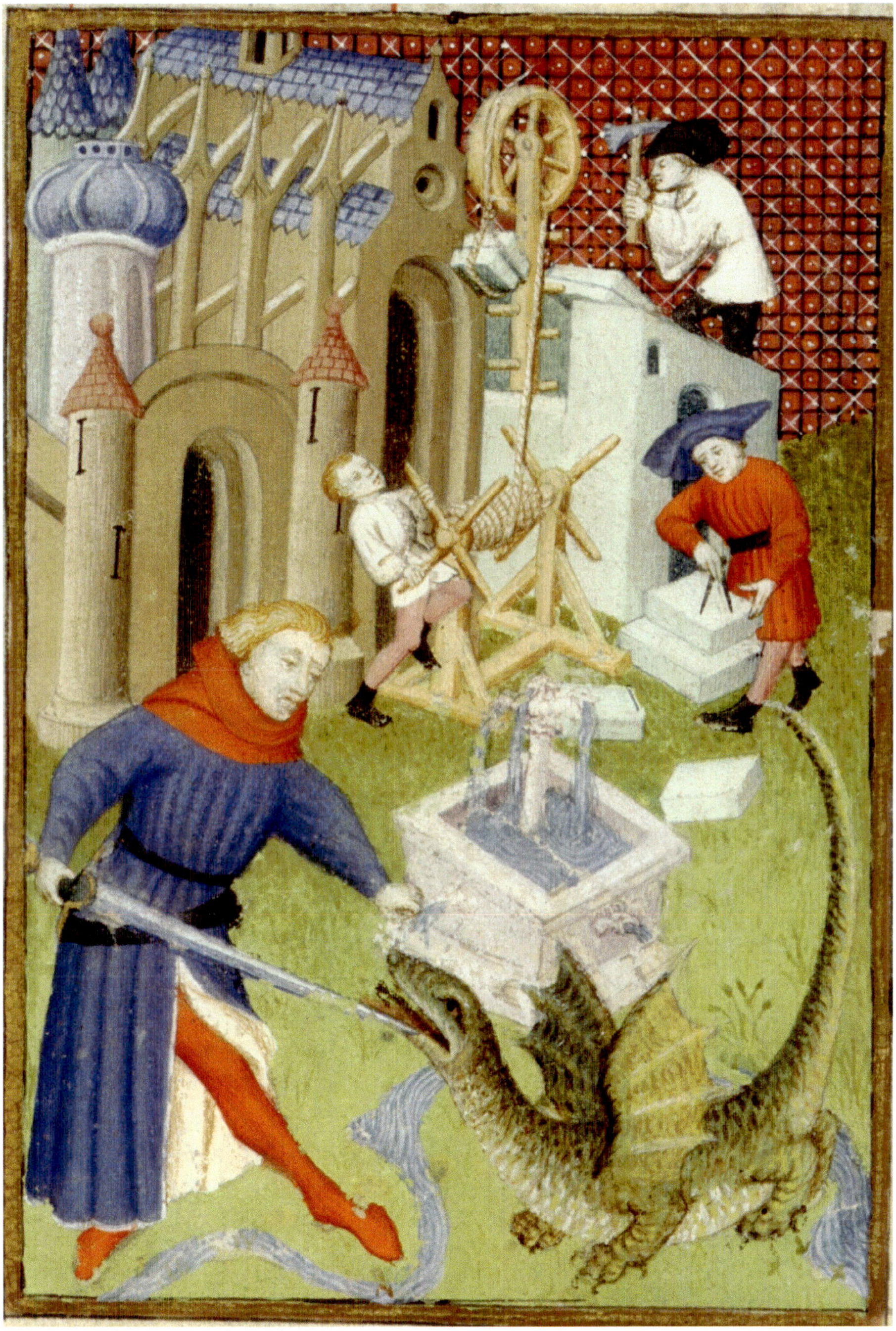

Figure 132: Master of the Cité de Dames, French, *Construction of Thebes and Cadmus killing the dragon at Ares's spring*, 1410-1414, from Christine de Pizan, Collected works ('The Book of the Queen'). © The British Library Board, London, Harley ms 4431, f. 109r

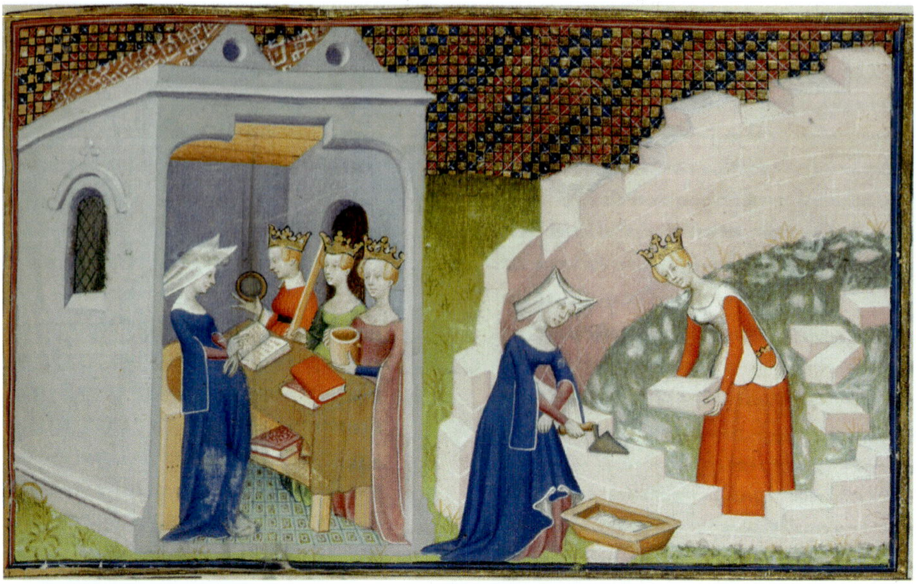

Figure 133: Master of the Cité de Dames, French, *Christine de Pizan before the personifications of Rectitude, Reason, and Justice in her study, with Christine and Reason building the 'Cité des dames'*, 1410-1414, from Christine de Pizan, Collected works ('The Book of the Queen'). © The British Library Board, London, Harley ms 4431, f. 290r

are discussed above on page 68. Christine de Pizan's collected works, known as *The Book of the Queen*, was produced for Queen Isabel of Bavaria who married Charles VI of France in 1385. Two of the works, *L'Épître Othéa* and *Le Livre des Épîtres sur le Roman de la rose* contain miniatures by the Master of the Cité de Dames, (Figs. 132 & 133) (*La Cité de Dames* is also a work in this collection, see Figure 73). In the first miniature building work is going on in the background (Fig. 132) with a mason using a pair of dividers, whilst another mason higher up uses an axe to cut a piece of stone; the crane is discussed on page 148. In the foreground Cadmus, King of Thebes kills the dragon at Ares' spring. The second image (Fig. 133) shows the personifications of Reason, Rectitude and Justice in a house. On the right Christine and Reason build the Cité des Dames; Christine has a masons trowel and a container of mortar at her side, and Reason is laying the stone.

The Master of Girart de Roussillon is named for the *Roman de Girart de Roussillon*, a manuscript he produced for Philip the Good, Duke of Burgundy in *c.*1460. Girart lived *c.*810-*c.*879 and he and his wife Bertha commissioned the building of twelve churches. These are shown in Figure 134; eight are in the main illustration and the remaining four are placed one on each border. Four of the churches appear complete while work progresses on the other eight. The church in particular which takes up much of the foreground is of some interest as it shows the walls to be constructed

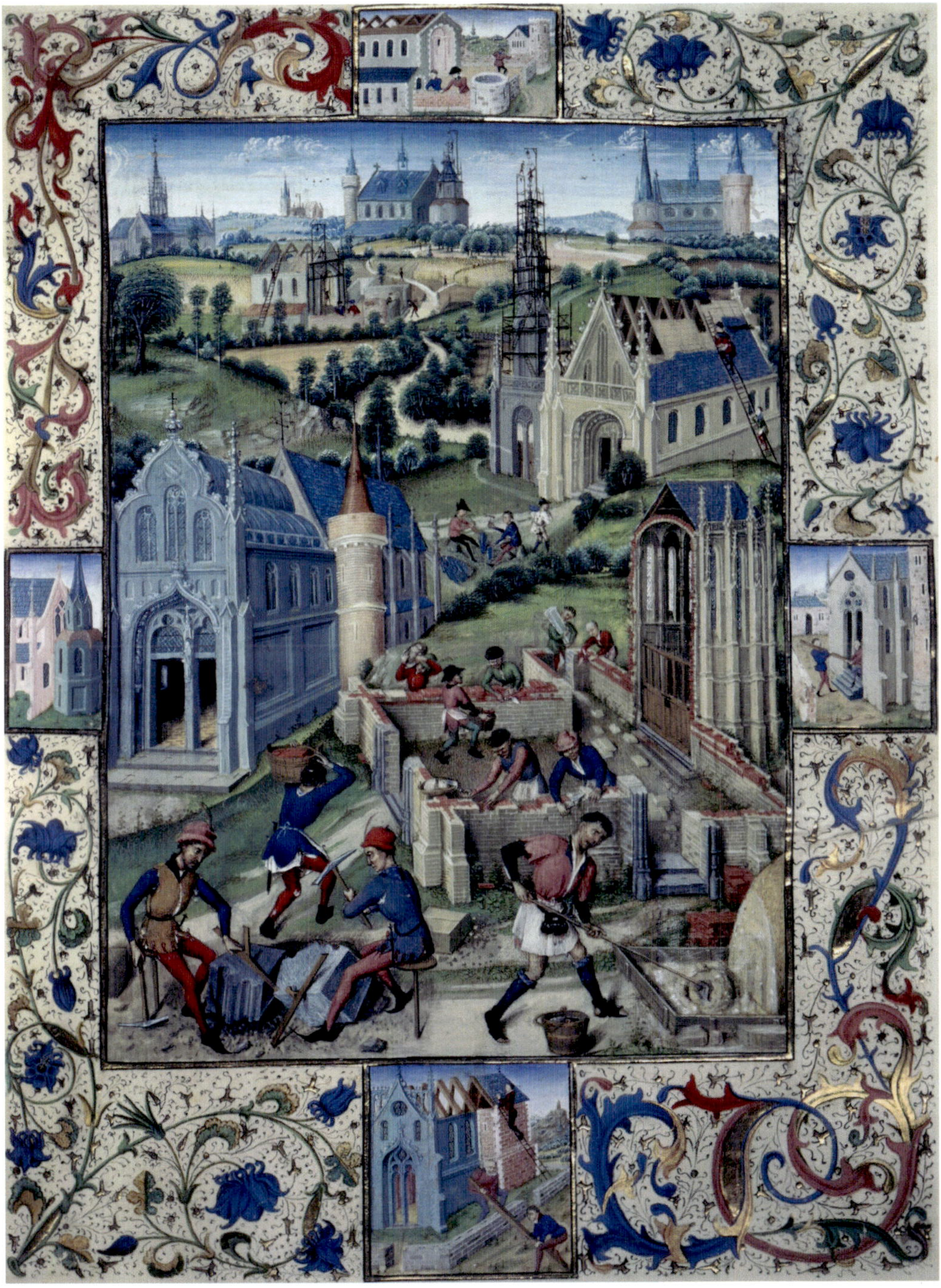

Figure 134: Master of Girart de Roussillon, Burgundian, active 1440-1465, *Construction of twelve churches by Girart and his wife*, c.1460. Österreichische Nationalbibliothek, Vienna, ms Cod. 2549, f. 164r (E 28.568-C)

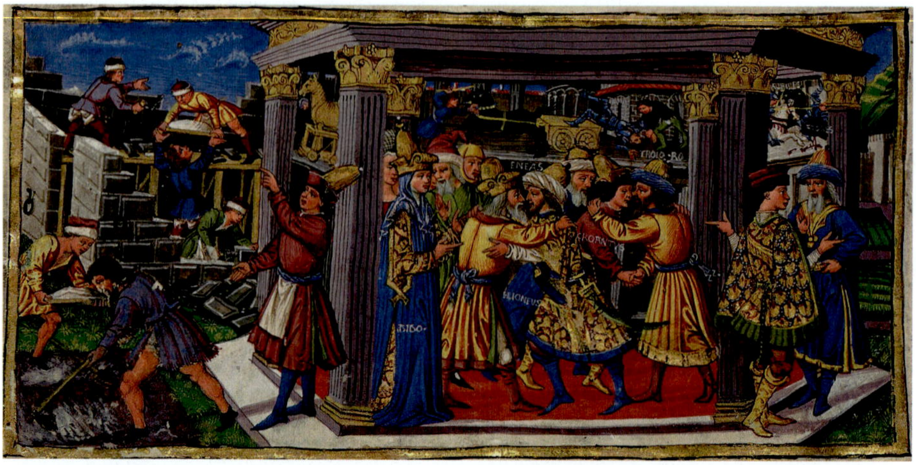

Figure 135: Apollonio di Giovanni, Italian, *Meeting of Dido and Aeneas and the Construction of the Medici Palace*, miniature from *Vergilius Publius Maro, Bucolicon, Georgicon, Aeneis*, 1460-1465. Palazzo Medici Riccardi, Biblioteca Riccardiana, Florence, ms Ricc.492, f. 72v. (©Photo Donato Pineider: 'Su concessione del Ministero dei Beni e le Attivita Cultural')

from stone facings with a brick core. The detail is compelling and very well observed, the brickwork is properly toothed into the stonework and on the interior brought forward to form decorative banding. This observation and depiction extends also to the masons and other workers who are all convincing in their tasks and tools. A scaffold will soon be required to take the building up to be next level. Two masons are working on mouldings for doorway or window jambs using axes with their well-crafted straight-edges, one of which is also a 45° angle-try which he is using to test the level and flatness of his work. They have made themselves stools to sit upon. These two have fine footwear in contrast to the labourer mixing mortar whose toes and heels are exposed; all of the workers in the foreground wear daggers. The chancel is complete and gated off so that services can be performed. In the middle distance are two slaters splitting slates with slate hammers, while another carries the finished slates to the roofers seen climbing the ladder on the church to the right. Here also very clearly on the roof are the sarking boards nailed to the rafters to which the slates are then nailed, an important detail of Medieval building and testament again to the Master's acute observation of what is happening around him.

Apollonio di Giovanni, 1415/17-1465 was a Florentine panel painter specialising in cassoni, as well as a miniaturist illuminating many manuscripts.[127] He favoured subjects relating to the *Aeneid* and Greek Mythology. In his two miniatures *The Meeting of Dido and Aeneas and the Construction of the Medici Palace* (Figs. 135 & 105) masons can be seen laying pre-prepared plain rusticated masonry blocks. In the left foreground mortar is mixed with a hoe and a shovel lies on the ground at the feet of

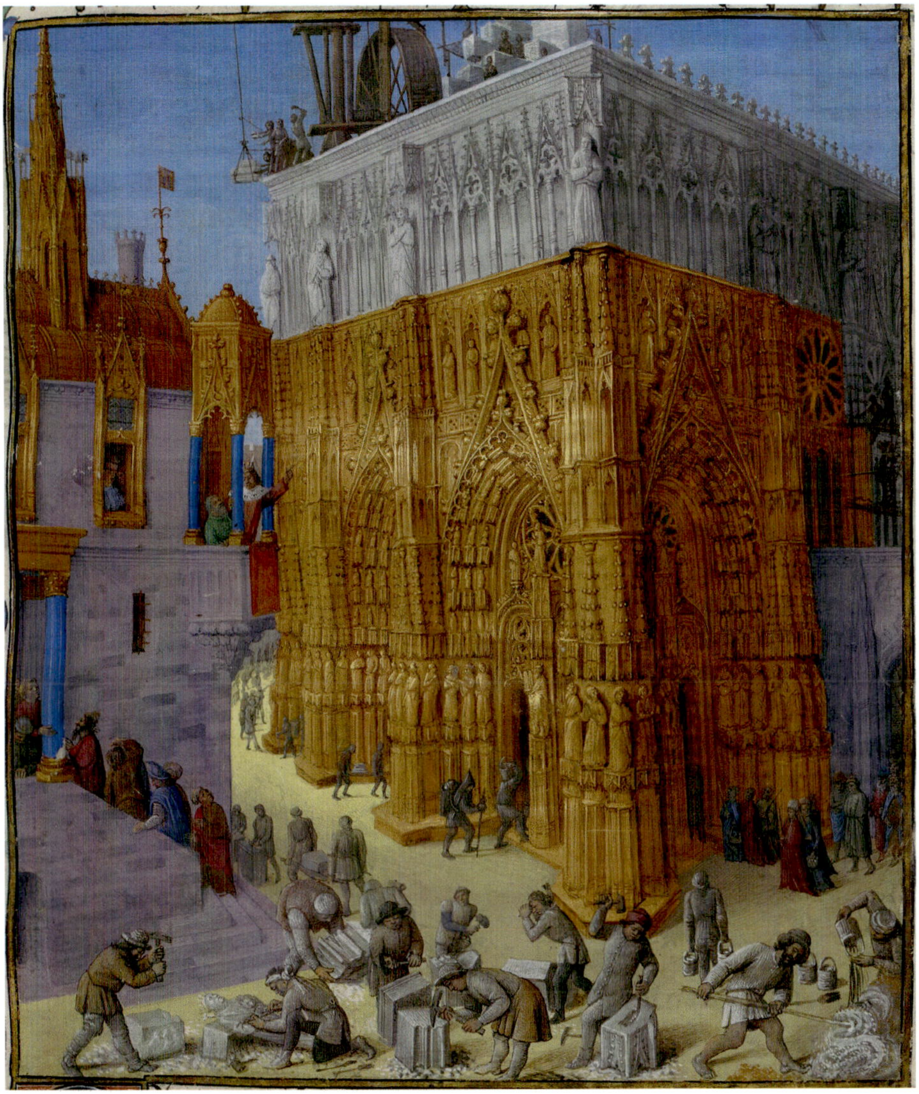

Figure 136: *c.*1465, Jean Fouquet, 1420-1481, or the Master of the Munich Boccaccio, active 1460-1480, under the former's supervision (probably one of his sons), *Solomon Supervises the Construction of the Temple,* from Josephus, *Antiquites Judaiques*. French, Bibliothèque Nationale, Paris, ms fr. 247, f. 163.

the labourer collecting a container of mortar. The actual rustication of the masonry on the palace is different to that depicted, being 'quarry-faced' or 'cyclopean', but the heavy iron rings that Apollonio depicts are on the real building. The crane seen in the second illumination (Fig. 105) is of particular importance and is discussed on page 152.

The Master of the Munich Boccaccio, active 1460-1480, is named after a copy of Boccaccio's *Des Cas des nobles hommes et femmes*, now in Munich. A manuscript

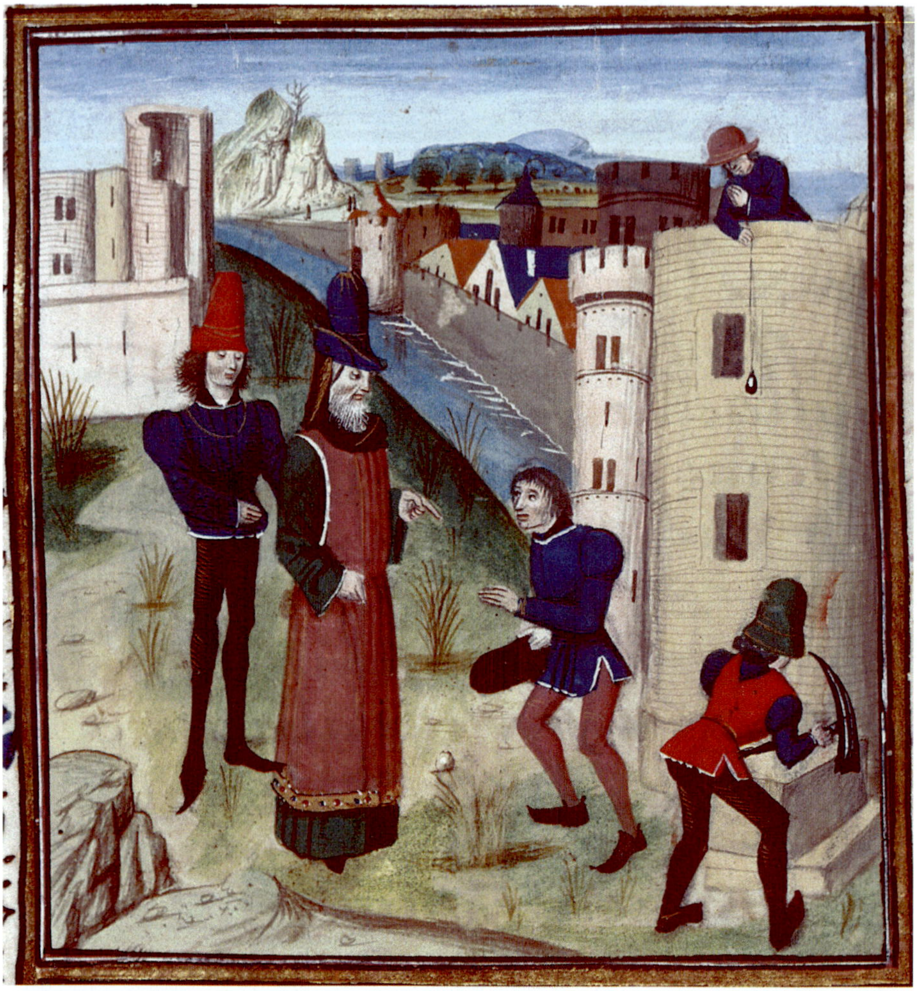

Figure 137: 1466-1467, Flemish, *The Construction of Rome*, from the 'Romuléon'. Bibliothèque Municipale de Besançon, France, ms 0850, f. 050.

he illuminated also contains a famous frontispiece by Jean Fouquet (1420-1481), possibly France's most important fifteenth century painter and quite possibly also the Master's father. So he could be either Louis or François Fouquet who both worked much under their father's supervision.[128] The scene in Figure 136 is another depiction of the *Construction of the Temple of Jerusalem*; here the building is a modified Tours cathedral and is therefore an unashamed capriccio, as mentioned above this was often the case for such representations of buildings. Tours was Fouquet's and, even if he were not his son, probably also the Master's birthplace, and Fouquet spent most of his life there, so the choice of the cathedral to reference for these purposes is no surprise. The building appears a second time in the same manuscript on folio

213v, where it is titled *Sack of Jerusalem and Blinding of Sedecias*.[129] It is a four square building with a gently domed roof; in the scene of construction masons work in the public square in front of the Temple, some squaring blocks, others running mouldings, and one carving a figure. Other workers are carrying materials towards and into the building and at the furthest doorway one emerges with an empty hod. In the right foreground two labourers are furiously mixing mortar. On the right through an archway the faithful enter the Temple through a side door. At the top workers haul in a load of stone lifted up by the crane, which is only partly visible, for the ongoing construction at the uppermost level. Parts of the illumination are unfinished with the upper story left uncoloured in grisaille drawing. Solomon is seen on the left on a balcony of the adjacent building where he is directing operations.

The *Romuléon* is a fourteenth century text compiled in Florence by Benvenuto da Imola which tells the story of Rome from its foundation up to the era of Constantine. The Flemish *Romuléon* of 1466-1467 was translated by Jean Miélot for Philip the Good Duke of Burgundy and shows two images of the construction of Rome. (Figs. 29 & 137) Although a less Roman city is difficult to imagine since what is depicted is thoroughly French Medieval. In the former image two masons are seen at the top of towers, one laying or fixing stone, while the other checks the vertical line of the wall with a plumb-line. In the foreground a mason addresses a block of stone with a large axe and a labourer climbs a ladder with mortar. To the left is a woman worker with an empty or partially filled basket on her back; she also holds a rod or stick over her shoulder. It is not clear what she is doing, and she may or may not be involved in the building work. The depiction of woman workers is very rare except much later and then only in works such as the 'Book of the Queen', already written about above. The second image shows possibly the same mason working stone on the ground and the verticality of a tower wall again being checked with a plumb-line. The patron of these works, although different in each image, may be a representation of Marcus Furius Camillus, the 'Second Founder of Rome'.

The mythical founding of Rome is represented in Figure 138 with all its battles, murders and suicide, including the suckling by the she-wolf of Romulus and Remus in the middle ground on the left. These need not concern us except for the construction of the ensuing city around some of these scenes. The city in this case is a fantastic pile of Gothic and Renaissance extravaganza illuminated by Jaques de Besançon in the French manuscript *Ab Urbe Condita* of 1467-1470; here the masons are convincing enough for his time with their trowels, plumb-line and axe.

Another illumination by the Master of the Rouen Echevinage (see Figure 130) is the *Construction of Venice, Sycambria, Carthage and Rome* in the manuscript *Chronique de la Bouquechardière*. This name is derived from the manor of Bourg-Achard in Normandy, within the lordship of Jean de Courcy, and the author of the chronicle, written between 1416 and 1422. It is a curious title as it is, in fact, a

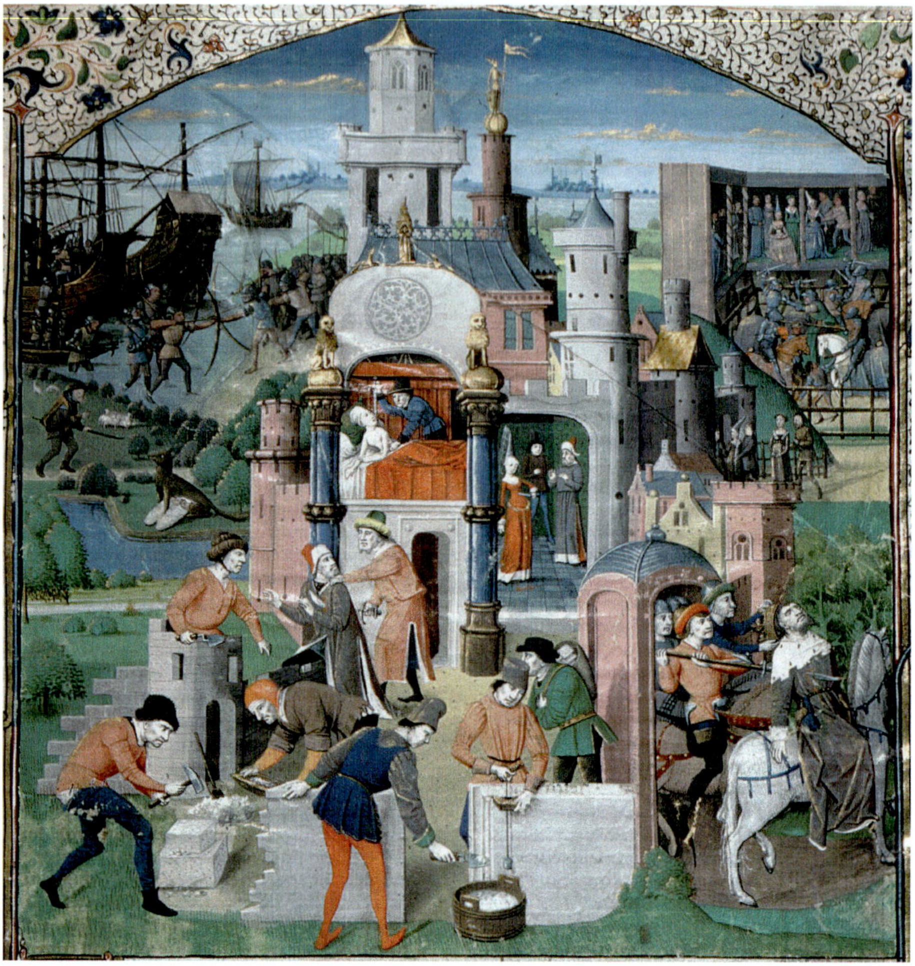

Figure 138: 1467-1470, attributed to Jacques de Besançon, French, active *c.*1460-1500, *The Origins of Rome*, from *Ab Urbe condita*. Bibliothèque municipale de Tours, ms. 0984, f. 001.

history of the world, or ancient times. Thus, the subjects in the depiction of the founding or construction of the four cities in Figure 139: At the top left are the Trojans observing the building of Venice; at top right is Helen and the Trojans watching the construction of Sycambria (spelt Cymmbre in the image), a city in Roman Gaul or possibly the city of Sicambria in Frankish mythology; at the bottom left is Dido and the building of Carthage and to the right Romulus oversees the construction of Rome. All four sites are busy with masons cutting and fixing stone.

Saint Augustine finished writing *City of God* in 427CE. Some thousand years or so later the text was translated by Raoul de Preses, copied down by an unknown scribe in Paris, and lavishly illuminated by an unknown artist in a manuscript

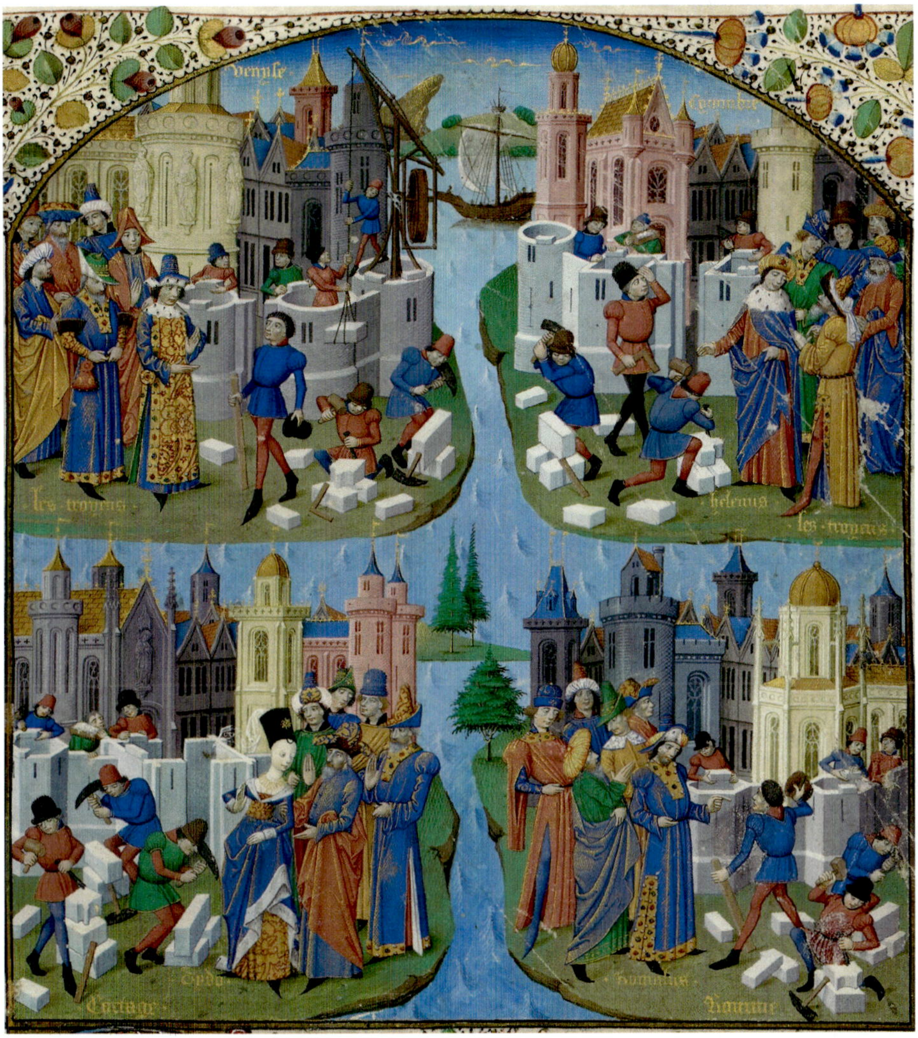

Figure 139: 1475-1500, Master of the Echevinage de Rouen (also known at the Master of the Geneva Latini), French, *Construction of Venice, Sycambria, Carthage and Rome*, from the *Chronique de la Bouquechardiè*. © The British Library Board, London, Harley MS 4376 f.150.

of 1480. The two images from the manuscript reproduced here (Figs. 140 & 141) are *Construction of the City of Enoch*[130] and *Construction of the Tower of Babel*. In the former the city is largely complete except for a tower and the city walls. The builders are so out of scale that it appears as if they are building a model rather than a real city. In the second image, the Tower of Babel, of course, will never be complete, although they have achieved five storeys. Enoch was a pre-flood city built by Cain who named it after his oldest son. In both images masons are using

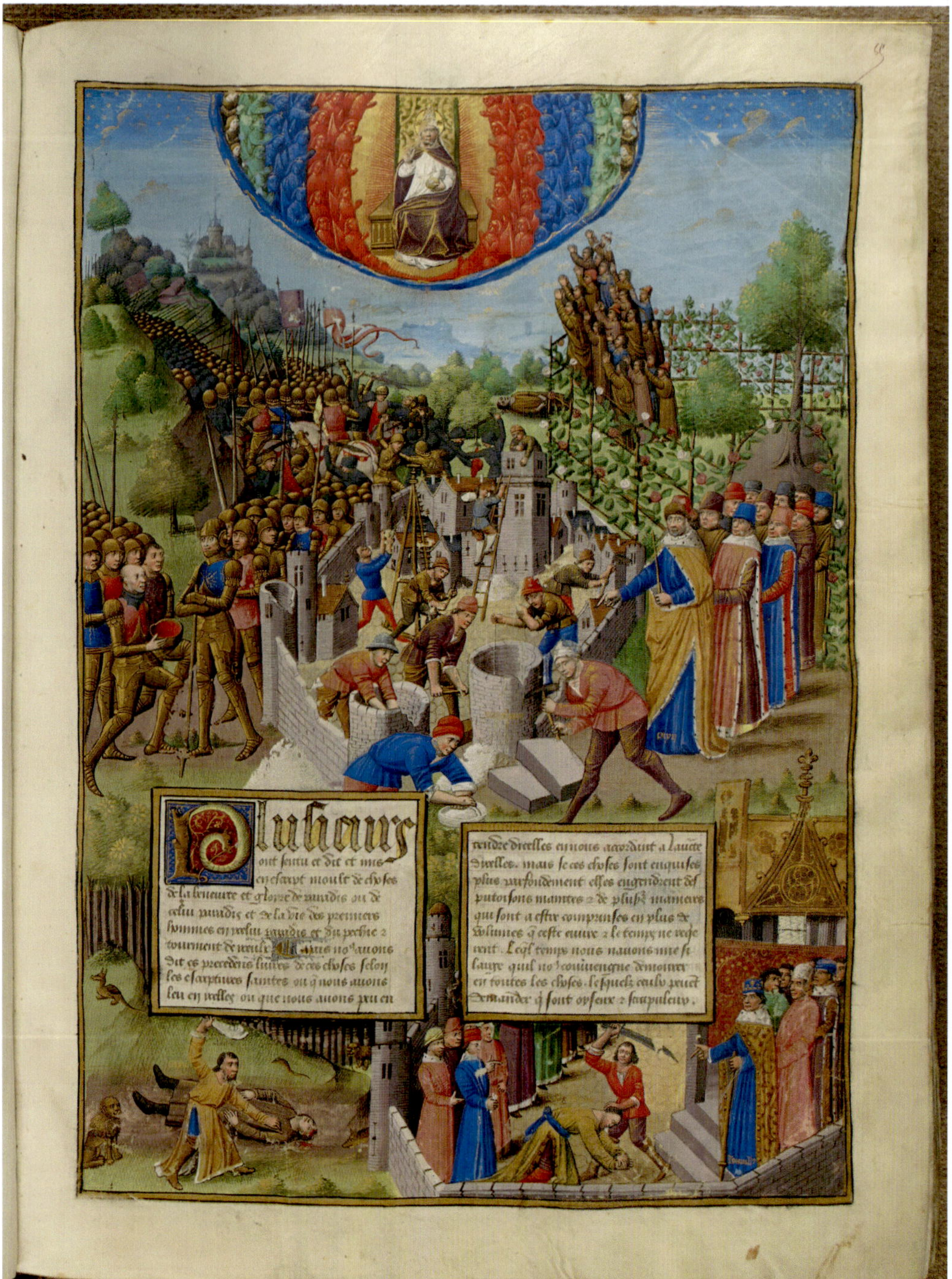

Figure 140: c.1480, French, *Construction of the City of Enoch*, illumination from 'La Cité de Dieu', Bibliotheque Municipale de Macon, France, ms 0002, f. 065.

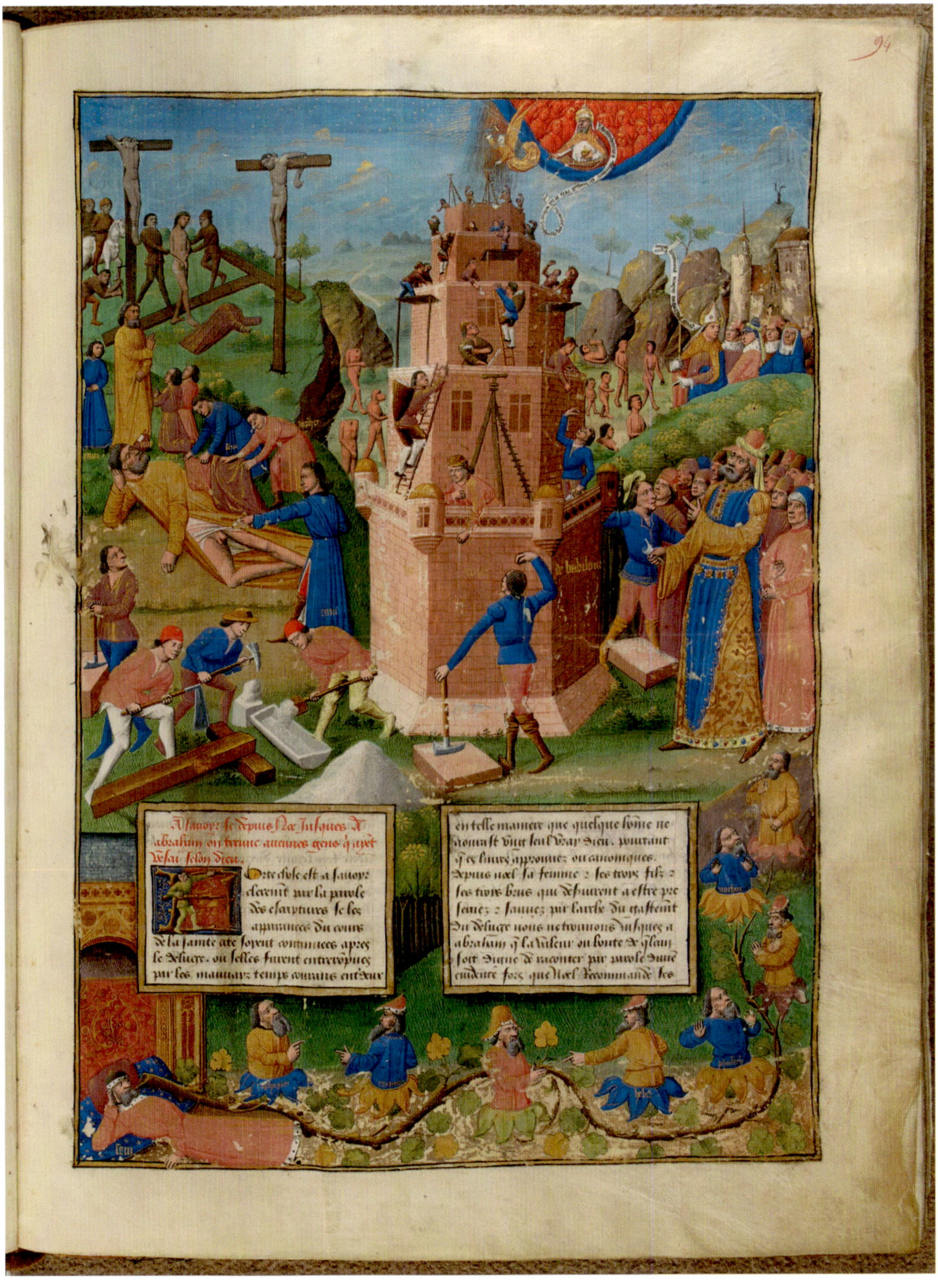

Figure 141: *c.*1480, French, *Construction of the Tower of Babel*, illumination from 'La Cité de Dieu', Bibliotheque Municipale de Macon, France, ms 0002, f. 094.

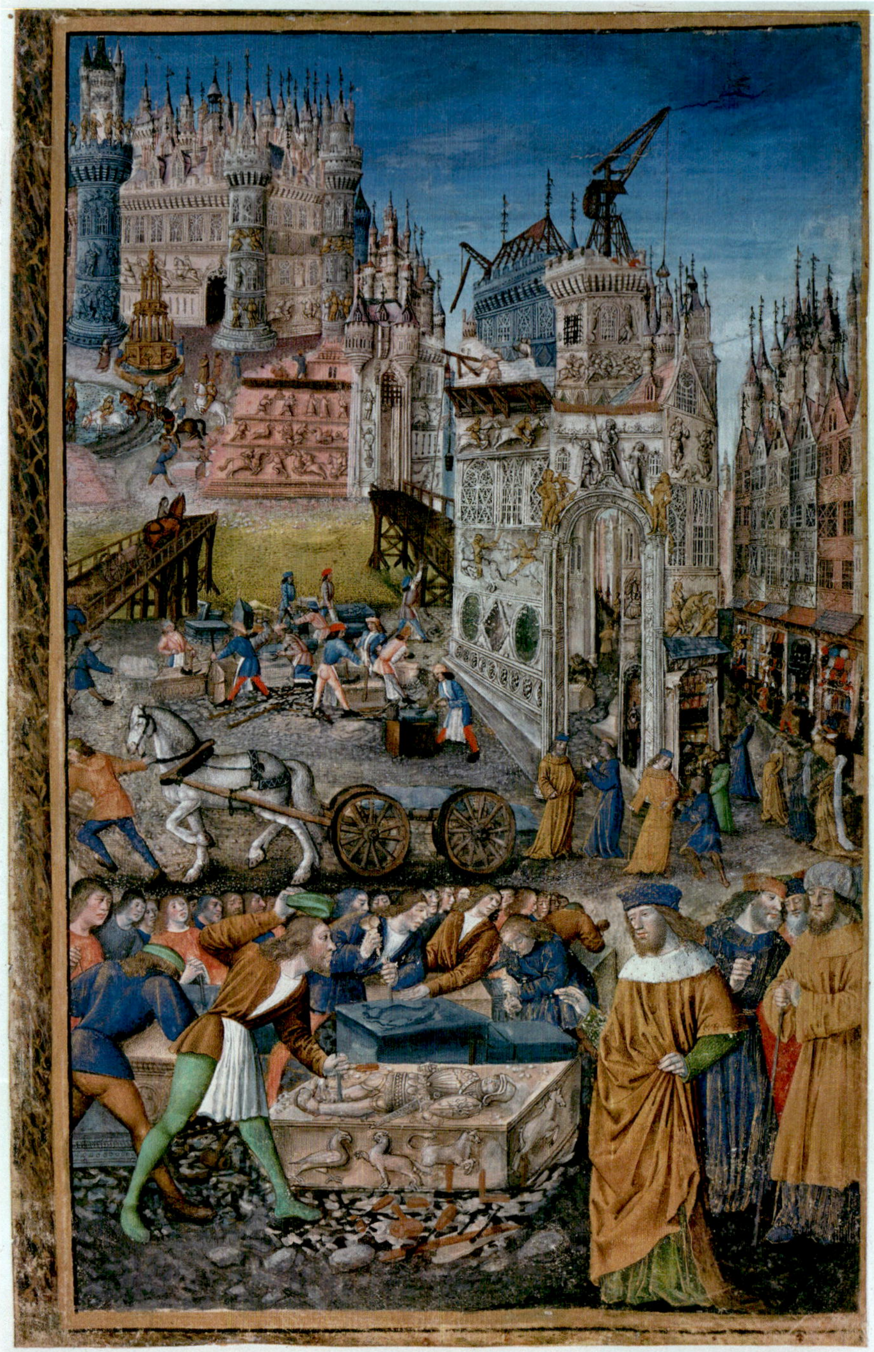

Figure 142: 1490s, Jean Colombe, French, *c.*1430-*c.*1493, *Rebuilding of Troy by Priam*, from *Recueil des Histories de Troie*, 0000649 bpk / Kupferstichkabinett, Staatliche Museen zu Berlin / Jörg P Anders.

axes, trowels and plumb-lines. The tripod or shear leg devices seen in both images, are discussed above on page 159.

One of the most fantastic cities depicted in these manuscripts is Jean Colombe's 1490 rendering of the city of Troy (Fig. 142) from the *Recueil des Histories de Troie*, of 1490. In the foreground a mason doffs his hat to King Priam who has ordered the reconstruction of the city. The mason has dropped his hammer or mallet but still holds his chisel in his left hand; his other tools are scattered around him on the ground including wooden handled chisels for fine work, a try-square, a straight edge, other chisels and a mallet. All these tools are mixed in with the chippings from his fine work, almost certainly in marble, perhaps a monument to a fallen soldier. Other masons behind him are at work on similar sized pieces. Another group of masons working on large pieces are in the middle ground of the image, which includes one, who in the foreground of the group, is sharpening a chisel on a whetstone, probably a slab of sandstone, with a bucket containing water to lubricate the process and obtain a good edge. Through the arch to the right are seen more masons at work whilst to the far right ordinary life goes on in the shopping district. High up on the scaffold and within one of the towers other masons are at work laying stone, their materials brought to them by crane and gantry.

In the charming, if naïve, image of the building of the Tower of Babel (Fig. 143) from the *Bible historiale complétée*, a translation by Guyart des Moulins of Peter Comestor's *Historia Scolastica* we see a mason carefully ruling a line on a piece of stone using a try-square and a metal tipped scribe.[131] Another mason cuts stone with an axe whilst a labourer carries materials up a ramp to a mason fixer who holds a trowel and stands on a wonderfully rickety scaffold. Both the ramp and scaffold are made from woven wattle hurdles as discussed in Chapter 1.

In the famous early fifteenth century French manuscript the *Bedford Hours* in the British Library[132] the artist known, from this particular work, as the Bedford Master shows us a frenetic late Medieval building site. (Fig. 144) This is also the building of the tower of Babel from Genesis: '*Come let us build ourselves a city, with a tower that reaches to the heavens*',[133] which we have seen much of. What did the master know of such building projects and where had he seen them? The client and the architect have their place, as nearly always; the lifting gear is reminiscent of Italian practice (see Chapter 1, Part 2); is this a minaret; had he been to Spain, and what of the camel? The masons and their tools are quite correct but the mouldings they are producing do not appear to relate to the building we see behind them. The masons on the ground have various hammers and axes and a particularly wide claw chisel in the middle foreground. To the left under cover mortar is being mixed, this is carried up the external staircase or ramp of the completed section in containers or hods carried on the labourers shoulders. Stone is hoisted up on winches from stage to stage. We do not know what is in the bucket being hoisted up in the centre; it may

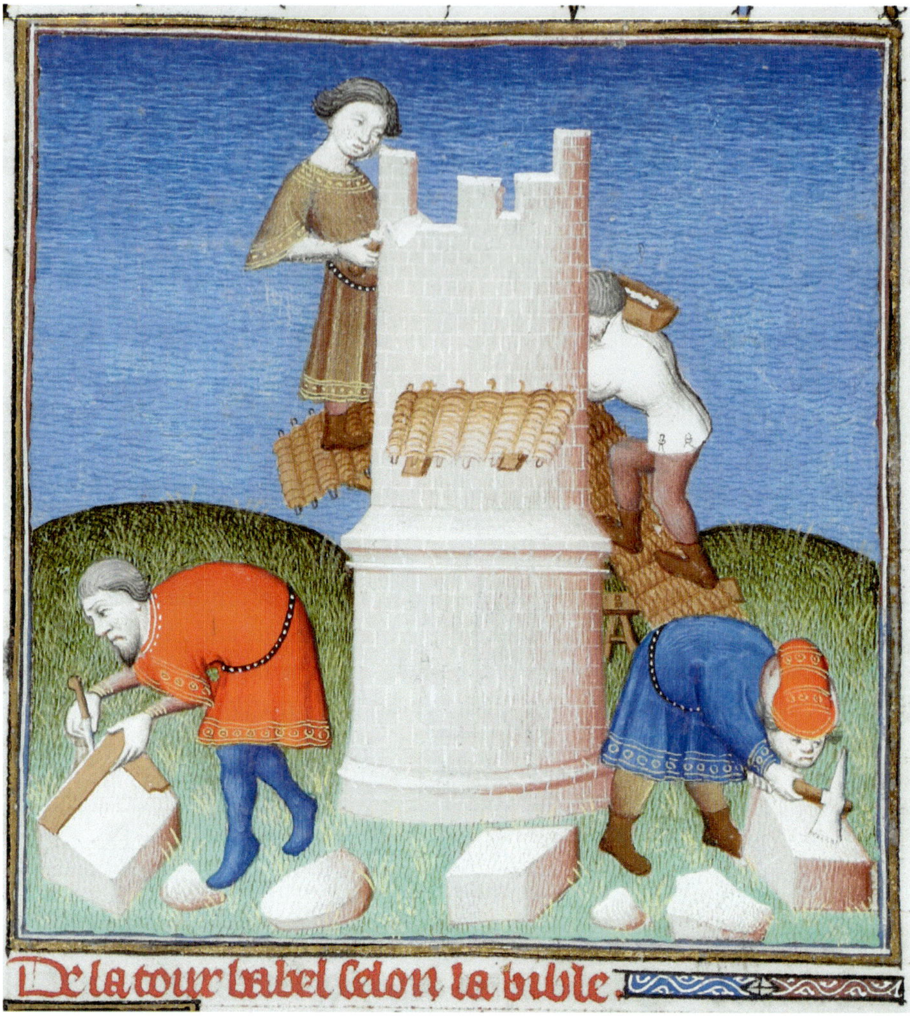

Figure 143: 1411, French, Bible Historiale, *Building of the Tower of Babel*. © The British Library Board, London, Royal ms 19 DIII, f. 16r.

be water for reworking stiff mortar. At the second storey all starts to go wrong with the wrath of God descending to bring such an audacious undertaking to an end.

The Grimani Breviary, one of the most lavish and elaborate Flemish manuscripts of the early sixteenth century, was produced in Ghent and Bruges *c.*1515-1520. This was illuminated by several of the leading artists and illuminators of the day including father and son Alexander and Simon Bening, Gerard David and here Gerard Horenbout in his highly detailed *Tower of Babel*. (Fig. 145) In this image we see almost everything there is to know in the construction process at the end of the fifteenth and into the early sixteenth centuries. In considering masons and

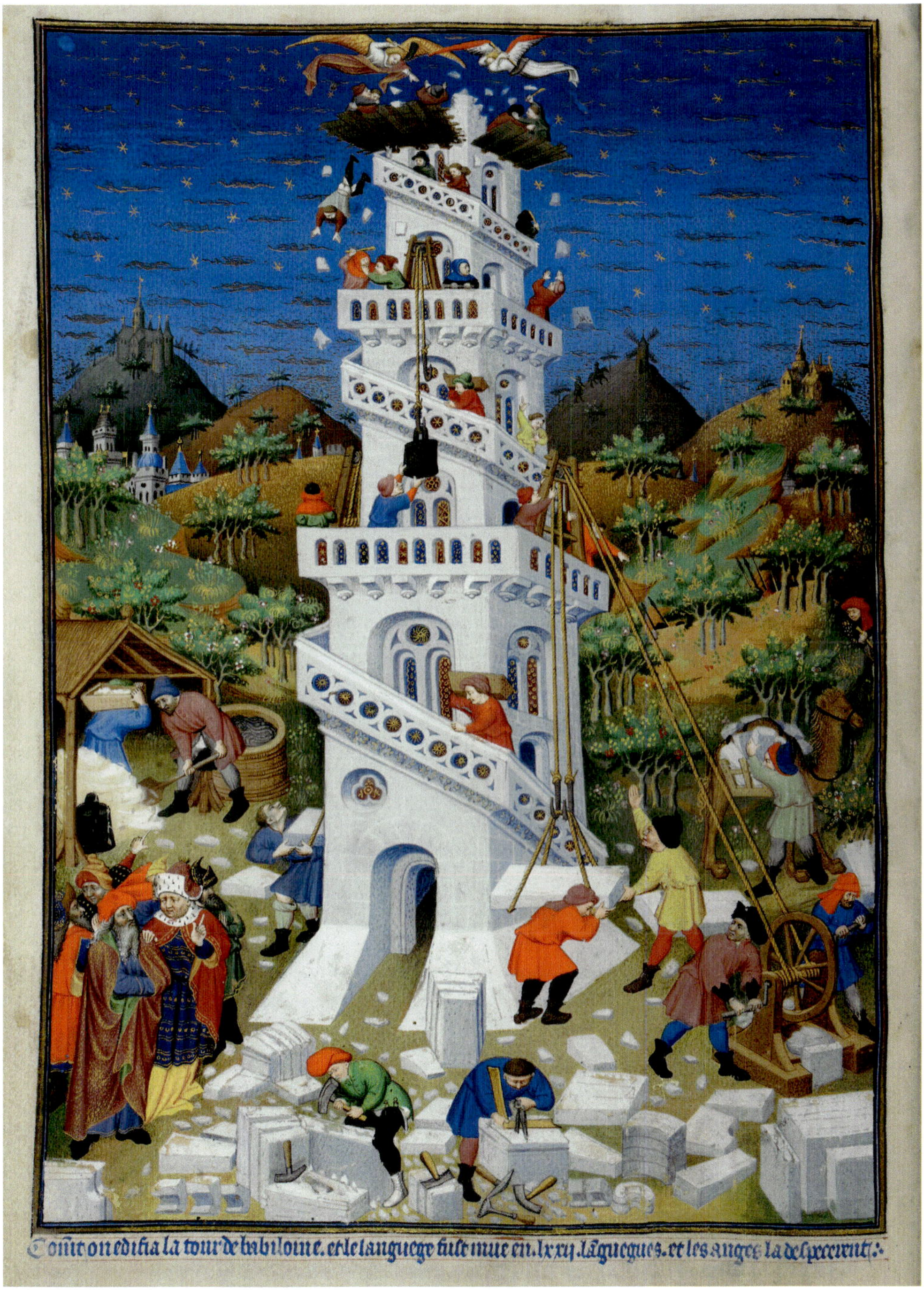

Figure 144: Bedford Master, *The Building of the Tower of Babel*, from a Book of Hours of the use of Paris (The Bedford Hours), © The British Library Board, London, Add. MS 18850, f. 17v.

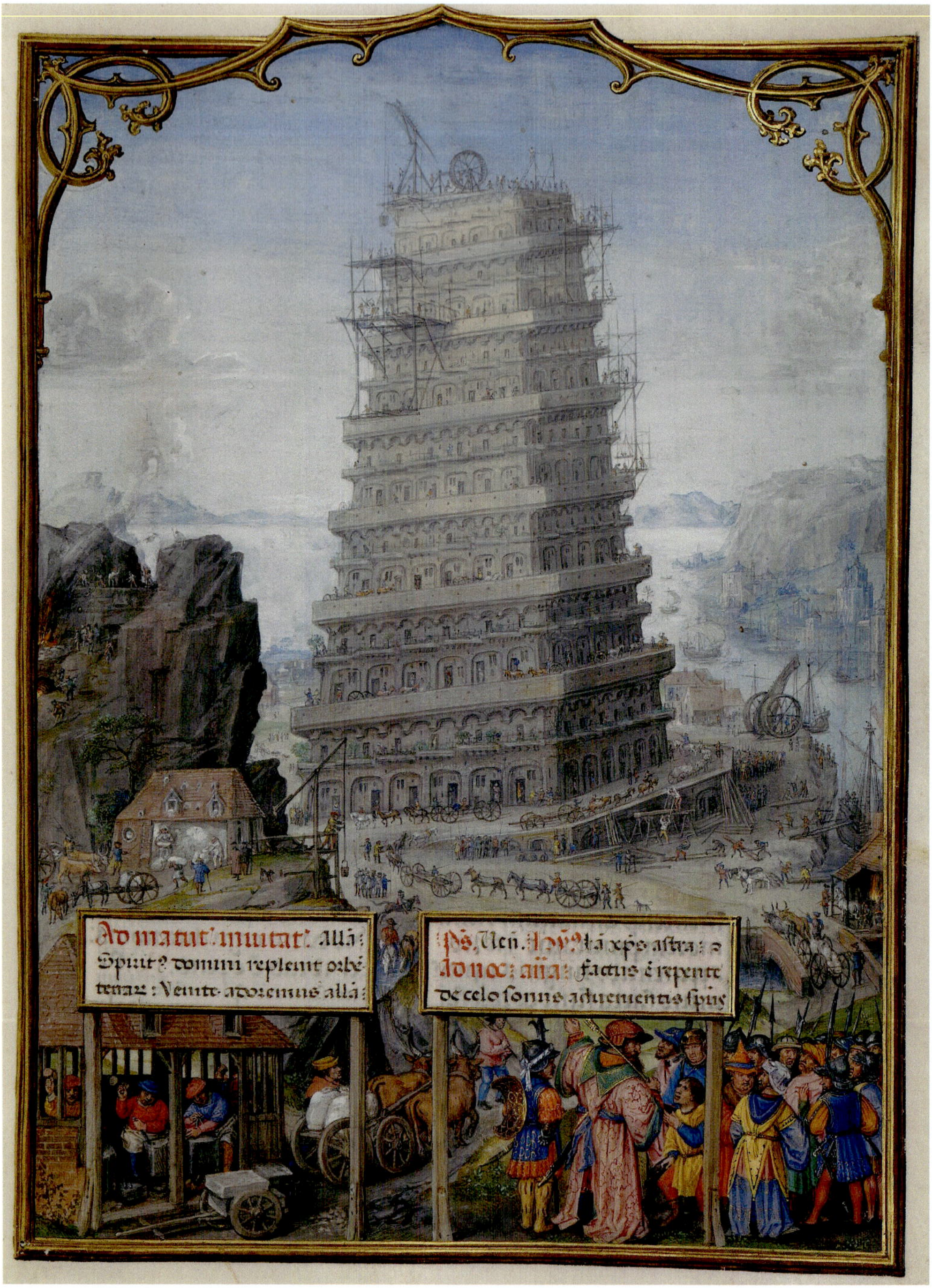

Figure 145: 1515-1520, Gerard Horenbout, or the Master of James IV of Scotland, Flemish, *Tower of Babel*, from the *Grimani Breviary*, ms. Lat. I, 99, f. 206r. Biblioteca Nazionale Marciana, Venice.

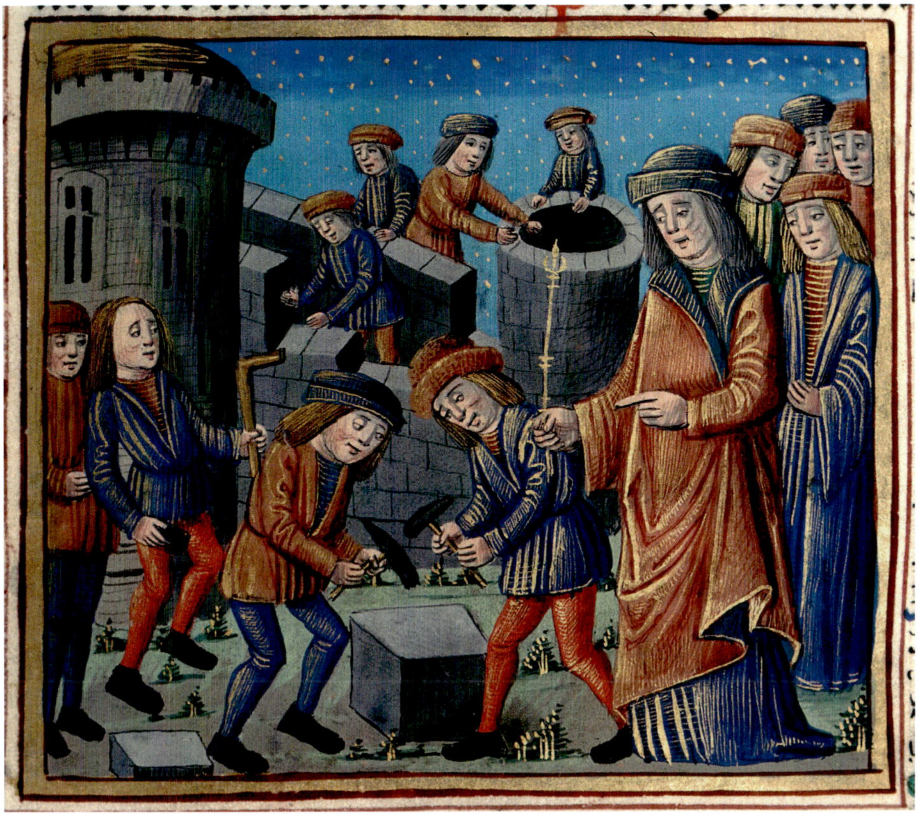

Figure 146: 1498, Attributed to (or Studio of) Master of the Echevinage de Rouen, French, *Construction of a town*, from the *Bréviaire of Charles de Neufchâtel*, ms. 0069, p.129, Bibliotheque municipale de Besançon, France.

the procurement of their materials we must look closely at the left foreground and middleground within the picture. At the top of the rock outcrop is the quarry for the building stone and mortar. Below this, indicated by fire, is a limekiln where limestone is burnt to produce quicklime. Further down is an open fronted brick building where the quicklime is processed into mortar by slaking with water, then mixed with sand and aggregate and bagged up for transport to where it is required.[134] As can be seen this is transported by horse or ox cart over the bridge and up the ramp part way up the building and then lifted by crane. Below is an open sided building or mason's lodge, where masons are working stone. Of the two most visible, one is cutting with a mallet and chisel and the other marks out a block with dividers. At the front of the lodge on the track is a cart of stone from the quarry awaiting further work. It was common that stone would be worked at the quarry into convenient blocks for the masons in the lodge or on the building. This was more economical and avoided carrying potential waste, and therefore extra weight, which could be left at the quarry. To the right foreground is Nimrod and his retinue, with possibly the architect or clerk of works on bended knee. Other noteworthy operations are the smith and farrier on the extreme right-hand side, just over the bridge, where horses could be re-shoed and tools for the masons and carpenters made and sharpened, and any ironwork forged. Above timber and other materials are unloaded from ships using gangplanks and a dockside crane. On the building are various scaffolds with winches for lifting and at the very top a crane. On the riverbank, by the dockside crane, is an Inn and what appears to be a troop of soldiers mustered for some possible or impending unrest. This image acts as a precursor, to Peter Bruegel the Elder's famous painting *The Tower of Babel*, of some forty or so years later, which is Figure 166, and is discussed earlier and below.[135]

In the *Breviary of Charles de Neufchâtel*, of *c*.1498 is a far more simplified and indeed naïve image of the building of a town. (Fig. 146) A mason holds a try-square and two others cut the same block of stone with axes whilst others work on the buildings behind. However, none of these figures is addressing the work with any conviction, and are merely actors in some unconvincing scene. The illumination is attributed to the Master of the Echevinage de Rouen, or his studio.[136] Comparing it to other paintings by the same Master, Figures 130 and 139, this seems barely credible, even for his studio.

In a late fifteenth century French book of hours or devotions with some twenty-six full page illuminations is a depiction of the building of a tower or fortress; probably a representation of the building of the Tower of Babel.[137] Many of the pages contain the arms of a family called Lipperkerke, so we may assume that the manuscript was made for a member of this family. The image (Fig. 147) forms the left and bottom border to a page of flower and leaf forms, themselves forming a full border to a small section of text. To the right is a 'babewyn' or 'gryllus' musician.[138] The building being constructed is a brick round tower, although where the vertical meets the horizontal at the base it expands rather shapelessly and abuts what appears to be the trunk of a huge tree-like form. The whole activity, until the top of the image, is associated with mortar and carrying it upwards. At the base a labourer mixes mortar with a very large hoe, while next to him another labourer shovels the mortar into a basket type hod on the shoulders of a waiting worker. Behind him another enters the Roman arched doorway dressed with stone to carry his load to the top. At the level of the first storey on a balcony two others, one with a basket hod and the other with a different type, make their way to the left; is this perhaps a spiral ramp, or are they scaffold lifts? Two others watch out of a two light window. At the next storey close to the top, a labourer passes his hod of mortar through a window to another on the inside. At the top where construction continues a crane is perched on the top of the wall with a load of bricks, two workmen haul in the load. A bricklayer works on the left outside the tower standing on an invisible scaffold.

In Figure 40 in Chapter 1, we see a mason's lodge on the ground with a mason cutting stone and a mortar preparation area, in this image from the Flemish *Spinola Hours*. Outside the mortar preparation area is a sieve for grading sand, or aggregate, for the mortar. This is an easel like apparatus on three legs on which the sieve, usually rectangular, is fixed vertically. The aggregate is thrown at the sieve with a shovel, the small particles pass through to form a pile behind, whilst the larger particles which do not pass the sieve fall in front. Here a pile of the finer aggregate required for the mortar lying behind the sieve is being used by the labourer mixing the mortar. In the foreground a mason checks a section of moulding before it is lifted to the work area further up the building. All around the two lodges are blocks of stone either awaiting work or finished, to be craned up or taken into the lodge for further work. A labourer disappears into the building with a hod of mortar whilst another on the scaffold lift above discharges his load of mortar into a large bucket to be lifted to the mason fixers by the crane.

In the Italian *Trattato d'Architettura*, by Antonio di Pietro Averlino[139], written *c.*1460-*c.*1465, for the Duke of Milan; on the first page decorating the capital letter 'P' is an image of two masons working on a square column, with possibly the author himself checking the progress. (Fig. 148) The standing mason demonstrates to the architect that the column, so far built, is vertical using a plumb-line whilst the mason who is crouching is tidying up the base with hammer and chisel. In a later translation into Latin by Antonio Bonifi of 1488-1489, and commissioned by Matthias I Corvinus, King of Hungary, is an illumination attributed to Bernardino Butinone (1435/6-*c.*1507).[140] On this particular page, seen in Figures 149 and 150, is a lavishly illustrated border on a lapis lazuli background including manticore and herm like figures, putti, birds, fish, rams heads, jewels, shells and foliage with much picked out in gold leaf. At the base is a scene, the construction of a palace, where masons work along the ground, including mortar mixing with a hoe, on the walls, masons fix stonework or lay bricks, one carries mortar and another checks with a plumb-line. There is rather a lot of direction from the ground and for good measure a grey cat. The lifting gear is discussed above in Chapter 1, part 2.

Moving from depictions of masons in manuscripts to larger paintings on canvas, boards and walls, where the scale becomes much larger, although the detail is not necessarily any greater. The magnificent altarpiece by Stefano di Giovanni, called il Sassetta, featuring the *Madonna della Neve*. (Figs. 151 and 152) The legend of the Madonna of the snow is celebrated throughout Italy on the fifth of August each year. The legend tells us that in 352 the Virgin brought snow to that part of Rome where the church of Santa Maria Maggiore now stands, on a sweltering hot day. The snow left on the ground formed the ground plan of the church. On the previous night this miracle was announced to both Pope Liberius, and to a rich patrician named Johannes in a dream. Johannes and his wife were childless and had prayed to the

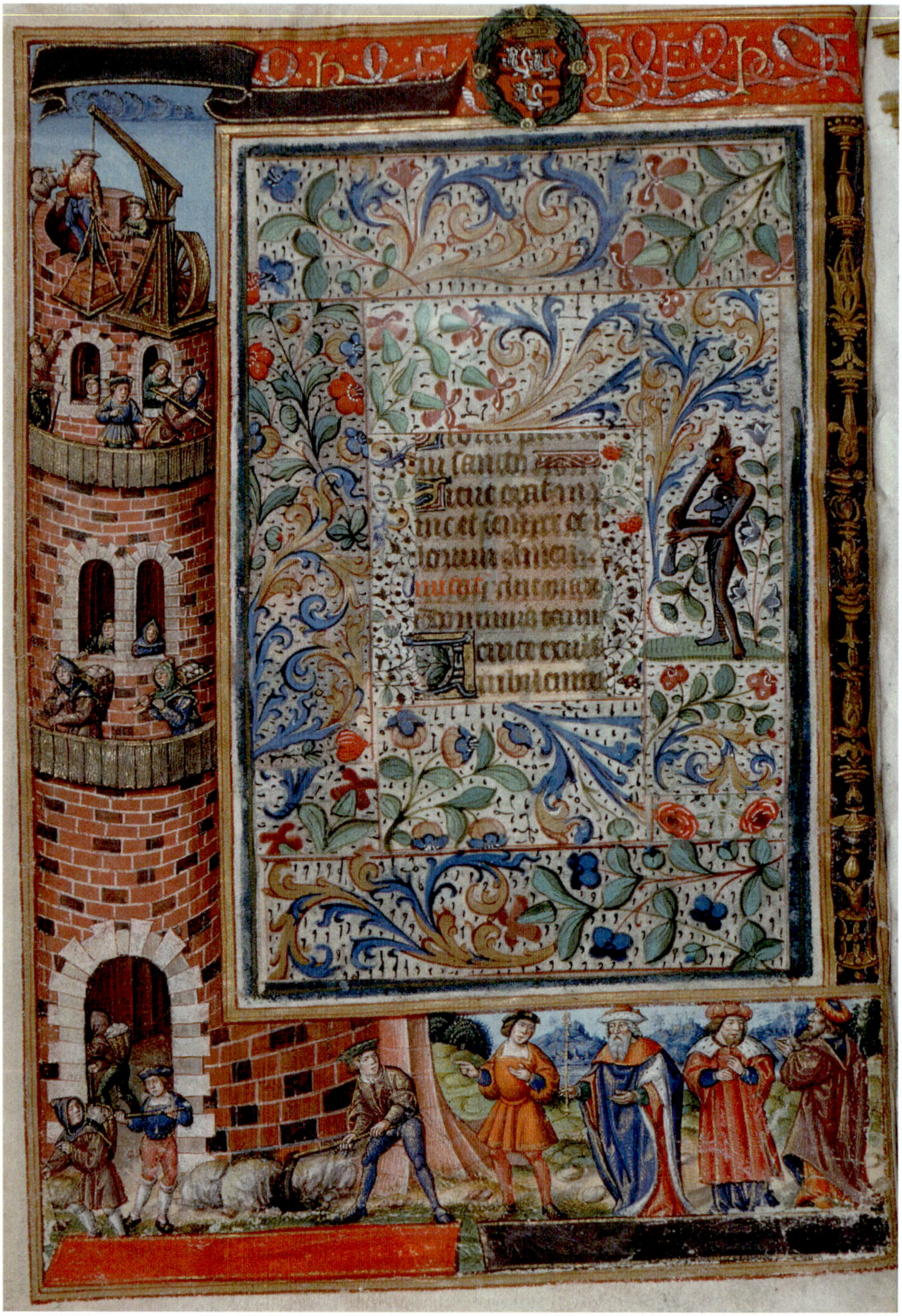

Figure 147: Late fifteenth Century, French, *Building a tower or fortress*, from a 'Book of Hours' ms. 0107, f. 020v, Bibliotheque municipale de Amiens

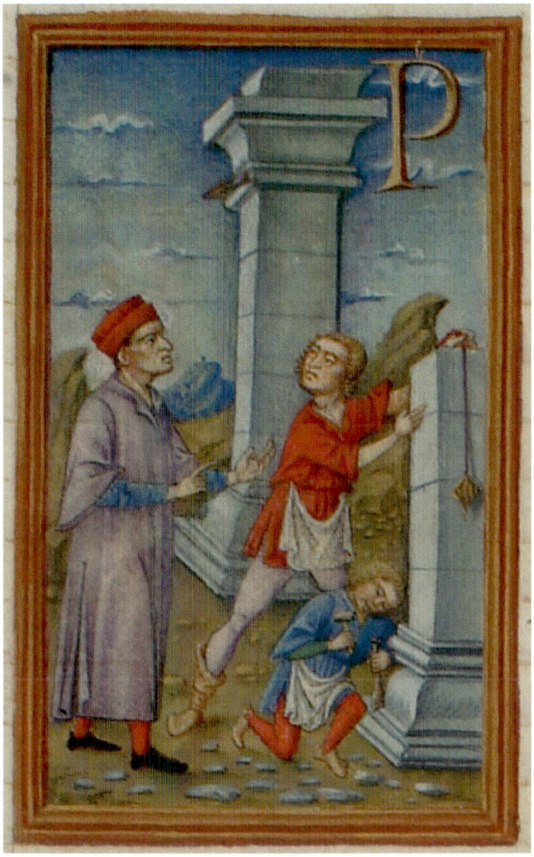

Figure 148: *c.*1465, Antonio di Pietro Averlino, known as Filarete, Italian, *c.*1400-*c.*1469, illuminated letter 'P' from his *Trattato d'architettura*. Biblioteca Nazionale Centrale di Firenze, ms II. I. 140 (giá Magl. XVII, 30), f. 1r.

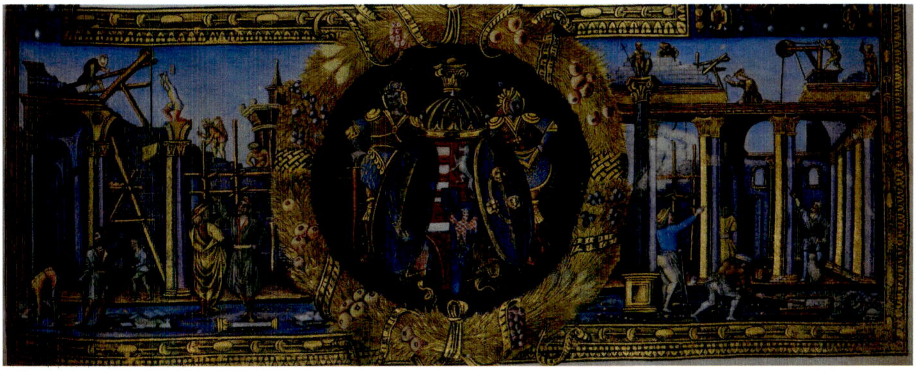

Figure 149: 1488-1489, Bernardino Butinone, Italian, 1435/6-*c.*1507, *A Palace under Construction*, detail of Figure 150, illumination from 'Trattato d'Architettura' by Antonio Filarete. Biblioteca Nazionale Marciana, Venice, ms. Lat. cl.VIII, 2 (-2796) f. 5r.

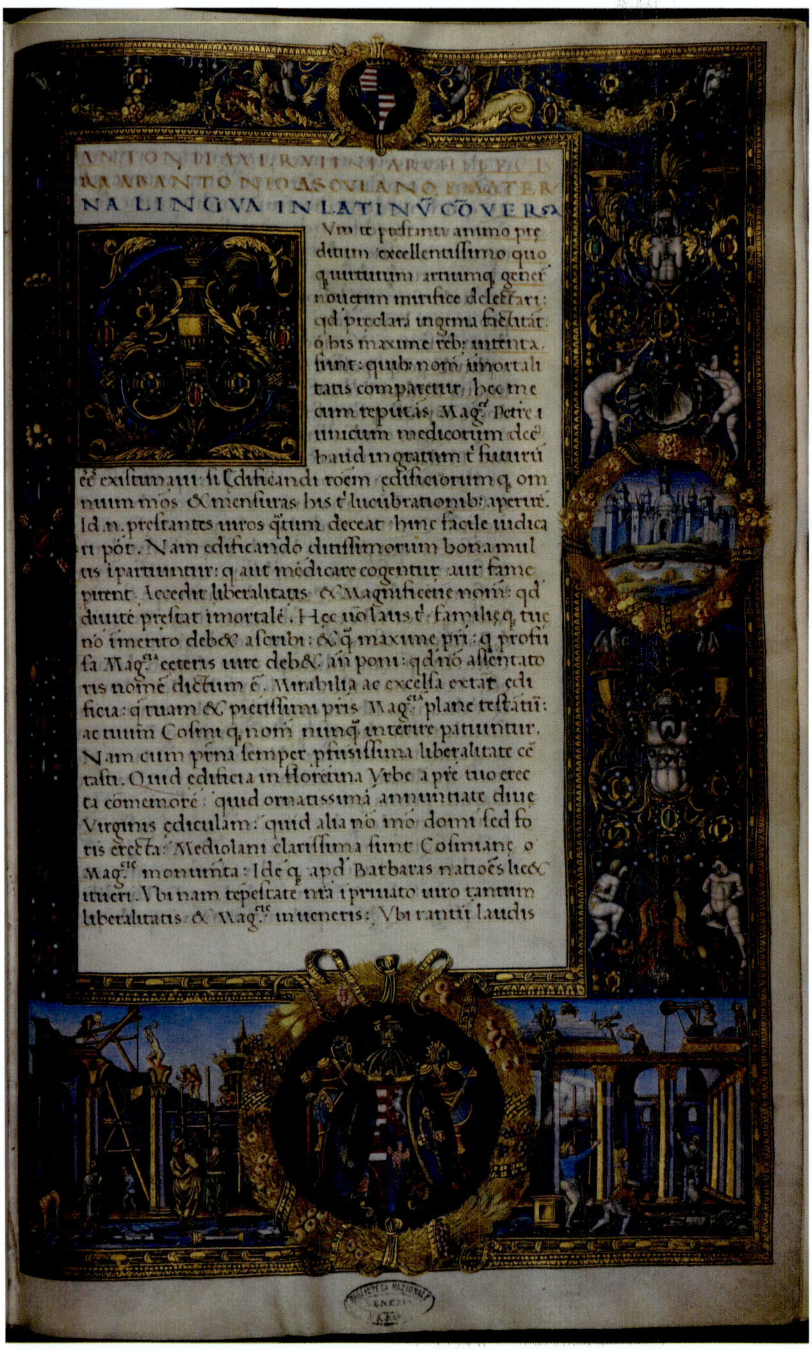

Figure 150: 1488-1489, Bernardino Butinone, Italian, 1435/6-c.1507, *A Palace under Construction*, illumination from 'Trattato d'Architettura' by Antonio Filarete. Biblioteca Nazionale Marciana, Venice, ms. Lat. cl.VIII, 2 (-2796) f. 5r.

Virgin to give them a sign as to how they could dispose of their wealth. The plan laid out in the snow was a sign to build a church on the spot, which was to become Santa Maria Maggiore. In 1430 a rich Sienese widow who was similarly childless commissioned Sassetta to paint the Madonna della Neve for her chapel in Siena cathedral.[141] This work is in the form of an altarpiece, with the main panel being the Virgin and Child enthroned with saints, and there is little reference here to the legend apart from angels bearing plates of snow and the one on the right kneading a snowball; the plate of snow on the left is barely visible in this archive photograph.[142] (Fig. 151) The depiction of the legend is reserved for the seven panels of the predella: from left, 1, the apparition of the Virgin to Johannes; 2, the miraculous snowfall; 3, the vision of Pope Liberius; 4, Johannes visiting the Pope; 5, Pope Liberius cutting out the ground following the plan in the snow; 6, the construction of the church; and 7, dedication of the finished church. All of the panels are damaged apart from panel 5, and panels 2, 3, 4 and 7 are seriously damaged with between approximately 30% and 75% loss.[143] Panel 6, representing the construction of the church (Fig. 152) shows two bricklayers, with one working on a scaffold and the other in the foreground checks his work with his trowel resting in the mortar on the board to his right. A mason sits in the foreground carving a stone detail and two labourers carry materials: Bricks in a basket, and behind the further wall, mortar on a board carried on the labourers head; the top of the ladder he is climbing is just visible. There are two other labourers, one who is grading sand into the pile just beyond the further wall and another who is unloading bricks from panniers on the two donkeys. A 'clerk of works' oversees the work. There is some damage to this panel and some detail is now lost.

Jan van Eyck's exquisite and minutely detailed drawing of Saint Barbara is both a portrait of the saint and a depiction of the building of the tower in which she was imprisoned by her father. (Fig. 153) The legend of Saint Barbara is briefly discussed in Chapter 1 on page 72. Although versions differ as to the place and manner of her martyrdom the association with a tower, her emblem, and confinement is a general one. There is some debate among art historians as to whether this is a finished drawing together with its integral frame, or if it is a painting for which only the under-drawing was completed. The colour in the background was added much later, by what the art critic Brian Sewell has called an 'idiot'.[144] The drawing on a gesso ground is fine brush stroke using ink and black pigment, and is signed and dated on the frame: 'JOH[ANN]ES DE EYCK ME FECIT. 1437.' The scene is of St Barbara in the foreground on a hillock; she reads from a book, a symbol of her learning, and holds a palm leaf indicating her martyrdom. In the middle distance the tower is being constructed and has reached the top of the second storey. The ground around the tower, although there some onlookers are present, is peopled almost exclusively by masons and building workers. Some of these manhandle large lumps of stone, with one carrying a piece on a wheelbarrow, while others to the extreme left are mixing mortar, and a

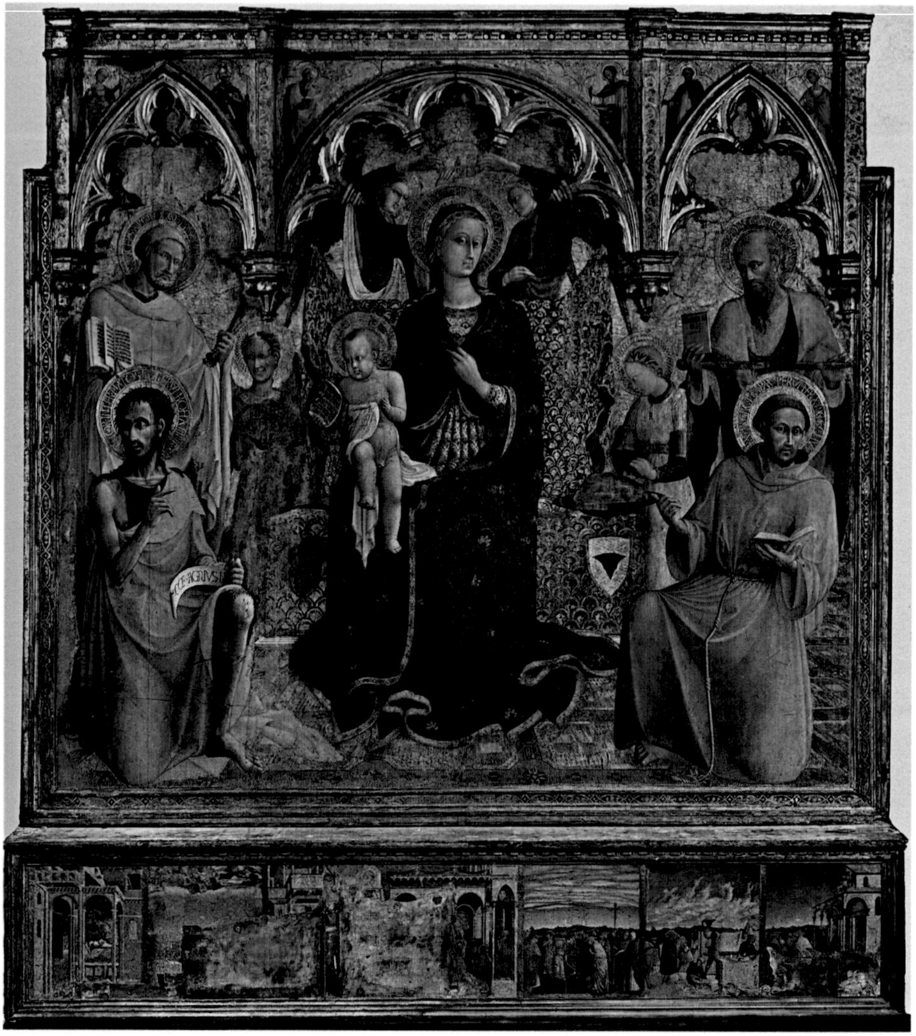

Figure 151: 1430-32, Stefano di Giovanni, called il Sassetta, Italian, *c*.1392-*c*.1451, Altarpiece *Madonna delle Nevi*, 240 × 216cm, panel, formally in Siena Cathedral, Uffizi, Florence. Anonymous archive photograph, *c*.1950, Fondazione Zeri, Universita di Bologna, Italy.

large tub of finished mortar is carried on a barrow by two workers to the rear of the building, to be lifted by the crane at the top of the building. To the right against the building is a mason's lodge or shelter where masons are cutting stone. Outside are finished column drums awaiting hoisting to the top, whilst a clerk of works looks on and encourages progress.[145] It is not known which church or cathedral Van Eyck drew on for his model. It may have been the cathedral of Our Lady, Antwerp which may have had more similarities in the fifteenth century than it has now.

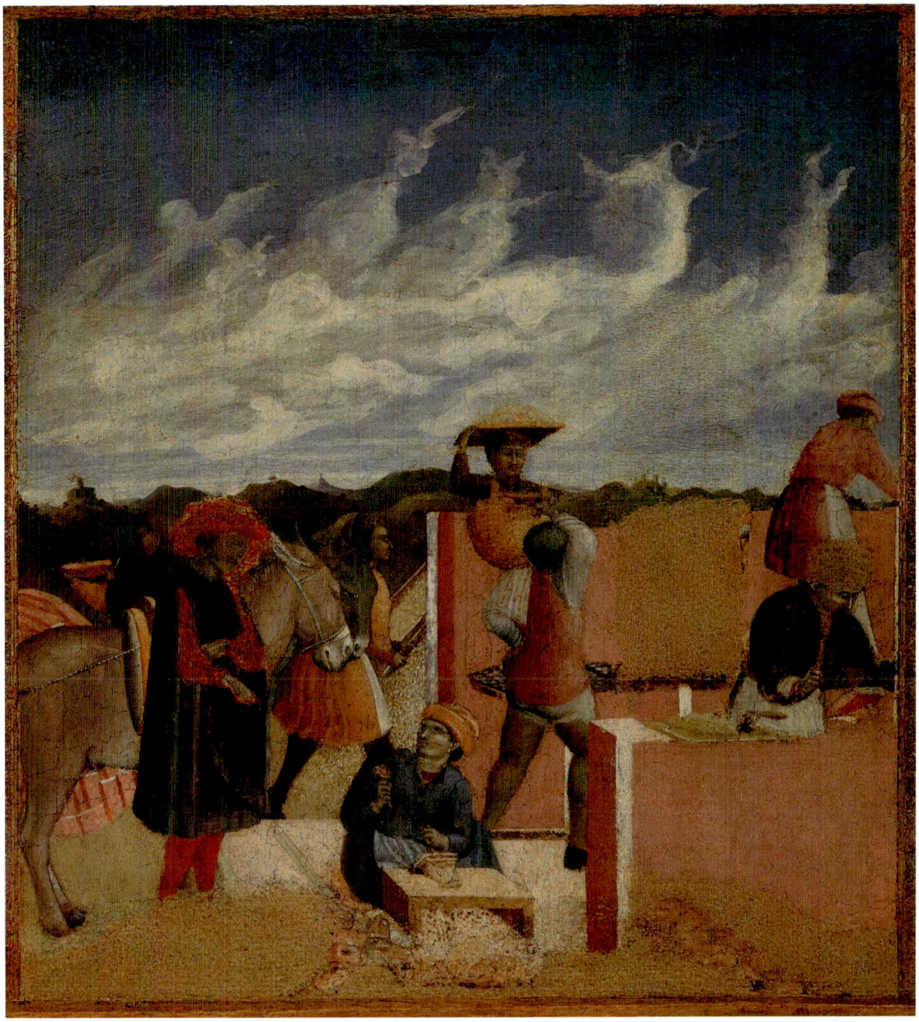

Figure 152: 1430-32, Stefano di Giovanni, called il Sassetta, Italian, *c.*1392-*c.*1451, *Construction of the Church which became Santa Maria Maggiore, Rome,* second panel from the right of the predella of the Altarpeice *Madonna delle Nevi,* 31 × 27.5 cm, panel, formally in Siena Cathedral, Galleria degli Uffizi, collezione Contini Bonacossi, cat. 00281712(7).

The painting of *The Victory of Alexander the Great and the Building of the City of Alexandria* is attributed to the Master of San Miniato, active 1460-1480, and so named for a number of paintings in the Basilica of San Miniato al Monte in Florence. This painting is significant both for the crane, which was discussed in the previous Chapter (Fig. 106) and, for this Chapter, the limekiln in the middle background. (Figs. 154 and 155) All construction sites required lime for building and kilns were essential for burning limestone to produce quicklime, the basic ingredient, once

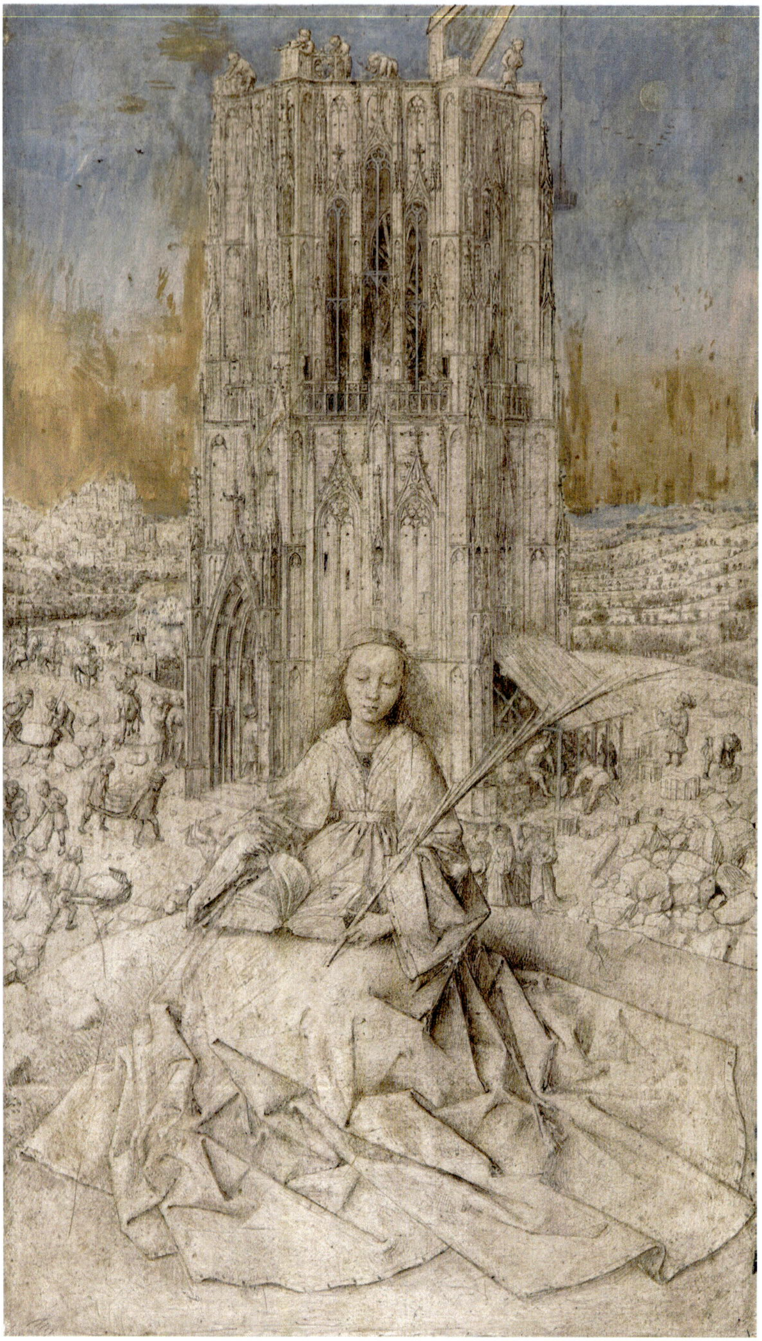

Figure 153: 1437, Jan van Eyck, Flemish, 1380/1400-1441, *Saint Barbara*, 41.2 × 27.6 cm (including original frame), black ink and oil on panel, signed and dated on frame, Royal Museum for Fine Arts, Antwerp © www.lukasweb.be – Art in Flanders vzw, photo Hugo Maertens.

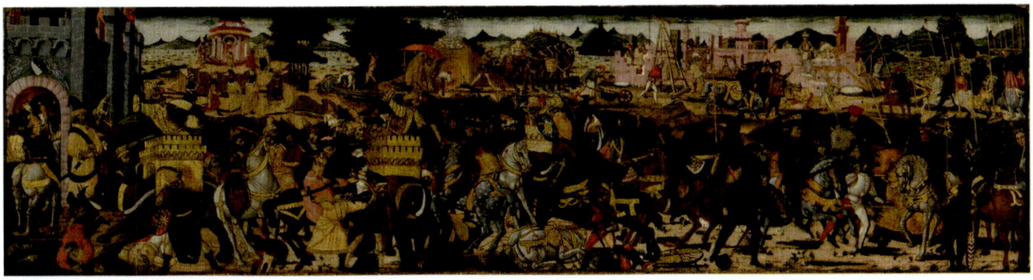

Figure 154: *c.*1460, Master of San Miniato (attributed), *The Victory of Alexander the Great and the Building of the City of Alexandria*, 51 × 160 cm, oil (?) on panel, originally produced for the front of a wedding chest, Ca' d'Oro, cat. D76. Galleria Franchetti, Venice.

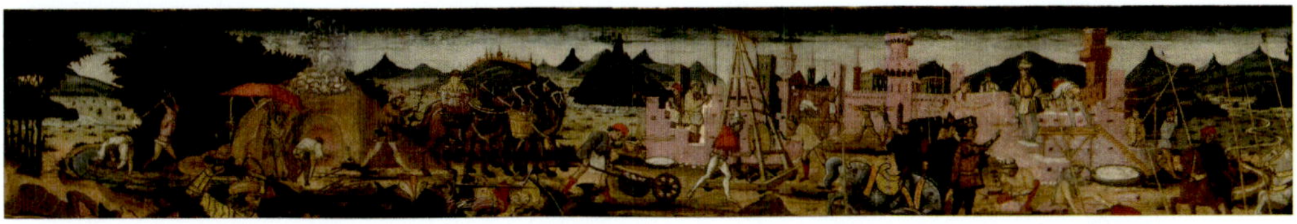

Figure 155: *c.*1460, Master of San Miniato (attributed), detail of Figure 154, *The Victory of Alexander the Great and the Building of the City of Alexandria*, 51 × 160 cm, oil (?) on panel, Ca' d'Oro, Galleria Franchetti, Venice.

slaked, together with sand or aggregate, for mortar (see above and Figure 145). Limestone of course was also used for the actual building if it was easily available. If construction was taking place in a geological area without limestone, either the quicklime had to be transported from distant kilns, or the raw limestone brought to kilns constructed locally for the purpose. Limekilns tended to be in rural locations, particularly by rivers (later also by canals), and in harbours,[146] as the product was also shared with agriculture. In this detail (Fig. 155) the limekiln is on the left, the limestone is brought in by horse transport with panniers; this is transported into the kiln at the top, with alternate layers of fuel, via a ramp which is hidden at the rear. The kiln is shown in the process of burning, but when this cycle is finished the fresh quicklime is raked out and, as is shown, transported to site in a wheelbarrow. It is then slaked on site in the large oval containers, one at each end of the site, and mixed with aggregate ready for the masons seen working on the building.

The lower left-hand panel of the closed view of the St Wolfgang Altarpiece, discussed earlier in Chapter 1 and Figure 34, depicts Saint Wolfgang and a helper or labourer building the church of St Wolfgang. (Fig. 156) Here are seen the tools required for mixing mortar in fine detail; a hoe lying in the foreground, a shovel standing in the mixed mortar, and two hods used for carrying the mortar. The helper pours mortar from a hod into a container being used by Saint Wolfgang to build the wall, and a mason's trowel sits in the mortar. Wolfgang taps a piece of stone into place using the iron head of his hammer. This would not normally be

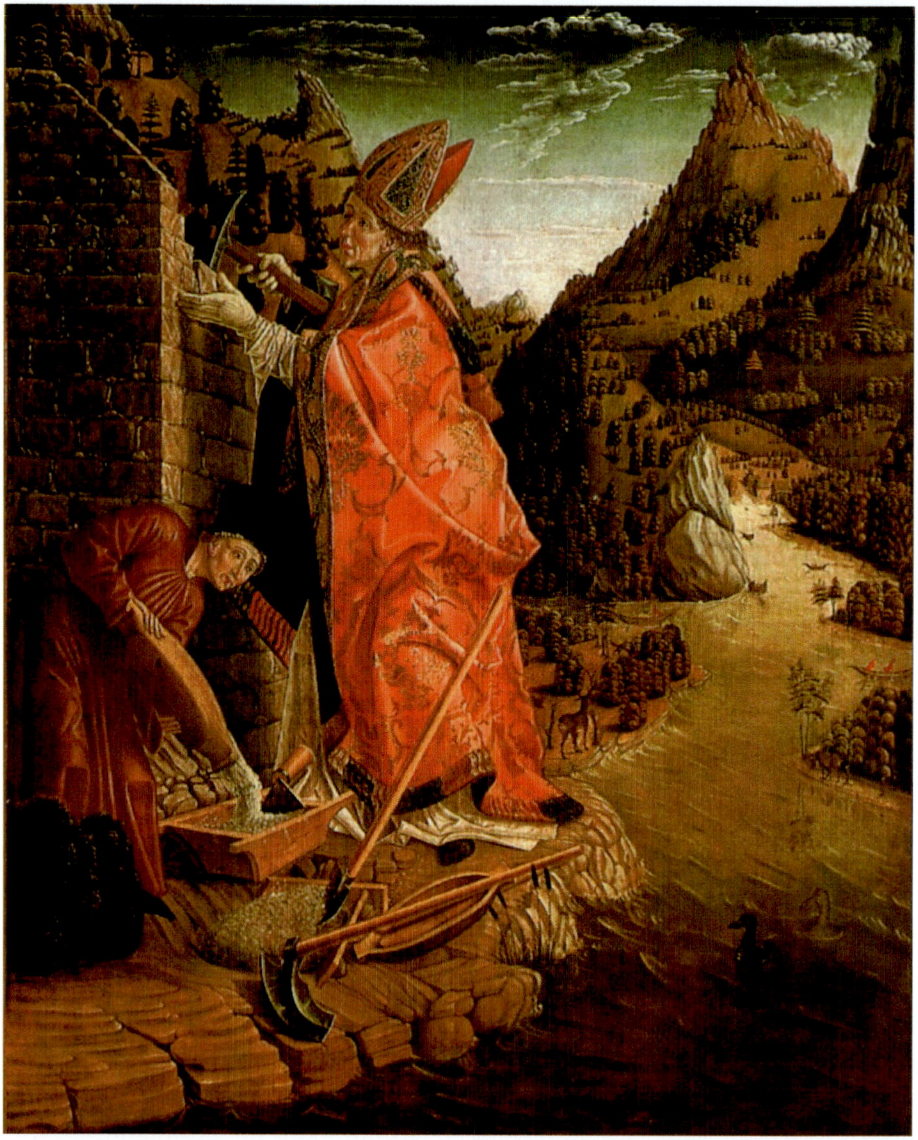

Figure 156: 1471-1481, Austrian, Michael Pacher, c.1435-1498, *Saint Wolfgang and Helper Building the Church of St Wolfgang*, Lower left-hand panel of the workday (closed) aspect of the *St Wolfgang Altarpiece*, 173 × 140 cm, oil on panel. St Wolfgang Church, Salzkammergut, Austria. Flickr, Jaime Antonio Alverez, with consent.

done this way as there is a danger of the impact of the metal on the stone splitting or otherwise damaging it, particularly as this is a heavy tool with both an axe and hammer heads. The normal practice would be to tap the stone into place with the wooden handle of the trowel, with which the mortar was laid.

The mason did not work only at the building site, they also often worked at the quarry. It made sense to work stone at the quarry, the master mason would send dimensions, and as the finished, or part finished, work was less weight and mass it reduced transport costs. The work would invariably be left only part finished to enable fine finishing on site, but the baulk of the waste would be left at the quarry where it was more easily disposed of. The finer waste on site would often be incorporated into mortar mixes as aggregate, or if it was a limestone quarry, burnt to produce quicklime. Andrea Mantegna's *Madonna and child of the Grotto* (*Virgin of the stonecutters*, or *Madonna delle Cave*), (Fig. 157) contains a scene, to the right of the seated Madonna, of masons working in a quarry. Two of them work on a column and two others work on what appears to be a length of cornice whilst an overseer looks on. Other pieces lie around including what looks like a capital and on a lower level, a sarcophagus is being hewn by two other masons. (Fig. 158) The quarry is within a fantastic explosion of a rock formation behind the virgin. Such landscapes are a recurring theme in Mantegna's work and Lawrence Gowing has said:

> Stone is everywhere in his pictures. It is cast up, stratified, fractured and fragmented [...] Mantegna's mountains are not those of the actual world: he seems rather to have invented, on the basis of the archaic Gothic formula, symbolic renderings of the character of the rock.[147]

A very similar scene appears in Mantegna's *The Man of Sorrows with Two Angels*, currently held in the Statens Museum, Copenhagen.[148] It also appears in his fresco *The Arrival of Cardinal Francesco Gonzaga* on the west wall of the Camera degli Sposi in the Palazzo Ducale, Mantua, also known as *The Meeting Scene*. (Fig. 159) In the background are marble quarry workings. (Fig. 160)

The Treasury doors to the now ruined Abbey of Saint Bertin, at St Omer in the Pas-de-Calais, northern France, now in the Musee des Beaux-Arts, Dijon depict scenes of the life of Saint Bertin, the second Abbot.[149] (Fig. 161) The bottom panel of the right-hand leaf shows construction of the Abbey, (Fig. 162) which commenced in the seventh century on the banks of the river Aa. Here a bricklayer is laying large bricks in Flemish bond. The more advanced building in the middle ground is also brick with a Romanesque arch, through which is a view of the cloisters. Or, is this bricklayer using precisely produced ashlar blocks? The extant ruins of the abbey are unequivocally stone, although they are a later building. The abbey was ruinous in the nineteenth century when Richard Parks Bonington painted it in 1823 and was further damaged in 1947 in World War II.[150] The bricklayer attends very closely to his task with his trowel. On his right is a tub of mortar and his level and hammer are seen in the very foreground. A labourer brings more bricks in a basket whilst Saint Bertin gives his blessing to the work. Behind are two column bases with a

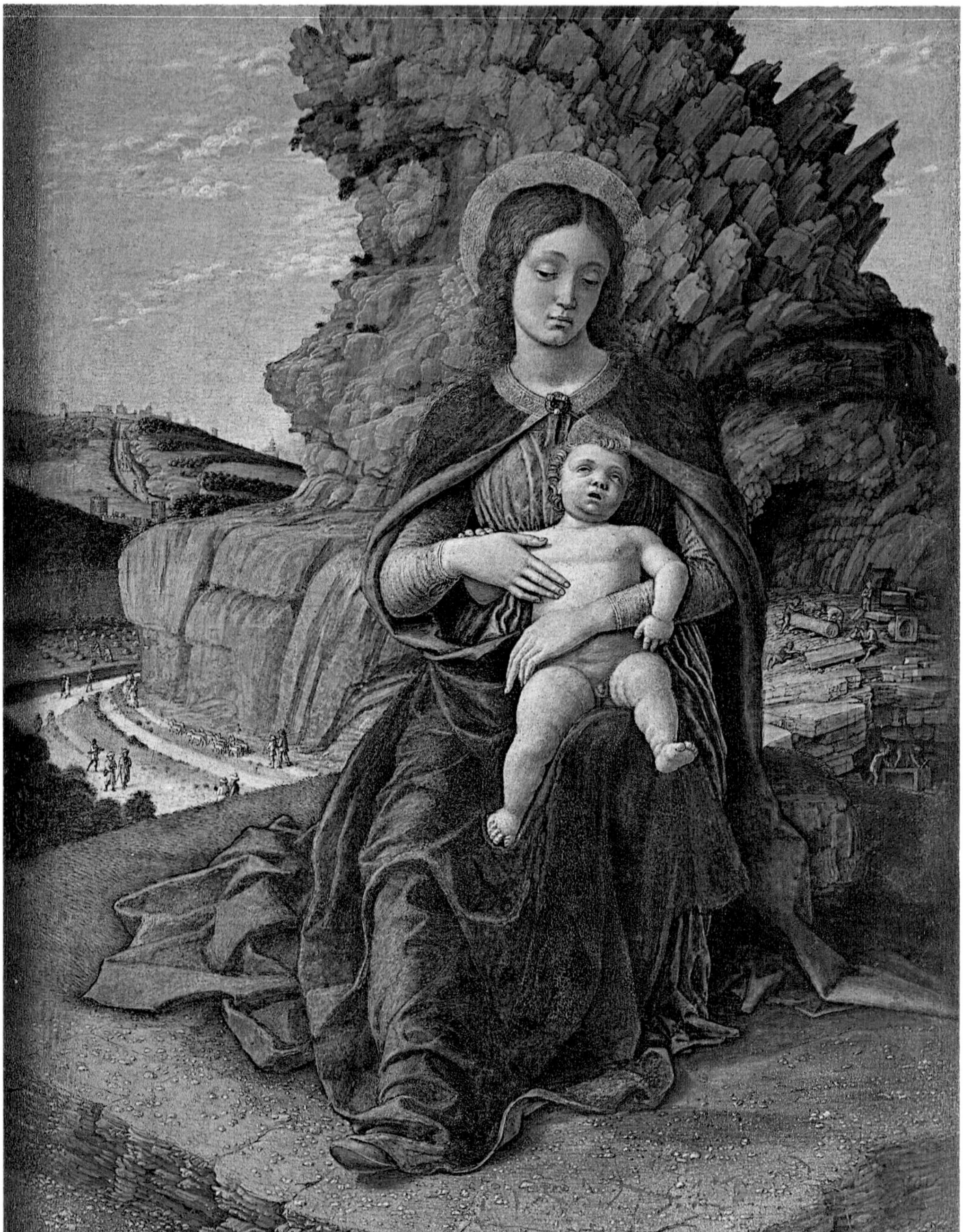

Figure 157: *c.*1489, Andrea Mantegna, Italian, 1431-1506, *Madonna of the Grotto*, 29 × 21.5 cm, tempera on panel, Galleria degli Uffizi, Florence.

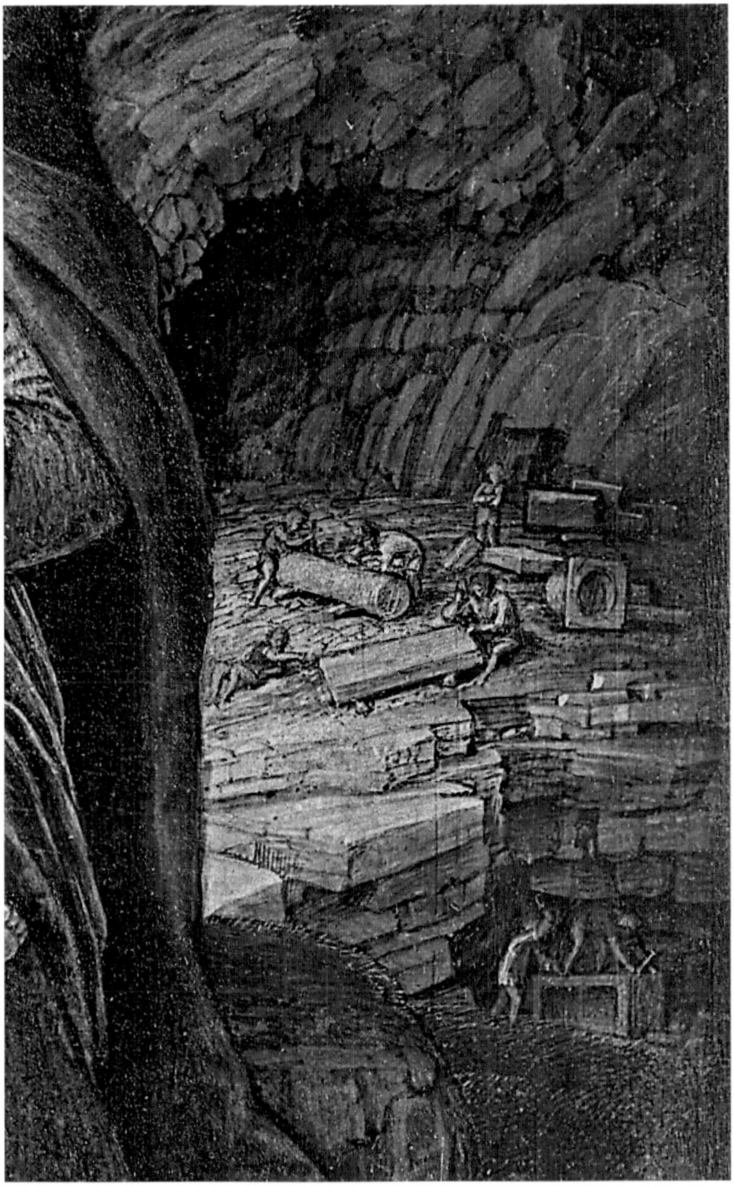

Figure 158: *c.*1489, Andrea Mantegna, Italian, 1431-1506, *Madonna of the Grotto*, detail of Figure 157, 29 × 21.5 cm, tempera on panel, Galleria degli Uffizi, Florence

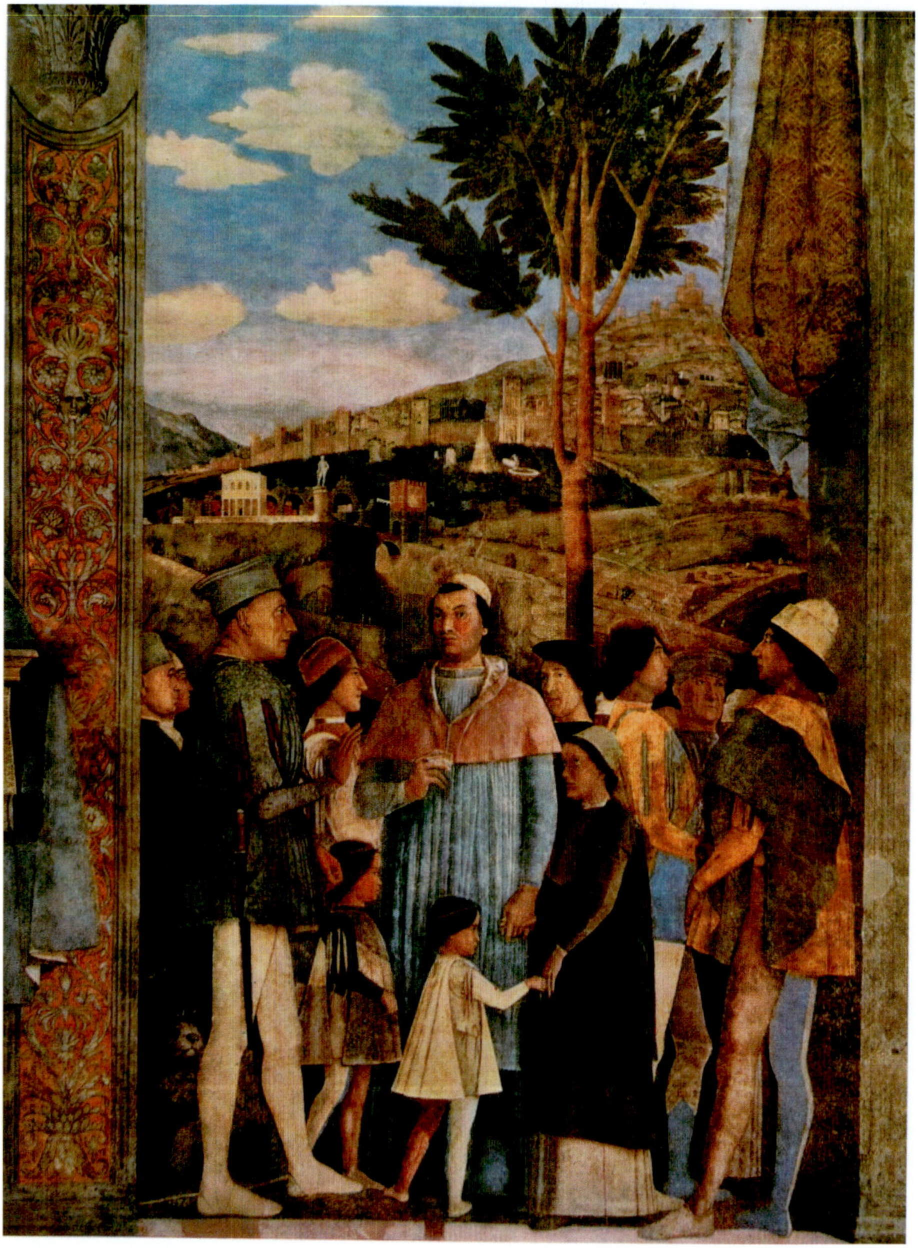

Figure 159: 1474, Andrea Mantegna, Italian, 1431-1506, *The Arrival of Cardinal Francesco Gonzaga*, fresco west wall, 'Camera degli Sposi', Ducal Palace, Mantua, Italy. GNU Free Documentation Licence.

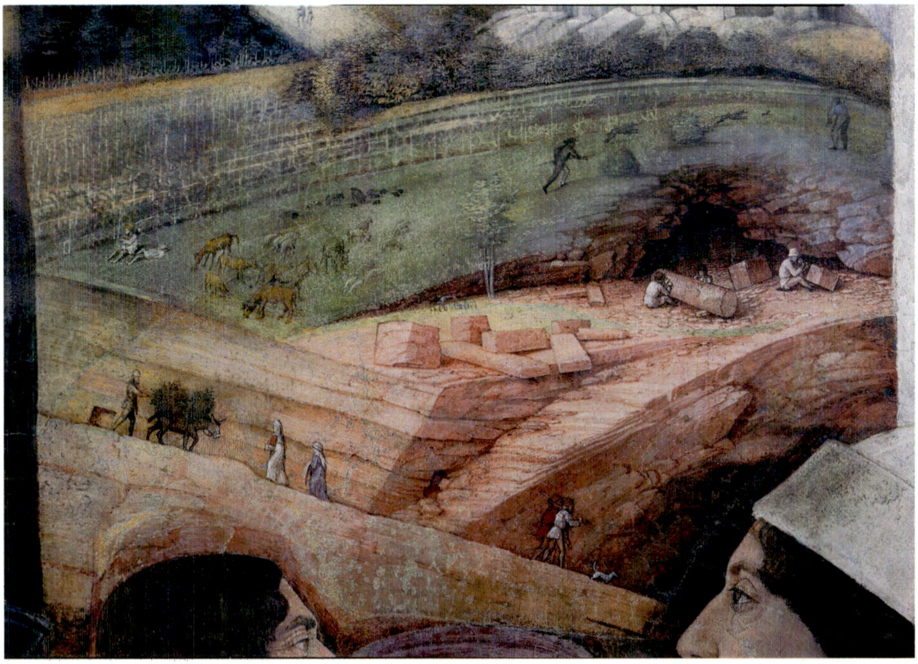

Figure 160: 1474, Andrea Mantegna, Italian, 1431-1506, *The Arrival of Cardinal Francesco Gonzaga*, detail of marble quarry workings, fresco, west wall, 'Camera degli Sposi', the Ducal Palace, Mantua, Italy. Palazzo Ducale, Mantua, Lombardy, Italy, Mondadori Portfolio/Electa/Antionio Quattrone/Bridgeman Images.

stack of column drums waiting to be set in place. Beyond that carpenters perform various tasks, for which see Figure 80.

The construction of another abbey, Klosterneuburg Monestry, founded in 1114 by Saint Leopold III, the Margrave and patron saint of Austria, is depicted in a panel painting by Rueland Frueauf the Younger. (Fig. 163) This is one of several scenes painted by the artist of the life of Saint Leopold for the abbey between 1496 and 1507. In the picture Leopold visits the abbey to view the building work, accompanied by his second wife Agnes of Germany, daughter of Henry IV, Holy Roman Emperor. The architect or clerk of works, on site for the occasion, greets the Royal party with bended knee and hat in hand. In the foreground a mason trims the side of a large block of stone with his axe, and other tools are by him: A straight edge, chisel and mallet. Hanging on a timber post is a try-square and beyond that a template leans against the wall. Such a templates were used by masons (and still are) to check the accuracy of a moulding profile, such as a cornice. Where large runs of the same moulding were required a template would be cut and shaped from a beaten iron sheet for longevity. The timber structure appears to be a temporary shelter for the altar, possibly erected for mass to celebrate the visit.

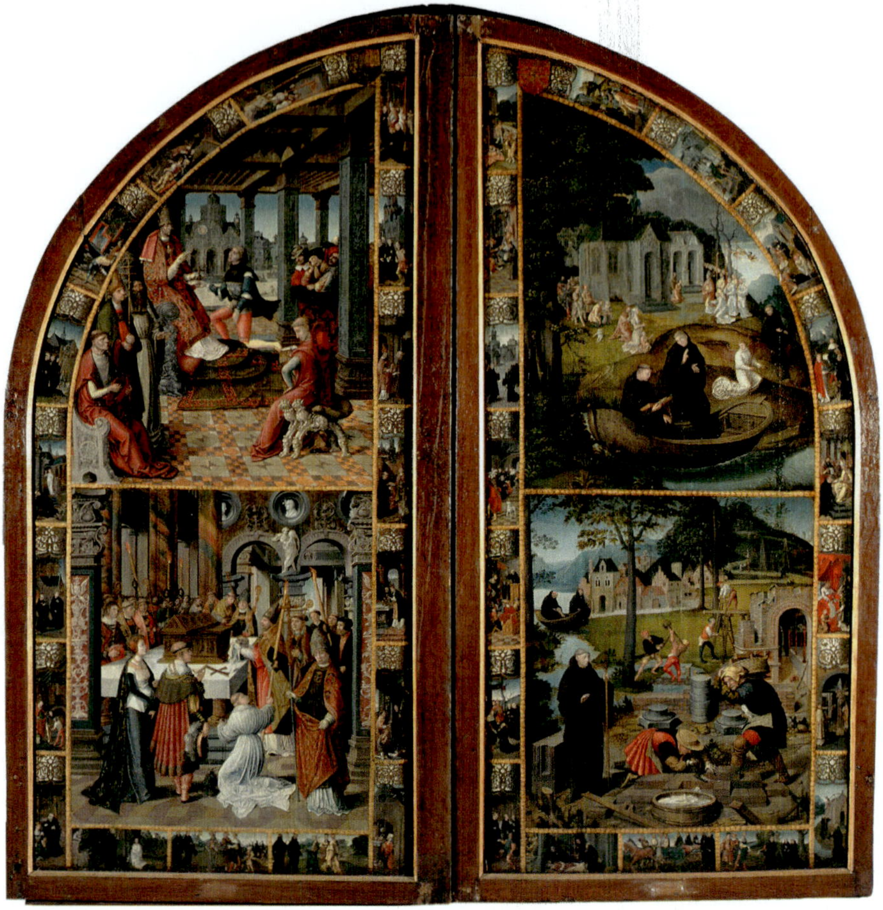

Figure 161: 1500-1533, Northern French, *Scenes from the life of St Bertin*, oil on wood panel, pair of doors to the treasury of the Abbey of St. Bertin at St. Omer, © Musee des Beaux-Arts de Dijon. Photo Hugo Martens.

Of similar date are the frescos of Giovanni Antotonio Bazzi, called Il Sodoma, and Luca Signorelli, on the cloister walls of the Abbey of Monte Oleveto Maggiore at Asciano, Italy. These illustrate the life of Saint Benedict of Nursia *c*.480-*c*.547. Signorelli's work is a few years earlier, of the late 1490s, than Il Sodoma's of *c*.1505-1508, who has also incorporated a self-portrait. Il Sodoma's work offers us two scenes of building work associated with St Benedict's founding of twelve monasteries. In Figure 38 in Chapter 1 a mason is carving the base for the column which lies beside him. Leaning against this are a pair of dividers and a drill of the type used for boring small holes. These are most likely to be drilled in the base of the column and top of the base for iron fixings; their position must be precise, hence the dividers for

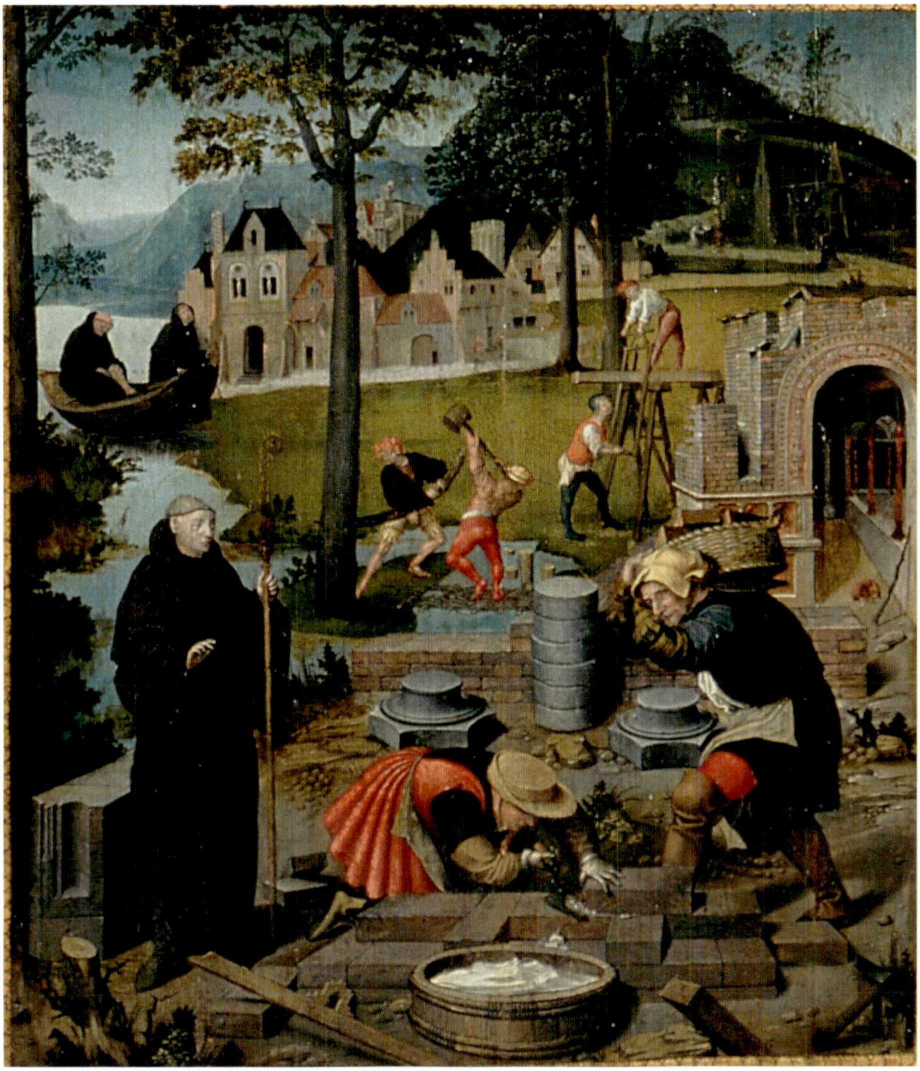

Figure 162: 1500-1533, Northern French, *Construction of the Abbey of St Bertin at St Omer*, oil on wood panel, lower panel of the right hand leaf of the doors to the treasury of the Abbey. © Musee des Beaux-Arts de Dijon. Photo Hugo Martens.

measuring. Behind the mason a monk kneels with a mason's trowel in his right hand. On the makeshift scaffold (see Page 78 above) another mason is working with a container of mortar and a bucket of water. He is plastering the ceiling of the vault where exposed brick can be seen to the left of the bucket, which he is yet to cover. He appears to be in conversation with Saint Benedict below him, who is pointing with a stick. To the right other masons work on the wall above the colonnade. A

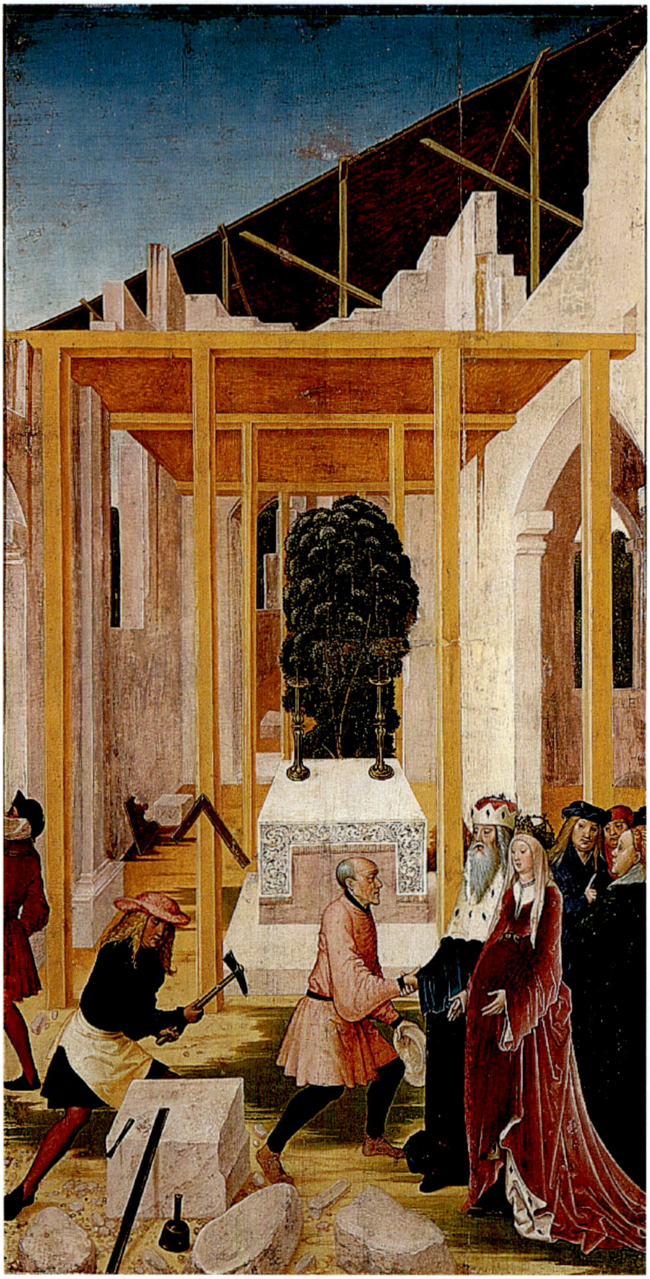

Figure 163: *c.*1505, Rueland Frueauf the Younger, Austro German, 1470-1547, *Margrave Leopold III (St Leopold) Inspecting the Construction of Klosterneuburg Abbey*, 76 × 39 cm, tempera on panel, Monastery of Klosterneuburg, Austria.

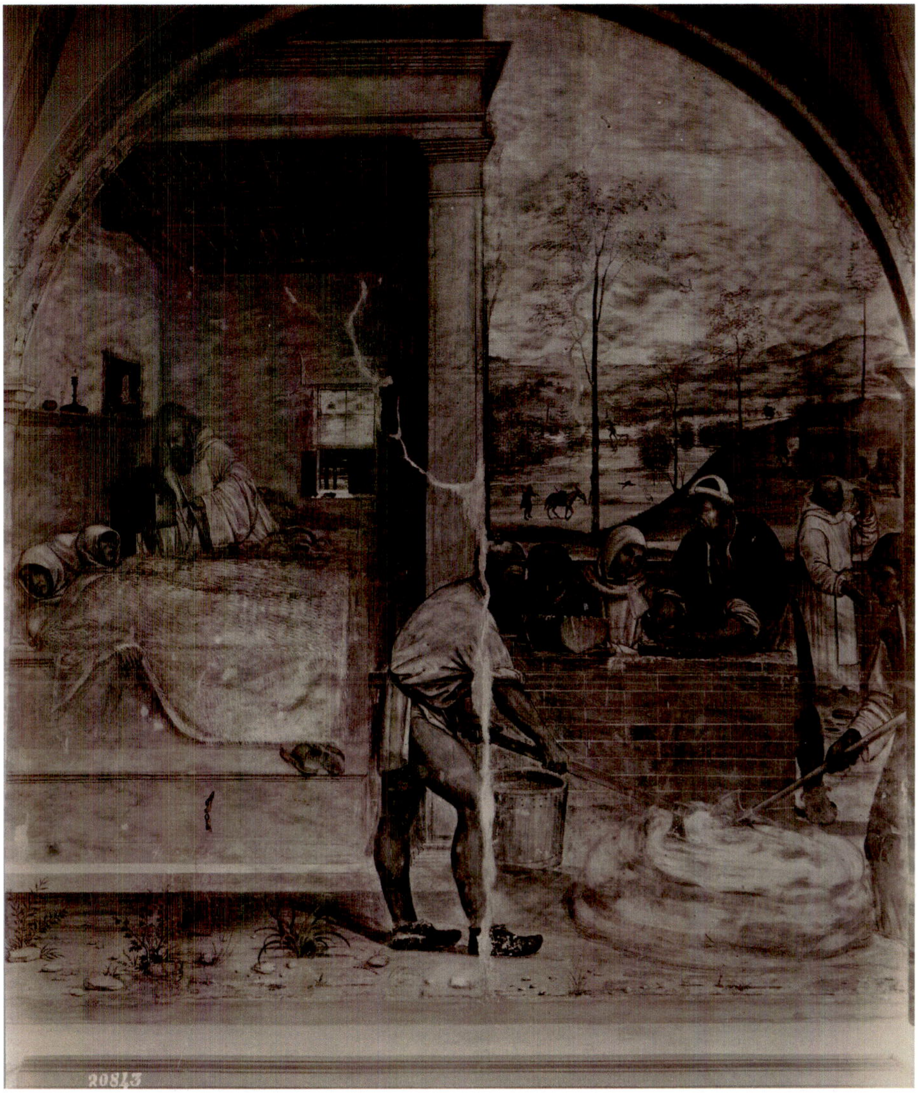

Figure 164: 1505-1508, Giovanni Antonio Bazzi, called Il Sodoma, Italian, *Life of St Benedict: Benedict Appears in the Dreams of Two Monks*, from a series of frescoes depicting the Life of St Benedict, Abbey of Monte Olieveto Maggiore, Asciano, Italy. Archive photograph *c.*1895, Fratelli Alinari. Fondazione Zeri, Universita di Bologna, Italy.

carpenter struggles with an upright timber post, and a worker climbs a ladder carrying a hod of mortar. In the second image (Fig. 164) a brick wall is being built. Two workers are mixing mortar in front of the wall, and behind it is a bricklayer with trowel in hand tamping the brick he is laying. A young monk places a container of mortar on the wall and another behind prepares to check the wall beyond with a plumb-line. To the left Saint Benedict appears in the dreams of two monks, bearing a model of the completed monastery.

Piero di Cosimo's strange and almost surreal painting of 1514-1518, *The Construction of a Palace* (Fig. 165) is broadly divided into two halves. The upper half consists of the palace comprised of two mirror blocks connected by a central courtyard and partly enclosed by the peripteral colonnade of square columns around both blocks. These support a continuous balcony with a balustrade. Above each column is a torch-bearing statue on a pedestal. The last of these is being set into place on the right-hand block using the hoist. Apart from this operation the building is complete. Beyond the central court the grounds have been formally laid out with mature trees to form an avenue. However the building has an empty shell-like appearance; there is no fenestration and no doors, lending to the surreal effect of the composition. To left and right a hilly landscape descends to parkland, which gives way to the flat arid landscape into which the building is set. On the right a group of mounted gentry have come to admire the building. Children lark around the base of the building including an exciting game of descending a rope rigged around the base of the statue at the far left. The arid landscape develops into the middle and foreground in front of the building; the lower half of the composition. Which is thoroughly surreal in character, as there is no coordination or any real purpose to the various construction processes going on and the scale of many of these is disturbing. In masonry terms, in the foreground are two masons working on an enormous block of dark stone, one of whom sits idly on the corner wielding his hammer. They are flanked either side by seemingly abandoned finished pieces. On the right is a composite capital perched at an angle on a lump of stone, and on the left a statue of a naked male figure lies flat on the ground. Another large block of stone is about to be offloaded from an oxcart. To the far right beyond two mounted figures on a horse, a circular column is being checked, and there is a box of chisels on the ground with a mallet. There are no circular columns on the exterior of this building, so is this for the interior? Although surely it would have already been erected? Or it may be intended as part of a landscape monument. In the middle ground two masons work a length of cornice and beyond them another naked statue is being transported on a hand truck. This is possibly intended for the interior, as indicated by the figure pointing to the open door of the right-hand block. Or is he maybe pointing to the garden beyond, indicating that these are garden statuary? All these figures are out of scale, particularly the two horsemen to the left. In front of the hoist is a strange base built of rubble stonework. On its left flank it is falling apart and to the right two workers carry what appears to be rubble in a barrow, although to where and for what cannot be discerned. At the left of the building donkeys with panniers are either removing materials from the site, or delivering them. Is this sand for mortar? But there are no other signs of mortar

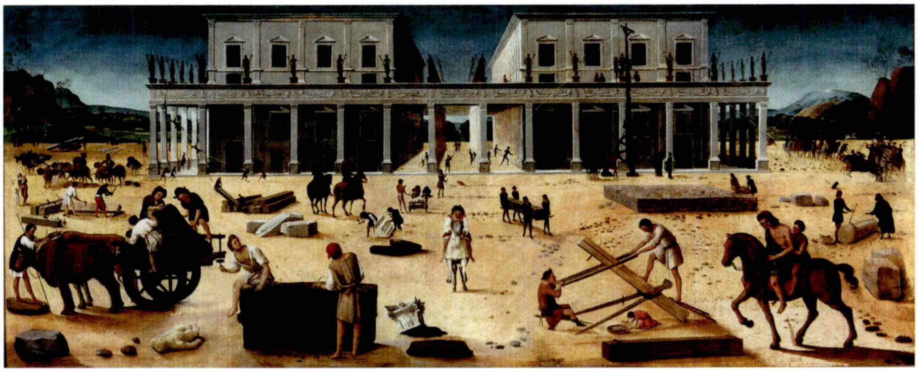

Figure 165: 1515-1520, Piero de Cosimo, Italian, 1462 -1521, *The Building of a Palace*, 83 × 197 cm, oil on panel, The Ringling Museum of Art, Sarasota, Florida, USA: Bequest of John Ringling, 1936, Collection of The John and Mabel Ringling Museum of Art, the State Art Museum of Florida, Florida State University.

preparation. Perhaps it is topsoil for landscaping. There is considerable depiction of carpentry here and this is discussed above on page 124. It is not clear whether the whole composition is some form of allegory or mythological representation. It is argued that the building may be a reference to designs by Piero's friend the architect Giuliano da Sangallo, but this is by no means certain.[151]

In Pieter Bruegel the Elder's, *The Tower of Babel*, of *c.*1525, we see the whole range of the construction process, depicted in compelling clarity, as it would have existed in the first quarter of the sixteenth century, and indeed for several centuries before that. The building is biblical, mythological and quite unreal. But its mode of construction is as it broadly would have been for any major building project, such as a cathedral, castle or palace, in Medieval and late Medieval Europe. However, contrary to popular belief, not all is entirely correct (see also page 142 above). When considering Pieter Bruegel the Elder's art Max J Friedländer says:

> There we find landscapes, biblical subjects and genre-like pieces. The categories cannot be clearly segregated. A biblical content is infused into landscape and all human activities are conceived as genre. Innumerable figures, a whole world, adventure, picturesque scenes, gaiety and edification, an endless spectacle unfolds, entertaining for the simple observer, a miracle for one who is following the course of Netherlandish art.[152]

This is all true of his *Tower of Babel*, a large painting some one and a half metres wide and just over a metre high, crammed full of the most amazing detail. (Fig. 166) Bruegel had travelled to Rome and had seen the Colosseum upon which the building is surely based. He presents us with a vast building still far from finished, which dwarfs its supposed model and is far bigger than anything on earth. We know that such a building is implausible, but did the ordinary person think that such a building could exist in foreign lands? Certainly the master builders of the time building cathedrals, palaces and fortifications knew such a thing was impossible. But Bruegel looked to the projects of these master builders, masons, carpenters, clerks of works and the like, for his details. The painting is greatly admired for its genre-like aspects, and accurate depiction of daily life. But there is much that is inaccurate, particularly the depiction of mason's work. Masons are everywhere, on the hill overlooking the site, on the dockside, at the base of the tower and all over the tower itself. It is not what they are doing, but it is the size of the blocks of stone that is troublesome, and beyond reality. On the hill overlooking the site where Nimrod and his retinue have come to view progress, the masons work with mallets and chisels squaring blocks, many of which are already worked and are enormous. They lie as if dropped from the heavens in utter confusion, as they are in many other locations, and what are they anyway doing there so far from where they are needed? The average weight of stone is 2,250 kilograms per square metre.[153] Calculating the size of four blocks from around the site using the size of adjacent figures the following results are startling: In this scene, with Nimrod, a worker starts downhill disappearing into the bushes on the long trek towards the site with a block of stone on his back, which is approximately half a metre wide, just below three quarters of a metre high (·7m) and 200cm thick; this gives a weight for this block of approximately 150 kilograms. Given the distance and terrain to be navigated, this is far too heavy, even for a worker of above average strength. The block to Nimrod's right which is being manhandled by four workers is approximately 1·4m high x 1m wide x ·4m deep, giving an approximate weight of 1,260 kilograms, or 1·2 tonnes. The block on the second tier of the ramp of the tower, immediately to the right of the small shed adjacent to the large crane, and probably the largest block on the site, is approximately 2m x 2·5m x 1m, weighing 11,250 kilograms or over 11 tonnes. Finally the block being lifted by the crane is 2m x 1m x ·2m and 900 kilograms, nearly a tonne. All these numbers are approximate, but virtually all the blocks shown in these higgledy-piggledy stacks around the tower are too large and too heavy for human or for mechanical handling of the time. Where masons are actually laying blocks, where work has been left with bonding gaps and where the bedding lines of the finished work are visible, much smaller blocks have been, or are being, used. These blocks of stone in size are anăchronistic and for their time could not have represented reality. They may be awaiting sawing to size, but there is no sign of this activity (sawing stone was common), but they would have had to be moved anyway to achieve this.

Bricks are being delivered to site by barge, from across the bay or further afield, where they are offloaded onto the dockside and foreshore into neat stacks. From there they are loaded onto horse drawn carts to be hauled up the steep track and disappear into the entrails of the building to emerge somewhere higher up, where

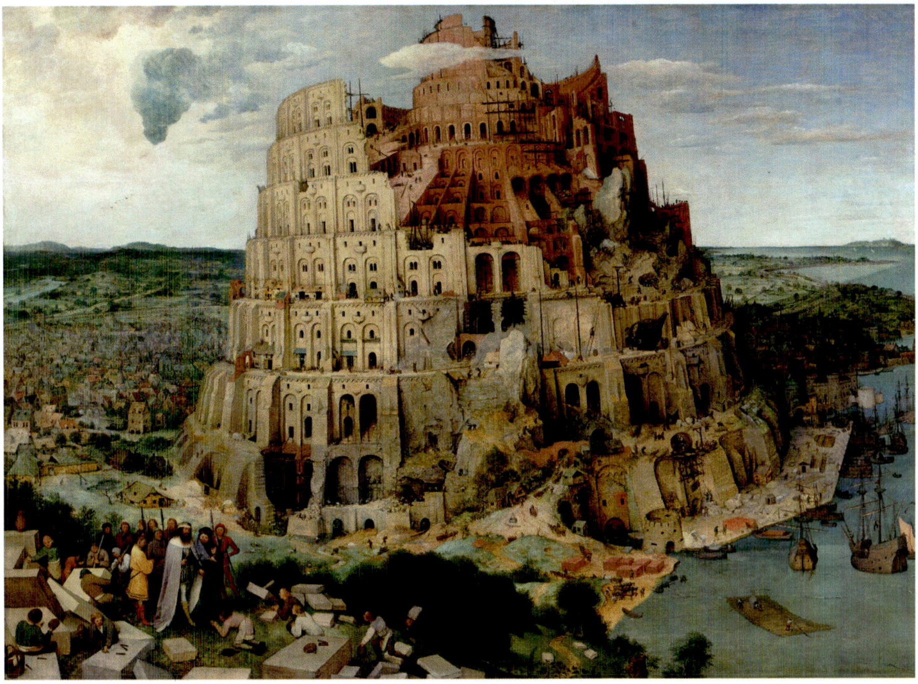

Figure 166: 1563, Peter Bruegel the Elder, Flemish, 1525-1569, *The Tower of Babel*, oil on Oak panel, 114 × 155cm. Kunsthistorisches Museum, Vienna: KHM-Museumsverband.

the inner core of the vast undertaking is all brick. Around the base of the tower and along the ramp, which rises up the outside, there are piles of sand and what may be prepared mortar. Sand is collected from the beach adjacent to the stacks of bricks, but this will have to be washed to get rid of the sea salts, which will otherwise have a devastating effect on the mortar. Lime is offloaded at the dockside and piled against the small tower, which is not a limekiln but appears to be a fortification. Apart from these signs there is no organised production of mortar, which would be required on a huge industrial scale. For such an undertaking at least a dozen or more limekilns would be necessary, working around the clock. But none are visible even in the landscape beyond the tower. They would have required huge quantities of raw limestone and fuel. There is a smithy however, for the production and maintenance of tools, located on land below the left shoulder of the tower, opposite a mill on the further bank of the river. The carpenters, cranes and other lifting gear are discussed in Chapter 1 (Figs. 44, 93 and 94)

The considerable extensions to Louis XIV's palace at Versailles were begun in 1678 under the direction of Jules Hardouin-Mansart. This work was recorded *c.*1680 in a painting attributed to Adam-François van der Meulen. (Fig. 167) In the background the palace itself is largely complete with the Alles des Ministres either side and the

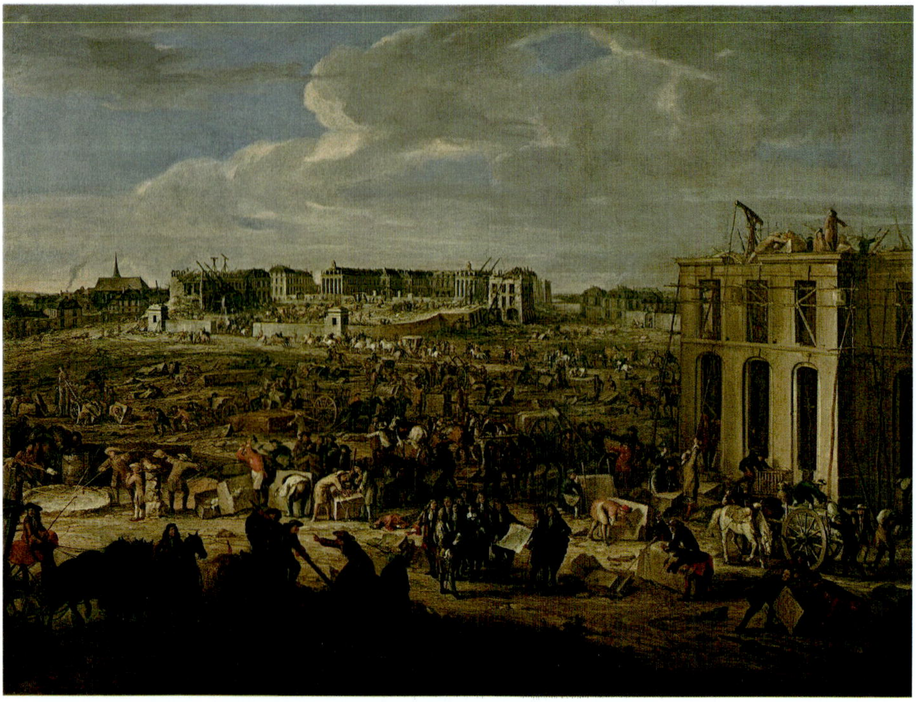

Figure 167: *c.*1680, Adam-François van der Meulen, Flemish, 1632-1690, *Construction of the Chateau de Versailles*, 103 × 138.5 cm, oil on canvas, Royal Collection Trust / © Her Majesty Queen Elizabeth II 2016.

Place d'Armes between under construction, with new ranges on the east side being added. Almost the entire area between the new east range and the Place d'Armes is taken up by a vast stone yard with dozens of masons working stone blocks and details. Intermingled with this work are transport bringing materials and taking away finished work. Mostly this is horse drawn, but in the middle distance on the left, six men haul a waggon containing a large section of cornice. In the left foreground is a lime-pit and mortar preparation. To the right Mansart with his assistants and Jean-Baptiste Colbert, the King's minister, show the plans to Louis. This is a very large and busy site, but there is a vague air of desperation; one can almost hear the King asking when it will be finished.

Stonemason contractors often had their own yards where some work would be carried out rather than on site; these may have been private, communal, municipal, or belong to a guild. Canaletto's *The Stonemason's Yard* is a good example of this. (Fig. 168) Here various orders would be carried out for sites within the city and then transported by barge; most sites would be adjacent to a canal. In Canaletto's yard one mason works on a section of moulding, whilst another addresses what is possibly a relief carving on a large slab. Leaning against the slab is a long rule and a large pair of calipers rest against a column drum. To the right a woman clears out the top of a well head, this may be an abandoned piece of work as a corner is

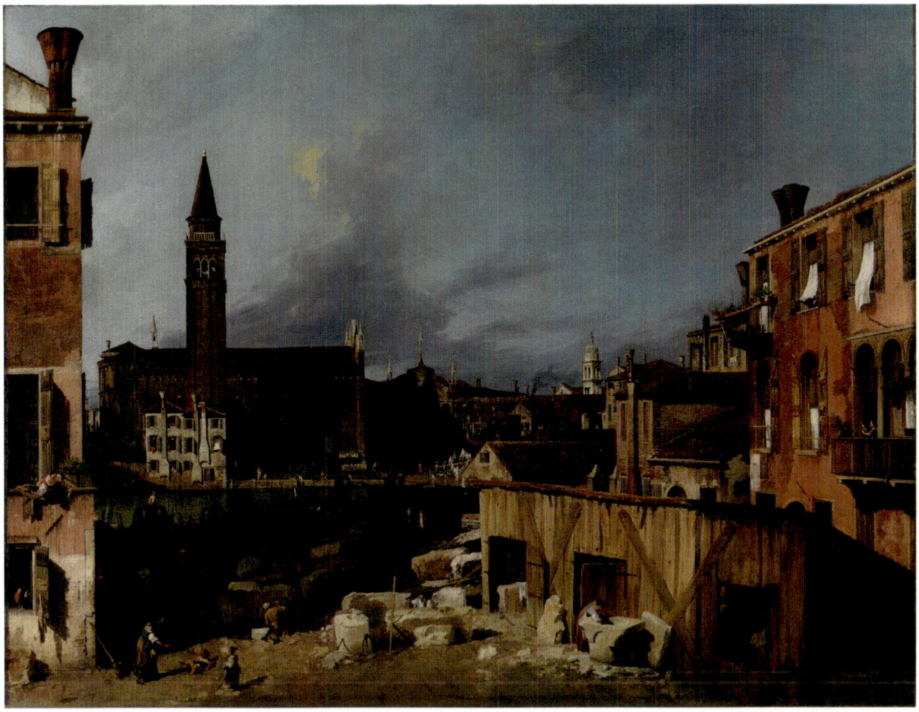

Figure 168: *c.*1725, Giovanni Antonio Canal, called Canaletto, Italian, 1697-1768, *The Stonemason's Yard*, 123.8 × 162.9 cm, oil on canvas, The National Gallery, London.

damaged. Work also progresses in the large shed. The site is at the Campo San Vidal, looking across the Grand Canal to Santa Maria della Carita. Today the church has lost its campanile and is part of the Accademia delle Belle Arti. The Accademia Bridge now rises in front.

A similar scene, although rural and some century later is George Jackson's watercolour *A View from Park Place across Whiteladies Road to the Royal Fort, Bristol.* (Fig. 169) Whiteladies road is now Queen's road and the whole area is now built upon. Two masons are sawing a large block of stone, in this yard these large blocks would be sawn and worked into smaller sizes for building. The stone is either carboniferous or oolitic limestone from the extensive quarry workings on Clifton Down, which was located a short distance away behind Jackson's viewpoint.

In Johann Christoph Frisch's 1802 painting of Frederick II King of Prussia and the Marquis d'Agens visiting the construction of the palace of Sanssouci at Potsdam, (Fig. 170) there is no grand vista of a vast construction site as in van der Meulen's rendering of Versailles. Instead there is a humble bricklayer and his two labourers, who appear to be working on a garden structure which requires some subterranean workings, or it might be an inspection chamber for drainage. The bricklayer spreads

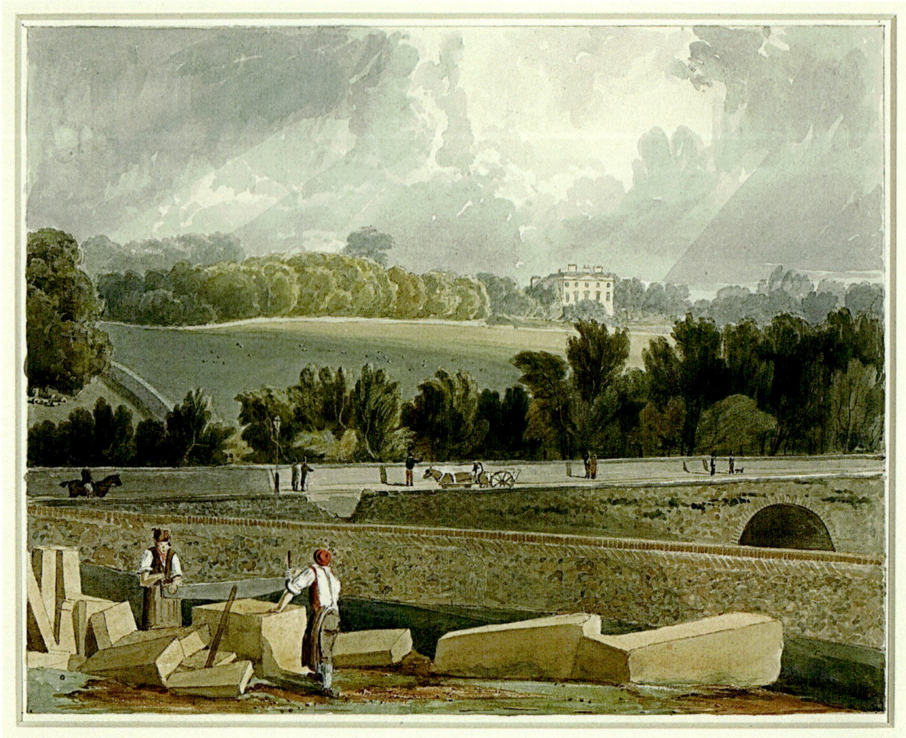

Figure 169: Samuel Jackson, English, 1794-1868, *A View from the Park Place across Whiteladies Road to the Royal Fort, Bristol*, 1824, 202 × 253mm, watercolour. Bristol City Museum and Art Gallery, Bequest of William J Braikenridge, 1908. Bridgeman Images.

mortar with his trowel for the next brick he is laying, which is held in his left hand. Next to him are his hammer and a tub of mortar, beyond on the other side of the opening are a plumb-line, the coiled cord of which hangs over the edge, and a levelling block. One labourer mixes mortar in a container with a shovel, while the other gathers bricks which he appears to have dropped out of his hod, and glances with some trepidation at the two noblemen in fear of being reprimanded. The bricklayer on the other hand is entirely engrossed in his work. The palace stretches away in the background seemingly complete and deserted of building workers.

Ford Madox Brown's *Work* (Fig. 171) depicts similar work to Frisch's, painting i.e., the construction of a sewer or drain. There are between six and eight workers depicted here, together with street vendors and passers by, but this is where the comparison ends. The scene is outside number 62 The Mount and Heath Street, Hampstead, north London in 1852, which is also the subject of an earlier study with just a single figure, a butcher's boy, walking down The Mount.[154] *Work* is a veduta painting, for it combines reality and the imaginary. The setting is realistic enough

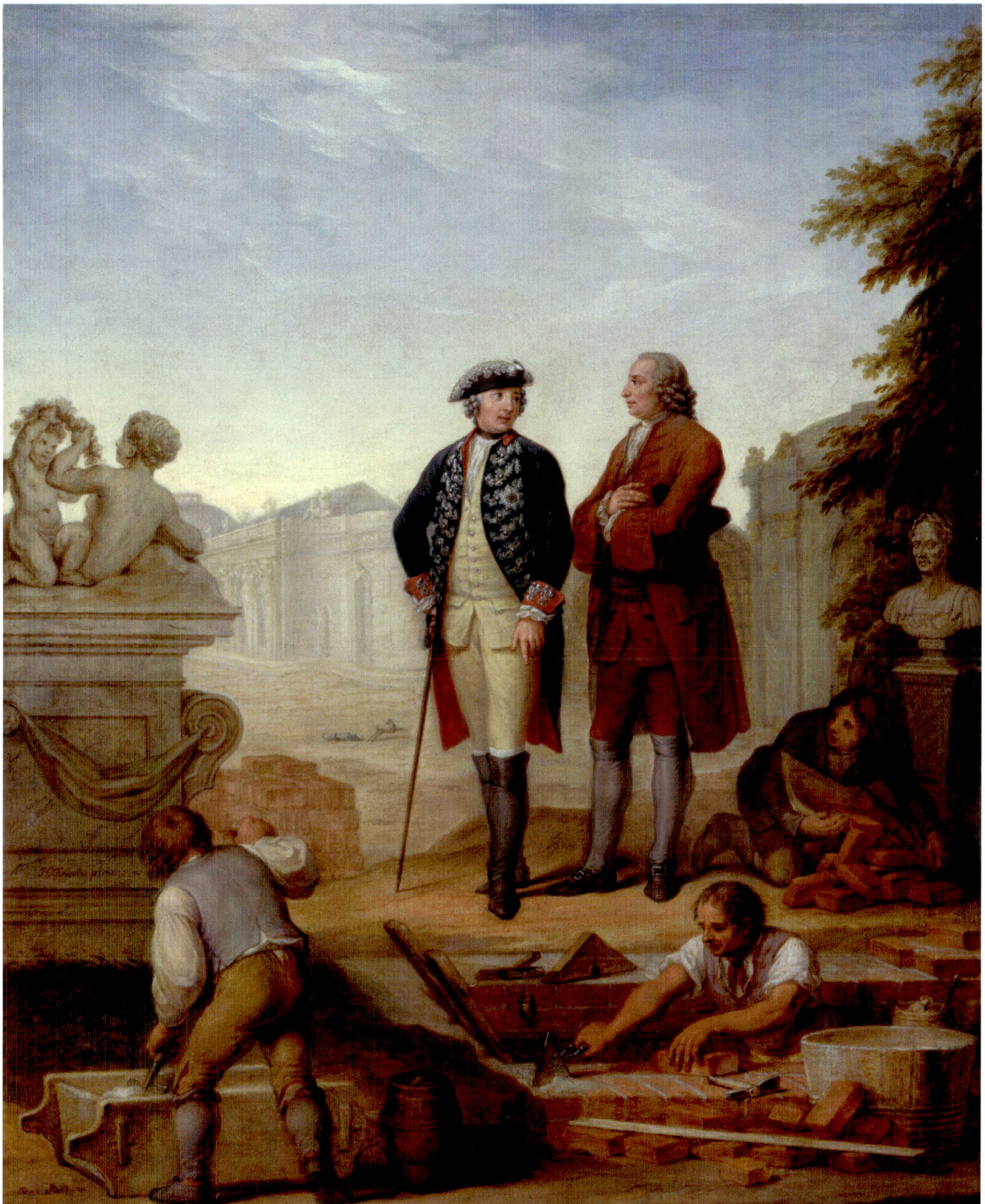

Figure 170: 1802, Johann Christoph Frisch, German, 1738-1815, *Friedrich II, King of Prussia and the Marquis d'Agens visit the Construction of the Palace of Sanssouci at Potsdam*, oil on canvas, 91 × 74.5cm. Stiftung Presußishe Schlösser und Gärten Berlin-Brandenburg/ Photograph: Gerhard Murza.

and the workers convincing, however the inclusion of all the other actors, realistic as they are, in such close proximity creates a fictitious narrative illustrating the moral of work. Even for the middle of the nineteenth century such work, dangerous work no less, would have been separated from the general public, if for no other reason than that the quicker the task is finished to the standard required, the sooner is payment made for it. These workers and indeed their employers, for this is a municipal undertaking, would not have countenanced such overcrowding and disruption to their site. Deep within the excavation are two navvies shovelling spoil onto the boarded platform at the top of the opening. All we see are the blades of their shovels clutched by a hand. Above them standing on the platform another worker, also the main focal character of the construction gang, shovels this spoil to a large pile beyond. Behind him a hodsman descends a ladder into the excavation with a hod of bricks on his shoulder, whilst another quenches his thirst before loading his hod. To his left a smiling lad with ginger hair is holding a ball in his right hand and carrying a bucket of water for the mortar mixing; this takes place to his left where a labourer mixes it with a hoe on the very edge of the retaining wall above a grass bank. Completing this circle of labourers or navvies, is the bent figure grading aggregate for the mortar by shovelling it at a sieve; the coarse unwanted material falls to the front whilst the smaller stuff passes through to pile up behind the sieve. Both these processes associated with mortar are too cramped for comfort and would have ideally been done a little further away, possibly in the square beyond; after all, they do have a wheelbarrow to transport it. The boy larking around with the barrow could conceivably be part of the gang given the date, however we are told that he is with the girl in the ragged red dress. He is her brother along with two other siblings, and she is there to drag him out of the way. In the wheelbarrow sits a lone small masons trowel of the type used for pointing and in the foreground lie various tools including a pickaxe, a broom and a try-square. Close by lying on a red rug is a small dog owned by one of the navvies and used when required for keeping rats out of the excavation. The other dogs belong to the other participants. The Whippet in the red coat possibly belongs to the beer and tobacco pipe seller who calls out his wares; being a sporting dog it suits him as he is also clearly a bookies runner and a pugilist. Brown worked on the painting for over ten years from 1852. In 1865 he exhibited it at a one-man show he arranged himself in Piccadilly, and wrote a catalogue entry in which he describes many aspects of the painting.[155] He tells us that the excavation was connected with water supply but this is no mere trench; an underground chamber is being built possibly for an important junction. At the front of the excavation is a block and tackle rigged over a stout piece of timber laid across the opening, from which something heavy is suspended. Given the bricks being taken down and the presence of the preparation of mortar there may also be bricklayers working underground.

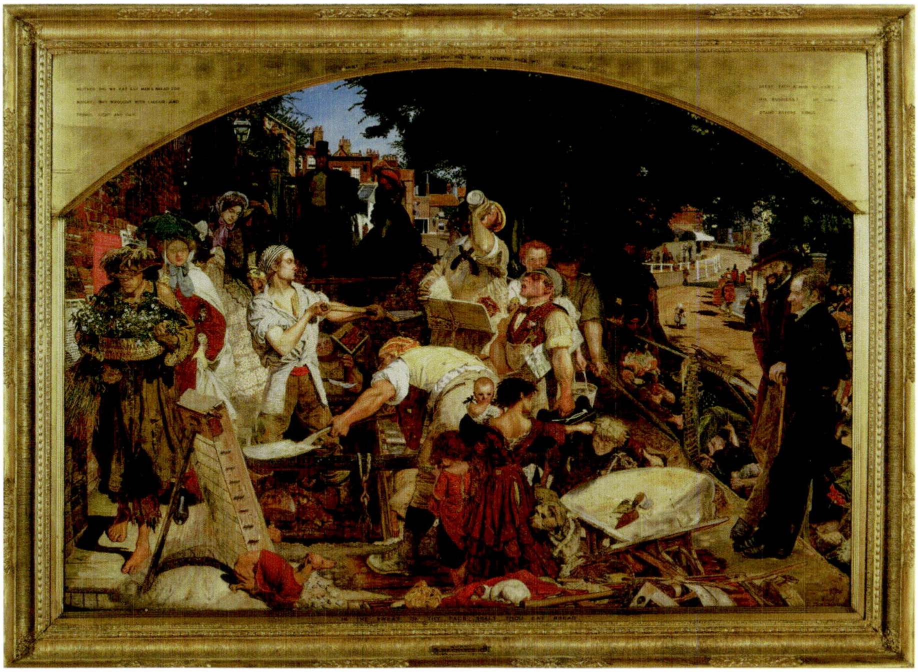

Figure 171: 1852-1863, Ford Madox Brown, English, 1821-1893, *Work*, oil on canvas, 137 × 197.3 cm, Manchester Art Gallery, UK/Bridgeman Images.

In J M W Turner's *Kirkstall Lock, on the River Aire*, (Fig. 172) we see a group of stonemasons and bricklayers at work on a building on the very edge of the river bank. In the background on the right are the ruins of the Medieval Kirkstall Abbey. In the middle the Leeds and Bradford turnpike's route over the Aire with the freely moving stage coach contrasts with the backlog of boats waiting to descend to the Leeds and Liverpool canal. The watercolour is essentially a 'Picturesque View' and is therefore, for our purposes, short on detail of the building site. This work is said to illustrate the interplay of 'modernism and history'.[156]

A coloured lithograph, in a book entitled *Méthode d'enseignement par l'aspect – les métiers et leurs outils*, by F. Streich and Dr K de Gerstenberg, published in Paris 1876, shows a large building site.[157] (Fig. 173) To the right are masons, some of whom are under cover working away at blocks and mouldings with their tools clearly delineated. In the foreground is a large timber box where mortar has been mixed, with a hoe in the box and a shovel to the side. Waiting to be taken to one of the buildings, for the fixers or bricklayers, is a hod of mortar with a mason's trowel. This is a book that educates through its illustrations. This image has also been discussed for its depiction of carpentry activities, see p 126 above and Figure 83, but it is worth reiterating that what it teaches us on the cusp of the twentieth century, is nothing substantial has changed, all is still human powered and the tools virtually identical to those of the preceding centuries.

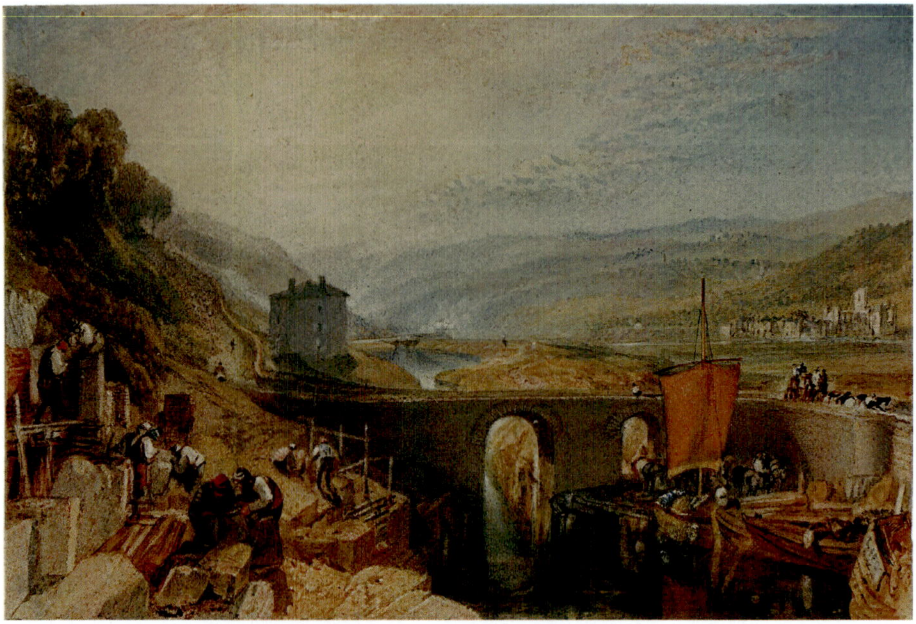

Figure 172: 1824-5, Joseph Mallord William Turner, 1775-1851, *Kirkstall Lock, on the river Aire*, watercolour on paper, 159 × 235mm. Turner Bequest, 1856, CCVII L, © Tate, London 2016.

Materials Production and Extraction

Masons working directly from a quarry are seen in Figures 157 to 160 in two works by Andrea Mantegna. In the small manuscript illumination of 1482 (Fig. 174) from *De proprietatibus rerum* (On the Property of Things) by Bartholomaeus Anglicus, a type of early Medieval encyclopedia, of great popularity at the time,[158] the miniature at the beginning of Book 16, 'On rocks, gems and minerals', above a decorated capital 'P', shows three workers all with pickaxes, two of whom, both kneeling, are handling shaped blocks.

J M W Turner's *View of Fonthill from a Stone Quarry*, (Fig. 175) depicts a quarry which was located in Wiltshire, southern England. The view is from the south and the quarry could be any one of a cluster in the area south of Fonthill Gifford where William Beckford built his enormous folly Fonthill Abbey between 1796 and 1813. The tower, which was some 90 metres high, and which is seen in the distance, collapsed twice during building. Finally, in 1825 Beckford commissioned Turner to paint several scenes of the Abbey. The stone, an oolitic limestone from this quarry, was probably used in the construction of the Abbey, although there were several other quarries in the immediate area, some of which were closer, and vast quantities of limestone were were required.

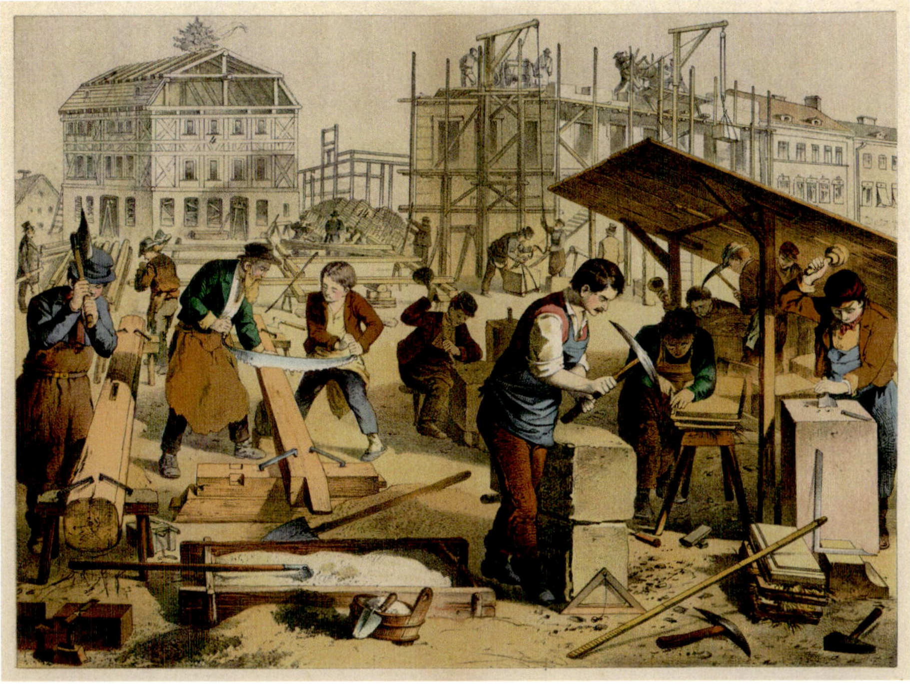

Figure 173: 1876, Illustration from F. Streich and K. de Gerstenberg, *Méthode d'enseignement par l'aspect les métiers et leurs outils*, chomolithograph, published 1876, Paris, editions Bonhoure. Mary Evans Picture Library.

In the pre-industrial period, the overland transportation of stone from a quarry to a building site was never easy. Quarries situated close to rivers and the coast utilised barge or ship transportation where possible, but some overland carriage was inevitable. Peter Paul Rubens' painting *Landscape with Stone Carriers*, c.1620, depicts the difficulties which were all too often encountered with poorly made and maintained tracks. (Fig. 176) Francisco de Goya's *Transporting a Stone Block* of 1786 (Fig. 177) portrays a much larger cargo on a heavy cart hauled by at least three pairs of oxen, over a better made roadway. The site, the construction of a fortress, is seen in the background. In the foreground a casualty from some nearby accident or confrontation is carried across the road on a stretcher.

Brick making was always located on or adjacent to clay deposits, where brick kilns were also found. In David Teniers the younger's painting *Brickmakers near Hemiksem*, c.1690, (Fig. 178) we see the whole process set out on the banks of the river Scheldt a few miles south of Antwerp. In the foreground clay is being dug out and transported to the brick-maker by wheelbarrow and bucket. The brick-maker is at the top of the track between two sheds and working at his bench. Bricks are

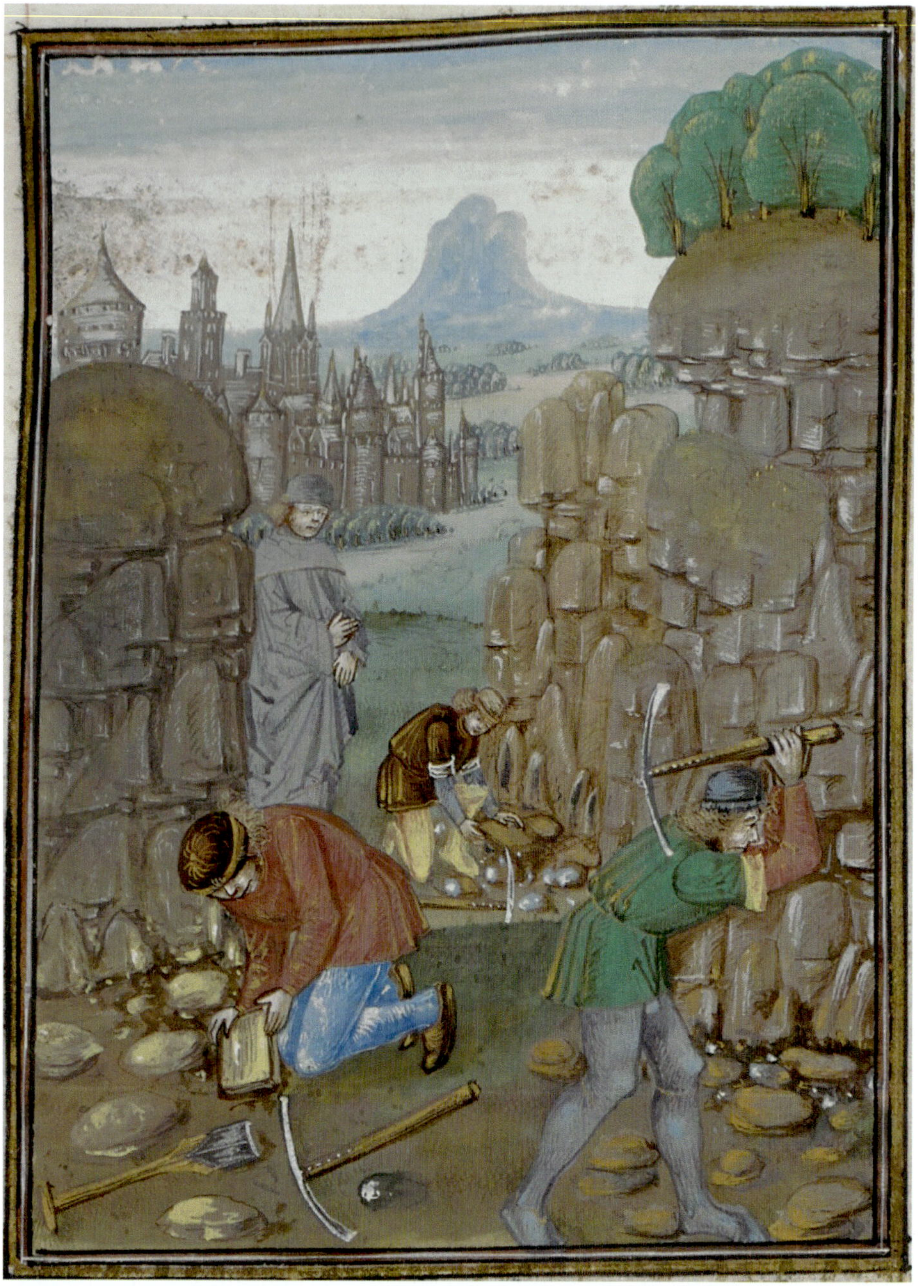

Figure 174: 1482, Assistant to the Master of the Flemish Boethius and/or the Master of Edward IV, South Netherlands (Bruges), *Miners at Work in a Quarry*, from Bartholomaeus Anglicus, translated by Jean Corbechon, *De proprietatibus rerum (Livre des proprietez des choses)*, MS Royal 15 E III, f. 102, British Library, London, UK © British Library Board/Bridgeman Images.

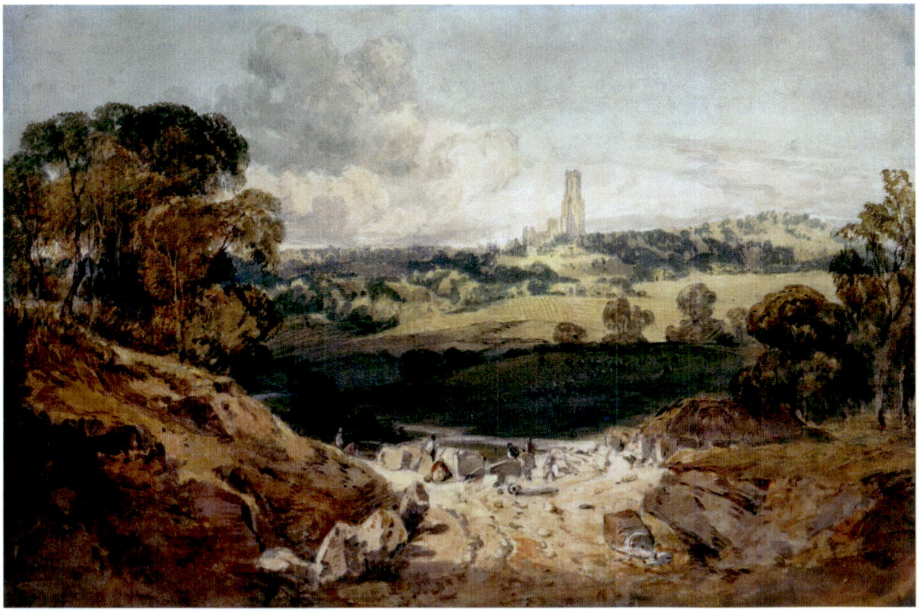

Figure 175: *c.*1799, Joseph Mallord William Turner, *View of Fonhill from a Stone Quarry*, pen, ink and watercolour, Leeds Museums and Galleries (Leeds Art Gallery), UK/Bridgeman Images.

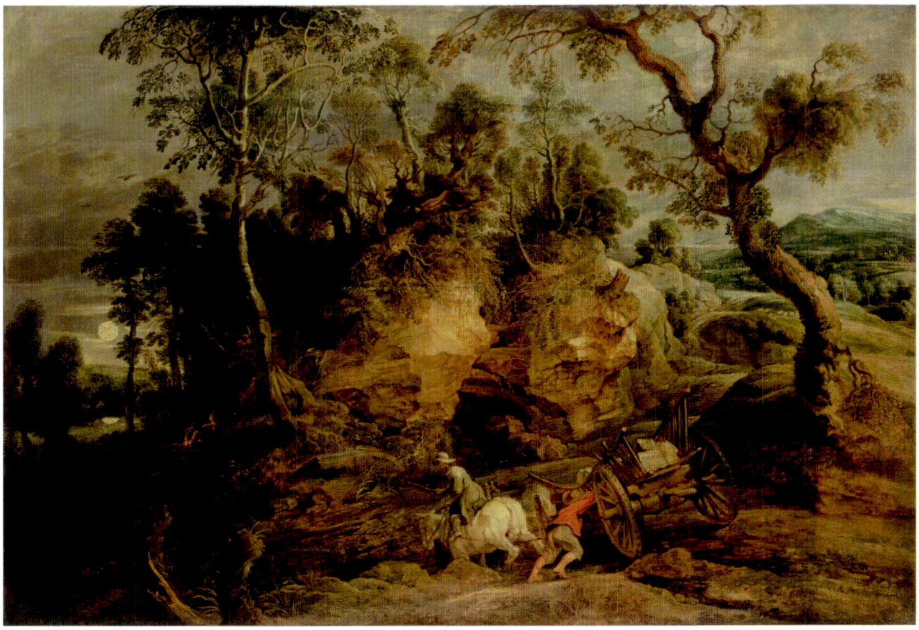

Figure 176: *c.*1620, Peter Paul Rubens, Flemish, 1577-1640, *Landscape with Stone Carriers*, 86 × 126.5cm, oil on canvas transferred from panel. State Hermitage Museum, St Petersburg, Russia/Bridgeman Images.

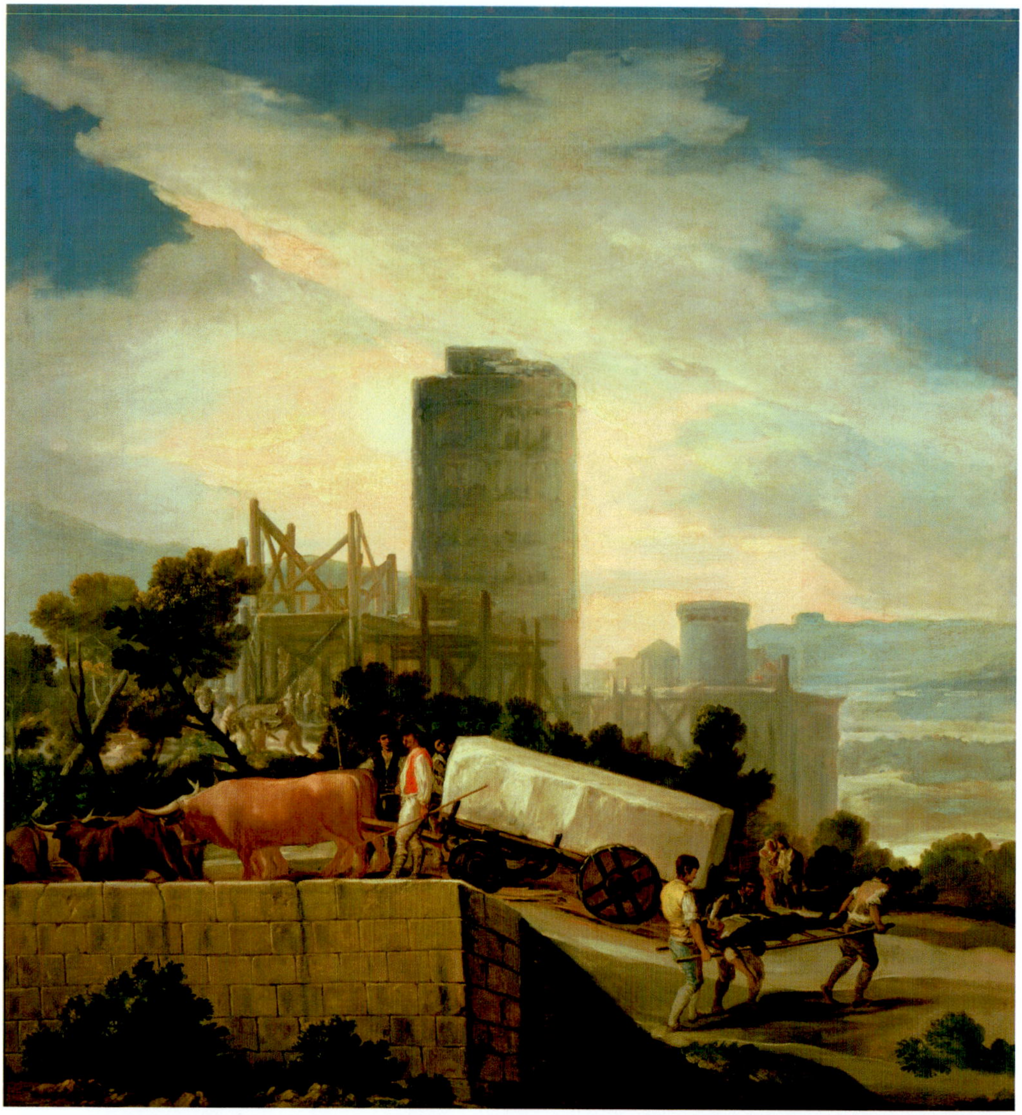

Figure 177: 1786, Francisco Jose de Goya y Lucientes, Spanish, 1746-1828, *Transporting a Stone Block*, 127 × 169cm, oil on canvas. Private Collection/Bridgeman Images

stacked for drying both in the open sided shed and on the ground in front, before being fired in the kiln seen at the bottom end of the shed. Various figures populate the scene, most of whom are workers, but others are standing in conversation. One worker, or a traveller, has rigged a wind shelter where he has settled down to rest. The large church is the Cistercian abbey of St Bernard founded in 1243, rebuilt in the late seventeenth century. The abbey established the brickworks in the thirteenth

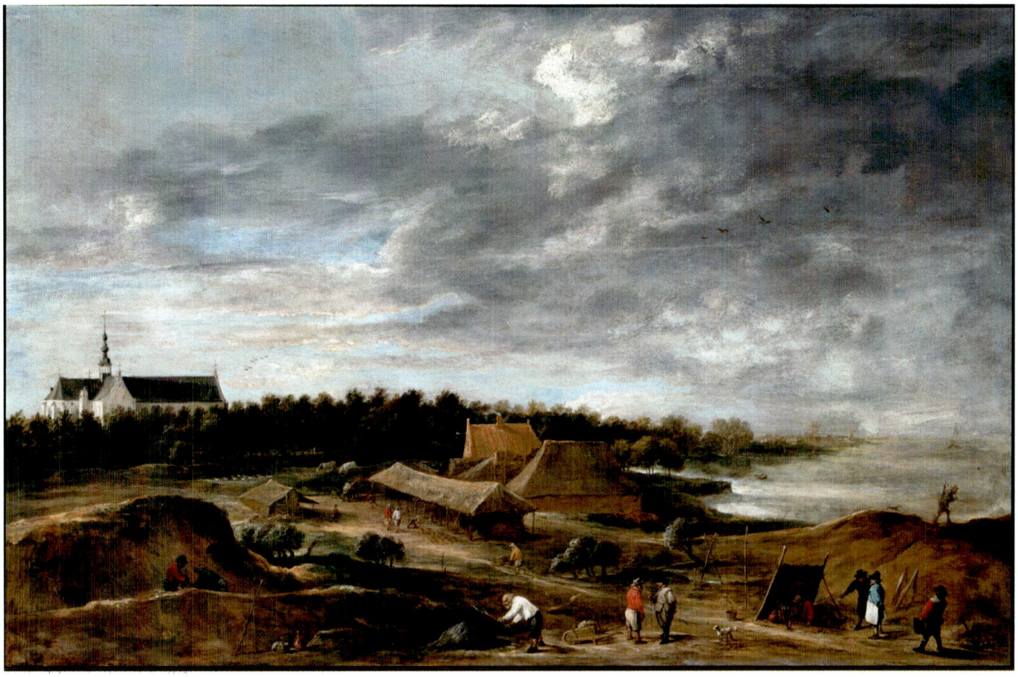

Figure 178: Before 1690, David Teniers the younger, 1610-1690, *Brickmakers near Hemiksem*, 43.8 × 67cm, oil on panel. DPG57, by permission of the Trustees of Dulwich Picture Gallery, London.

century, then one of the oldest brickyards in the Netherlands although the site is now in Belgium.

A number of large brick kilns are seen in Alfred Clint's painting of *Hampstead from the south-east*, 1852-1855. (Fig. 179) These utilised the extensive clay deposits around Hampstead in north London, which was growing at the time with the addition of many grand brick built town houses. Also beyond the kilns in the middle distance work is progressing on the Camden Town to Willesden railway opened in 1855.

Lime for mortar also required burning in a kiln. Figures 145 and 166 show lime-kilns in use associated with specific building sites. First limestone or chalk are quarried, then transported to the kiln, combined or layered with fuel and then burnt to produce quicklime (see page 206 above).[159] This process has existed since antiquity and is multi industrial, that is, it is intimately associated with both construction and agriculture. David Teniers the Younger's painting *A Lime-Kiln with Figures*, (Fig. 180) depicts a lime kiln, probably in the Lowlands. Here two men take limestone in wheel barrows to the top of the kiln, with the worker on the left just mounting the ramp which climbs to the top. Two men in the left foreground are breaking fuel, probably coal, which is also to be taken to the top. The quicklime would be raked out at the base from the opening on the left where a worker is emerging carrying

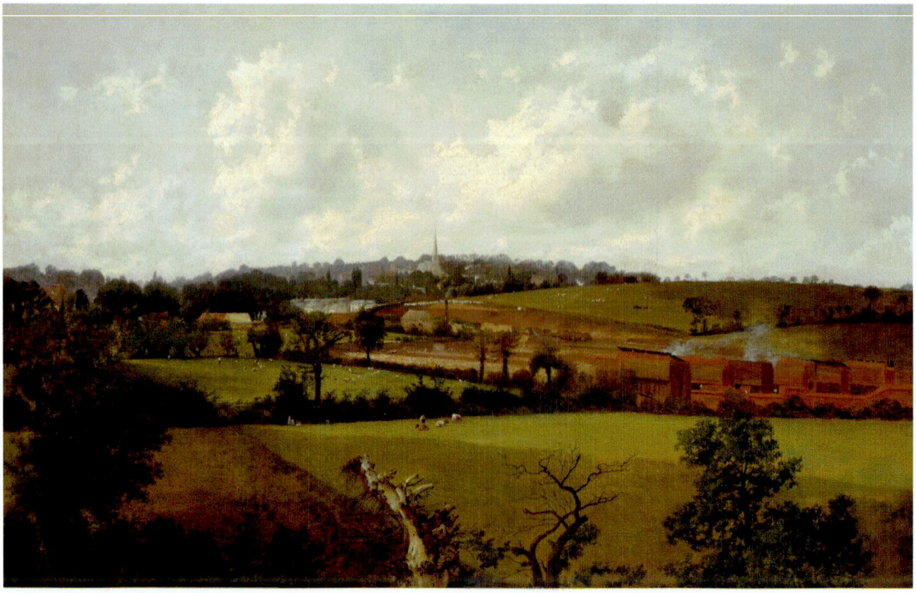

Figure 179: 1852-1855, Alfred Clint, English, 1807-1883, *Hampstead from the south-east*, oil on canvas, 610 × 914mm. Purchased 1935, N04809, © Tate, London 2016.

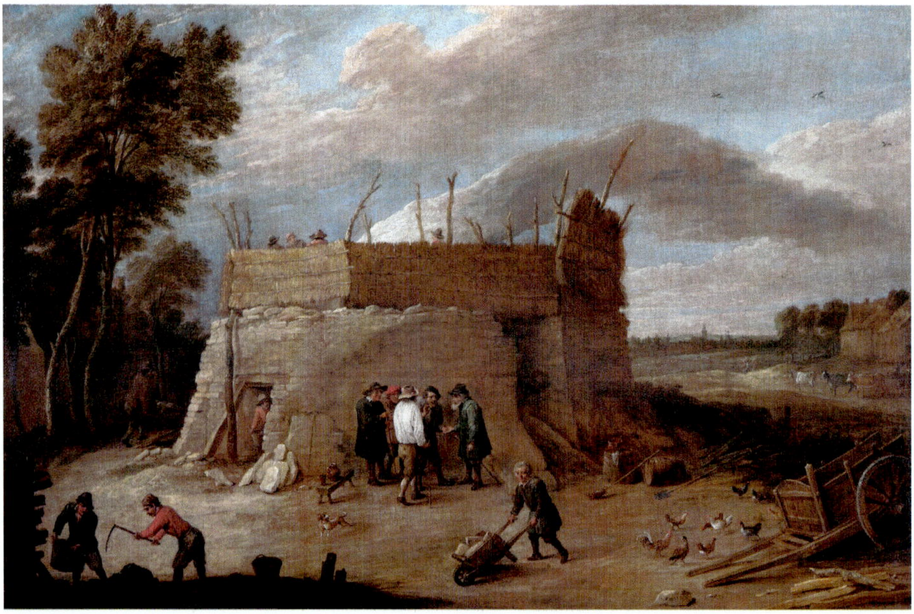

Figure 180: Seventeenth century, David Teniers the Younger, Flemish, 1610-1690, *A Lime Kiln with Figures*, oil on canvas, 58.5 × 88cm. Apsley House, The Wellington Museum, London, UK © Historic England/Bridgeman Images.

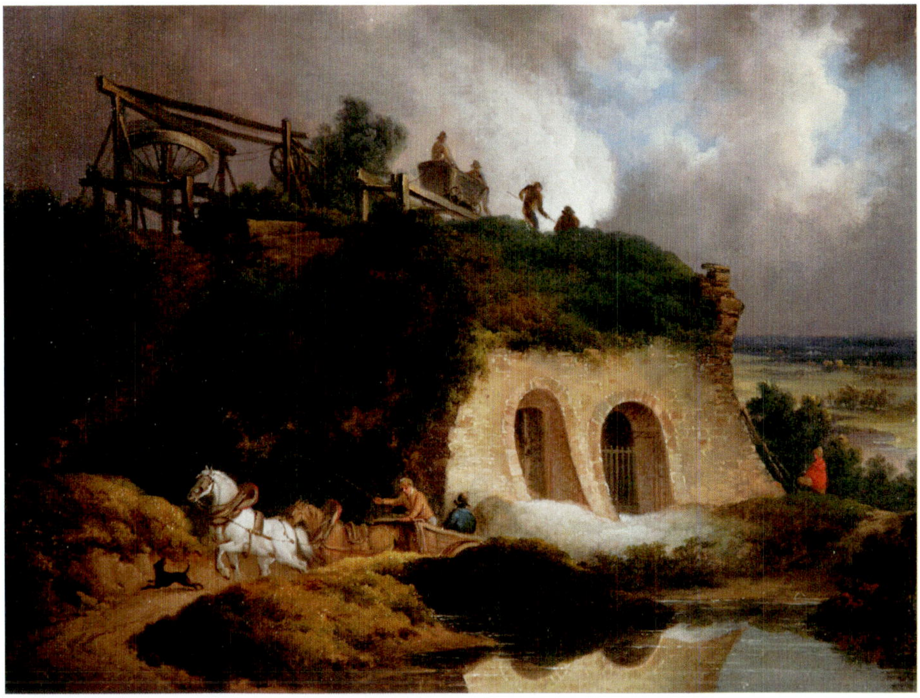

Figure 181: Late eighteenth century-early nineteenth century, Patrick Nasmyth (attributed) 1787-1831, or Charles Towne, 1763-1840, *Lime Kilns*, oil on panel, 23.3 × 30.9 cm, Museum of Barnstable and North Devon, UK.

a basket. The group in front of the kiln are striking a deal, between a man who is perhaps the kiln owner and a local farmer or builder. The kiln is being charged or loaded and is not actively burning yet. Another painting, *The Kilns*, (Fig. 181) from a century or so later, shows a more mechanised system for loading the kiln at the top. A horse gin is being used to haul up trucks on a primitive railway on an inclined plane from the base of the kiln.[160] Two workers ride the truck on its return downward journey whilst two others tend to the charge of limestone and fuel at the top. A horse drawn cart leaves the site; it is either empty from having delivered the raw materials, or else carrying a consignment of quicklime.

The other constituent mixed with lime to produce mortar is the aggregate or sand. This was extracted or quarried from pits such as those seen in Figures 182 and 183. John Linnell's *Kensington Gravel Pits*, (Fig. 182) shows workers digging out gravel, which is then sieved into various grades by the workers close to the wheelbarrows. Adjacent to them are various piles of graded aggregate for different uses, such as for mortar for bricklaying, render, or concrete. William Turner's *Gravel Pit on Shotover Hill near Oxford*, is a much less populated scene with only a single worker with a wheelbarrow. These hills above Oxford had provided building materials since Medieval times.

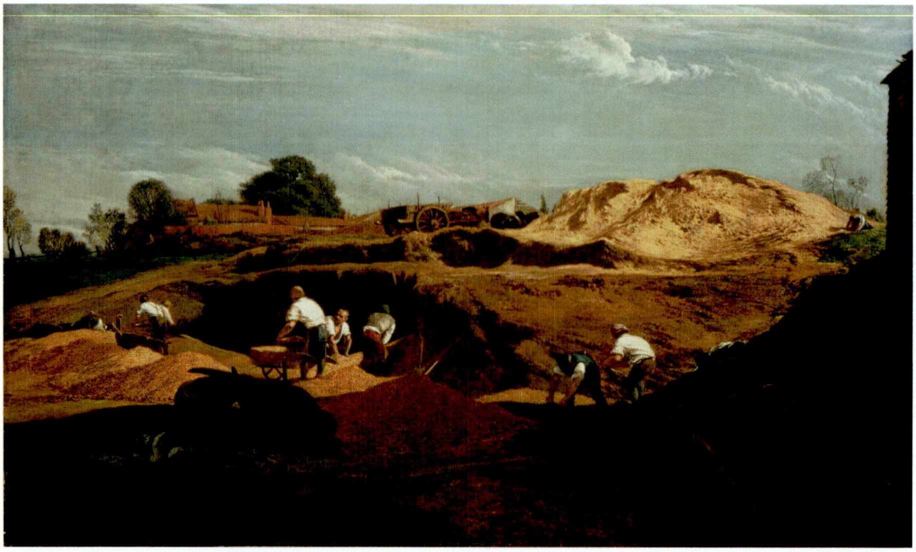

Figure 182: 1811-1812, John Linnell, English, 1792-1882, *Kensington Gravel Pits*, 711 × 1067 mm, oil on canvas. ©Tate Gallery, N05776, London, UK.

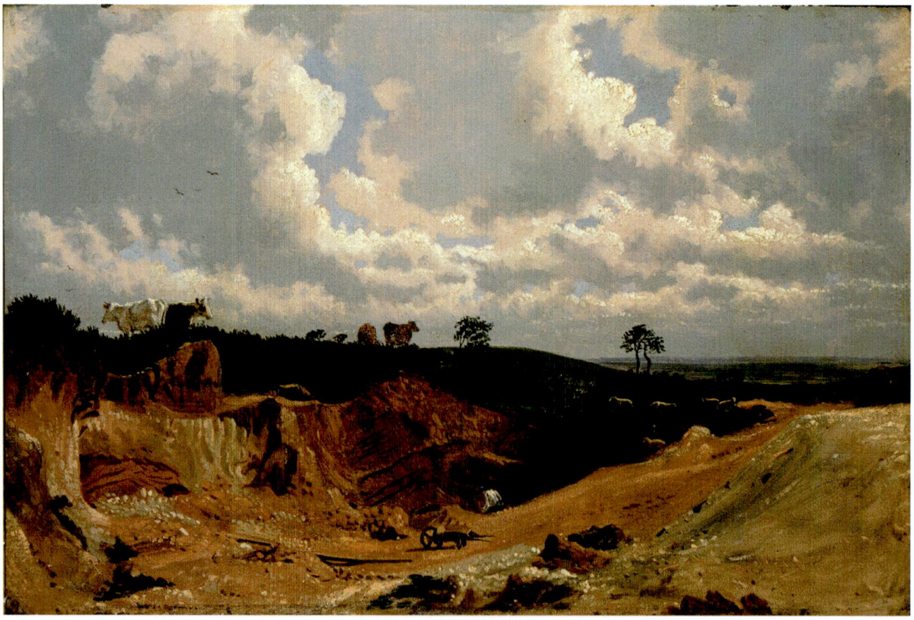

Figure 183: 1816, William Turner of Oxford, English, 1789-1862, *Gravel Pit on Shotover Hill, near Oxford*, 17 × 40 cm, oil on board, ©Ashmolean Museum, University of Oxford.

3. The Smith

No building site was without iron and none could operate without a blacksmith, for both the mason and carpenter relied on the smith for their tools. Furthermore virtually all buildings were full of accessories made of iron by the blacksmith down to the industrial period of mass-production in the late eighteenth and nineteenth centuries.[161] Smiths are shown producing various goods to illustrate the environment in which they worked.

The mining and winning of ores to produce metals dates back to *c.*3000-2000 BCE and the early Iron Age to 1200-500 BCE. In the bible, in the Book of Job, is a poem, a fascinating lyrical description of ancient mining techniques:

> There is a mine for silver
>   and a place where gold is refined
> Iron is taken from the earth,
>   and copper is smelted from ore.
> Mortals put an end to the darkness;
>   they search out the furthest recesses
>   for ore in the blackest darkness.
> Far from human dwellings they cut a shaft,
> in places untouched by human feet;
>   far from other people they dangle and sway.
> The earth, from which food comes,
>   is transformed below as by fire;
> lapis lazuli comes from its rocks,
>   and its dust contains nuggets of gold.
> No bird of prey knows that hidden path,
>   no falcon's eye has seen it.
> Proud beasts do not set foot on it,
>   and no lion prowls there.
> People assault the flinty rock with their hands
>   and lay bare the roots of the mountains.
> They tunnel through the rock
>   their eyes see all its treasures
> They search the sources of the rivers
>   and bring hidden things to light.[162]

Having produced the ore it was necessary to smelt it in a furnace in order to extract the metal. This was not a straightforward process as ores consist of the required ore minerals and many unwanted impurities. Smelting took place at a relatively low temperature, as early furnaces did not have the later technology of the blast furnace. The result was a spongey mass or lump called a bloom comprising the iron and impurities, hence the name bloomery for these early furnaces. The impurities were driven off by hammering, as the lump cooled it was reheated and the resulting near pure iron became known as worked or wrought. This wrought iron was the raw material of the blacksmith who was supplied by the furnace operators. In the medieval period virtually every village and town had a smith. The smith was a powerful person who enjoyed high status, equal to, or greater, than the mason or carpenter. It must be remembered that not only did a smith make the tools for building and the various iron artefacts for domestic living, but also arms and armour for war.[163]

The Blacksmith on the Building Site

There are very few images in art of the blacksmith employed directly on construction sites. There are two only in Chapter 2, Figures 145 and 166, and here we will take a closer look. In Figure 145, on the extreme right-hand side, just over the bridge is a smithy with a farrier shoeing a horse in front just off the roadway. Within the smithy the fire from the forge is clearly seen. (Fig. 184) The farrier is a branch of blacksmithing, making iron shoes for horses, which was the only mode of locomotion for land transportation in the period. This accorded them huge importance since a horse cannot travel very far over rough terrain without good shoes. We can imagine within the smithy blacksmiths making, repairing and sharpening tools for the many workers on site, and also producing hardware and fixtures for doors and windows etc. There must be 100 doors visible on the two elevations of this building. Although it is not a real building, this number is commensurate with a large fortress or cathedral if one includes all internal features that would require such fixtures.

Similarly in Peter Bruegel's painting of the *Tower of Babel*, seen in Figure 166, there is a smithy to the left of the building situated just below the river. (Fig. 185) Here there are two forges, one inside and the other outside to the right. Between this and the smithy is a circular whetstone for sharpening masons chisels. This is on a stand and would need to be rotated for use. There must be more on the site because it is a long way to come if you are working high up. However most of the masons would have with them a flat piece of finely grained sandstone and a bucket of water for the care of their chisels.

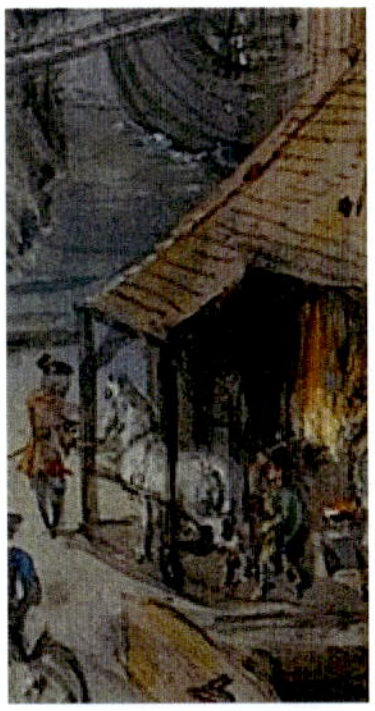

Figure 184: 1515-1520, Gerard Horenbout, or the Master of James IV of Scotland, Flemish, *Tower of Babel*, detail of Figure 145, from the *Grimani Breviary*, ms. Lat. I, 99, f. 206r. Biblioteca Nazionale Marciana, Venice.

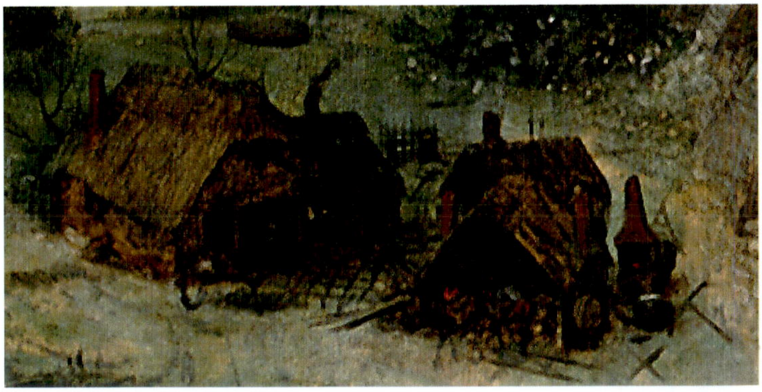

Figure 185: 1563, Peter Bruegel the Elder, Flemish, 1525-1569, *The Tower of Babel*, detail of Figure 166, oil on Oak panel, 114 × 155cm. Kunsthistorisches Museum, Vienna: KHM-Museumsverband.

Smiths in Mythology

On the portals of the door to Hylestad stave church in Norway, *c.*1200, demolished in the seventeenth century, and now in the museum of the University of Oslo, are scenes from the *Saga of the Volsungs*. The Norse Epic of Sigurd the Dragon Slayer, made universally famous by Richard Wagner in his Ring cycle.[164] The scenes on the right-hand jamb depict, in roundels or medallions at the base, the smith Regin forging the sword Gram for Sigurd, (Fig. 186) and at the top Sigurd slaying the dragon Fafnir. On the left jamb, in scenes that run into each other, are Sigurd roasting Fafnir's heart and subsequently slaying Regin; so the smith dies by the very sword he has forged. All part of a long and intricate saga woven around a treasure hoard of gold and its ability to bring about family feud and murder. It must be emphasized that the smith occupied a position second only to the warrior in the Viking world. For it was they

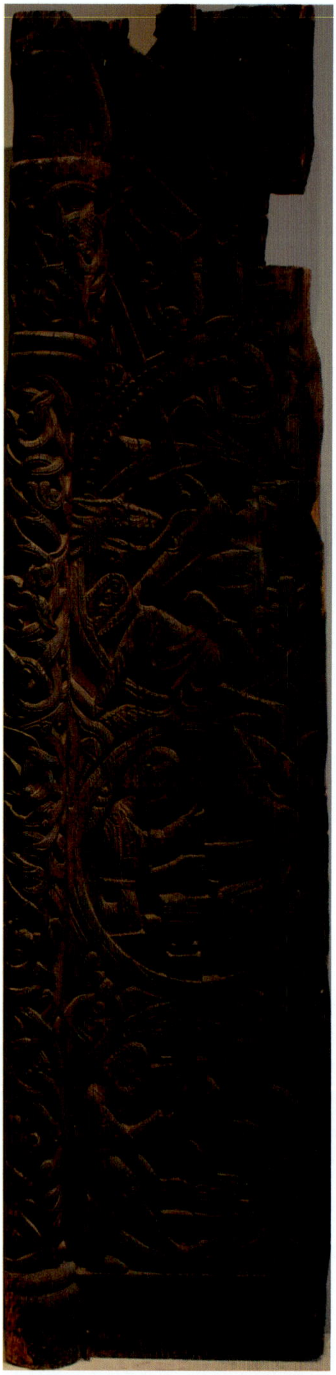

Figure 186: *c.*1200, the smith Regin forges the sword Gram (*Grani*) for Sigurd son of Sigmund. Carved wood door portal from Hylestad Stave Church, now demolished, University of Oslo Museum, Oslo, Norway. Photo: Author.

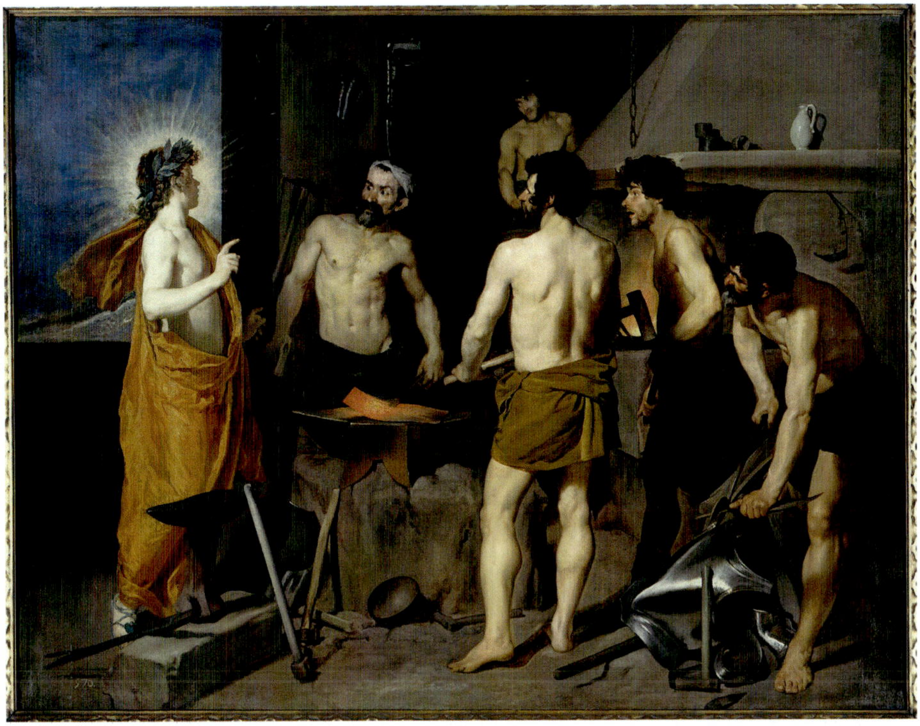

Figure 187: 1630, Velazquez, Diego Rodriguez de Silva y, Spanish, *The Forge of Vulcan*, oil on canvas, 223 × 290cm, Prado, Madrid, Spain. Photo © Photo Josse/Bridgeman Images

who made the weapons including that most highly esteemed personal possession of any warrior, the sword, often endowed with a proper name. The smith was the prince of craftsmen for not only did he forge the sword of the warrior, but also the axe-head for the woodman and the chisels and other iron tools of the carpenter and mason.

In mythology Vulcan is the God of fire including volcanoes, metalworking and the forge, he is often depicted as a blacksmith. In Velazquez's painting *The Forge of Vulcan* (Fig. 187) Apollo visits Vulcan's workshop to inform him of the infidelity of Vulcan's wife Venus, hence the look of utter fury from him and his assistants. Here they are making armour; the blacksmith holds a sheet of red-hot iron on the anvil with tongs, where he will strike it with the light hammer in his right hand. The two strikers to his left will then beat the area he has indicated with their heavy hammers giving shape to the sheet. Behind the group is the forge with an assistant in the background working the bellows, while another assistant cuts a piece of sheet iron with a pair of shears. In the foreground and hanging next to the window are various tools, mostly hammers and tongs. This is a mythological scene and Vulcan is a God. In reality these smiths would not be bare-chested or without footwear, otherwise the potential for injury is certain.

Two Fourteenth-Century Images

In the 'Holkham Bible Picture Book', *c.*1327-1335, at folio 31r we see Christ on the road to Calvary.[165] (Fig. 188) In the lower half of the image holes are being drilled in the cross with an auger. This is somewhat counter-intuitive as pre-drilling may give insufficient grip to the nails and they may pull out. Following this is an image depicting the nails for the crucifixion in the process of being forged by the smith and his wife, although the smith is in conversation with an attendant and his wife is doing all the work. She holds a red hot nail in tongs and beats it into shape on a small anvil on top of a tree trunk. Immediately to her left is the handle controlling the bellows, which give blast to the forge and bringing the fire to sufficient heat to bring the iron to the red heat that allows shaping and forming.

A similar forge is shown in the *Liber Astrologiae* of the second or third quarter of the fourteenth century, at folio 5r.[166] (Fig. 189) Here we see a side-on view showing the bellows and the fuel store behind the forge, with a pair of tongs and a hammer lying beside it. The blacksmith is to the right holding the piece which is being forged in a pair of tongs and hammering; the striker is to the left, his two-handed operation indicating that his hammer is heavier.

Adam Dürr, a Nuremberg Blacksmith

Adam Dürr, a Nuremberg blacksmith, *c.*1531-1606, is seen forging what look like railing finials (fig. 190), so this work could be associated with construction. Several lie around the base of the anvil, and some are on the ground and on the bench behind him. Railings for a house, church or public building could mean a requirement for several hundred. Lying at the edge of the fire are blanks in the process of heating up and awaiting their forging into shape, but there is also what appears to be a chisel with a wooden handle. This may be a carpenters tool that he is tempering. Also hanging on the wall and in the window opening are various tools he has made; a large carpenters saw lies at his feet. The inscription at the top of the folio reads:

> *Adam Dürr, seines Handtwerckhs ein Neberschmit, Burger alhie ist den 16. July A(nn)o03 ins Zwölff Brüederhaus Eingenommen worden, seines alters 72 Jar, und ist hernachen den 30. May A(nn)o 1606 darin gestorben, hatt also in diesem Allmusen gelebt 2 Jar 10 Monath und 16. Tag. D(em) G(ott) G(enad). 16*

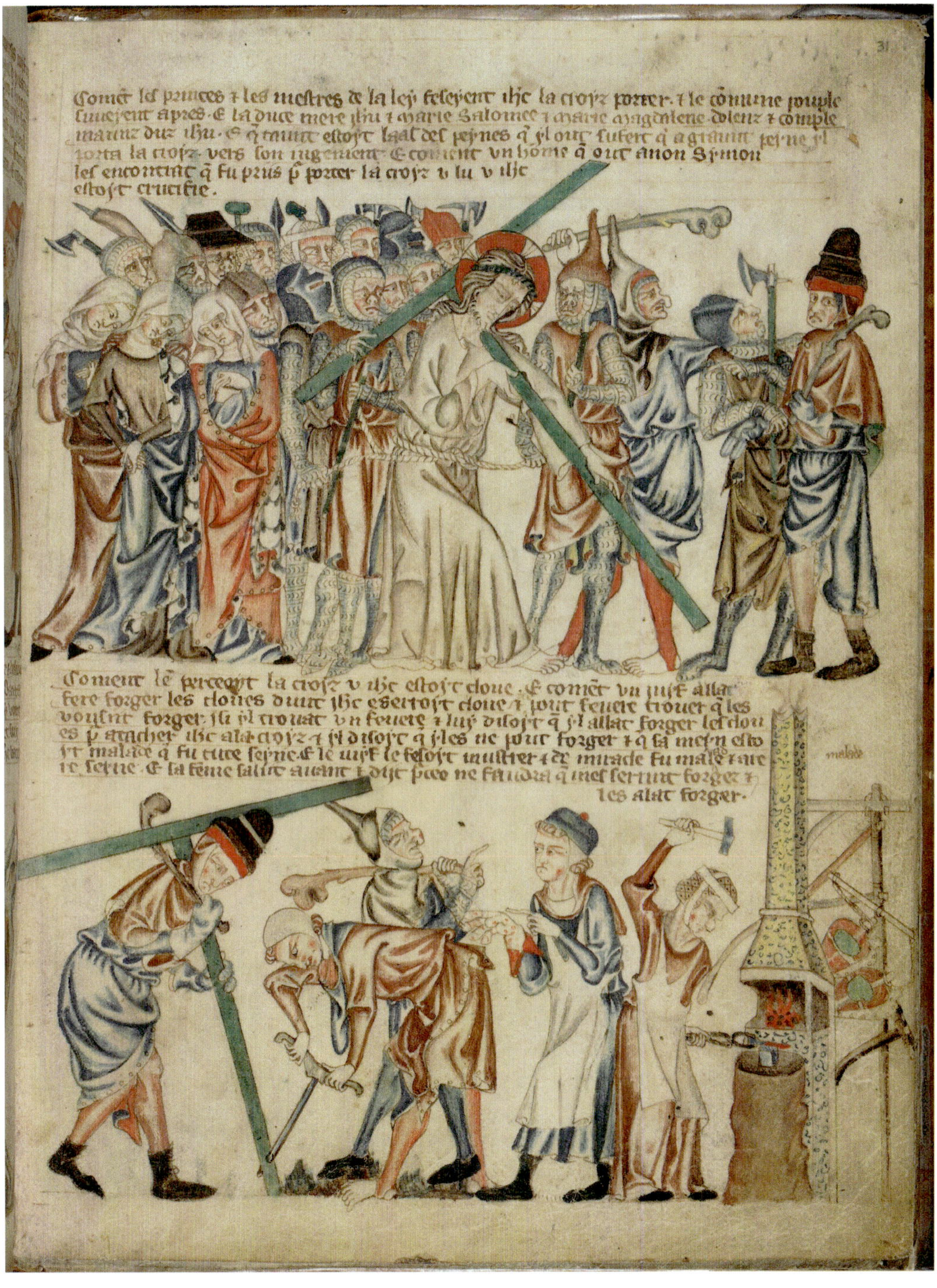

Figure 188: 1327-1335, *The Smith's Wife Forges Nails for the Crucifixion*, © The British Library Board, Add Ms 47682 f. 31r.

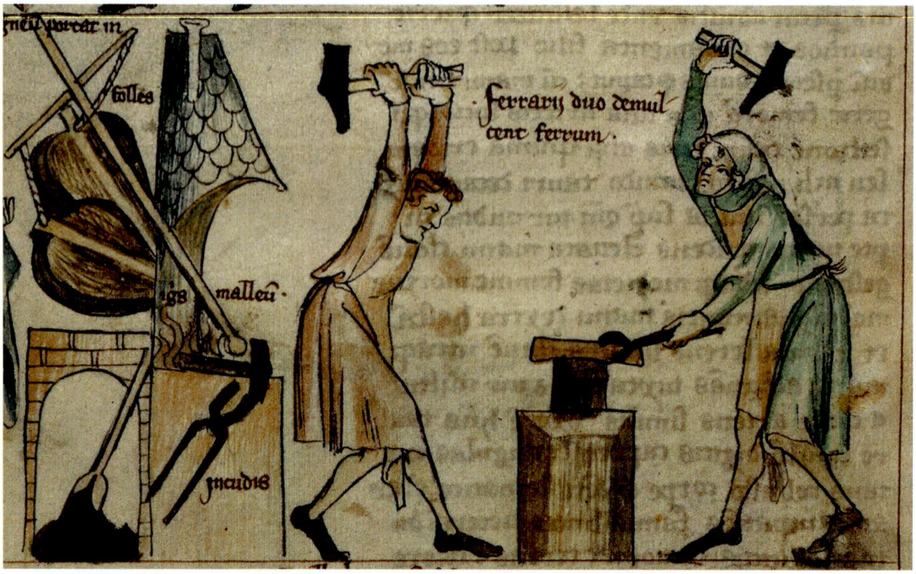

Figure 189: Second or third quarter of the fourteenth century, *Blacksmith and Striker Forging a Piece of Iron*, Georgius Fendulus, *Liber Astrologiae*. © The British Library Board, Sloane Ms 3983 f. 5r (detail).

Broadly translated, this states:

> Adam Dürr, a man of his craft [smith], Burger [Citizen] alhie was taken into the Zwölff Brüderhaus [Twelve Brothers House Foundation] on July 16, Anno 1603, his age 72 years, and after that died May 30, Anno 1606 in it, so he had in this Allmusen lived 2 years 10 months and 16 days. D (em) G (ott) G (enad).[167]

The Striker

The main actor in Goya's painting *The Forge* is the blacksmith's striker. (Fig. 191) He has his back to us and is swinging a heavy blow with the sledgehammer he wields above his head. The other assistant holds the red-hot sheet of iron firmly in place on the anvil with tongs whilst the blacksmith himself indicates where the striker's blow should fall. This will continue until the iron has been formed into the shape required. Further heating in the forge will certainly be required before the final form of the metal is achieved. The composition is robust and the heavy work of the smith and his workshop is clearly illustrated. it is reinforced by showing just these three workers in isolation; the forge itself is not visible, and nor is the usual paraphernalia of the smithy.

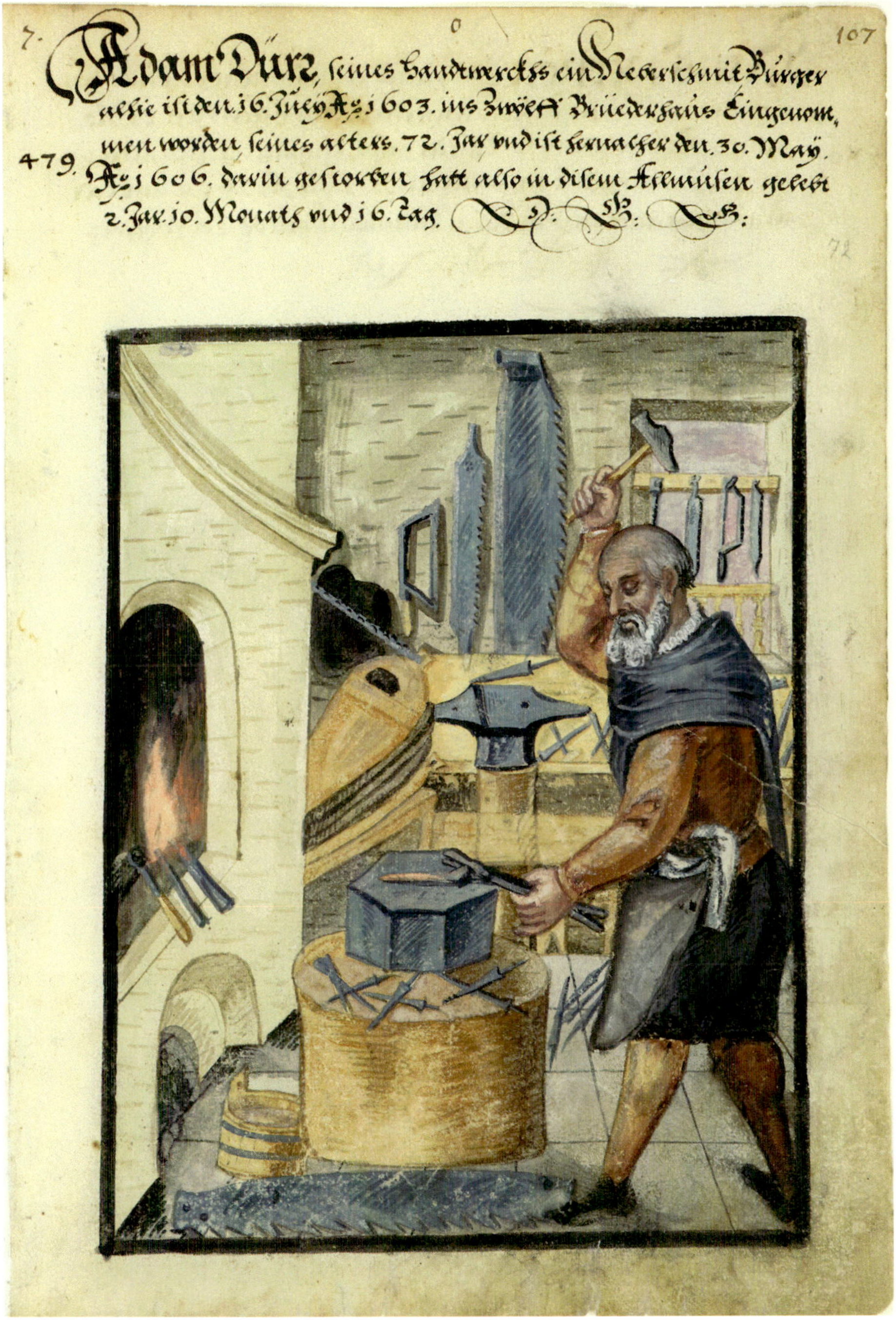

Figure 190: 1550-1791, House book of the Mendelian Twelve Brothers Foundation, Vol. 2, Nuremberg, 75-Amb-2-317b, f. 72r. Stadtbibliothek im Bildungscampus Nürnberg, Amb. 317b.2°, f. 72r.

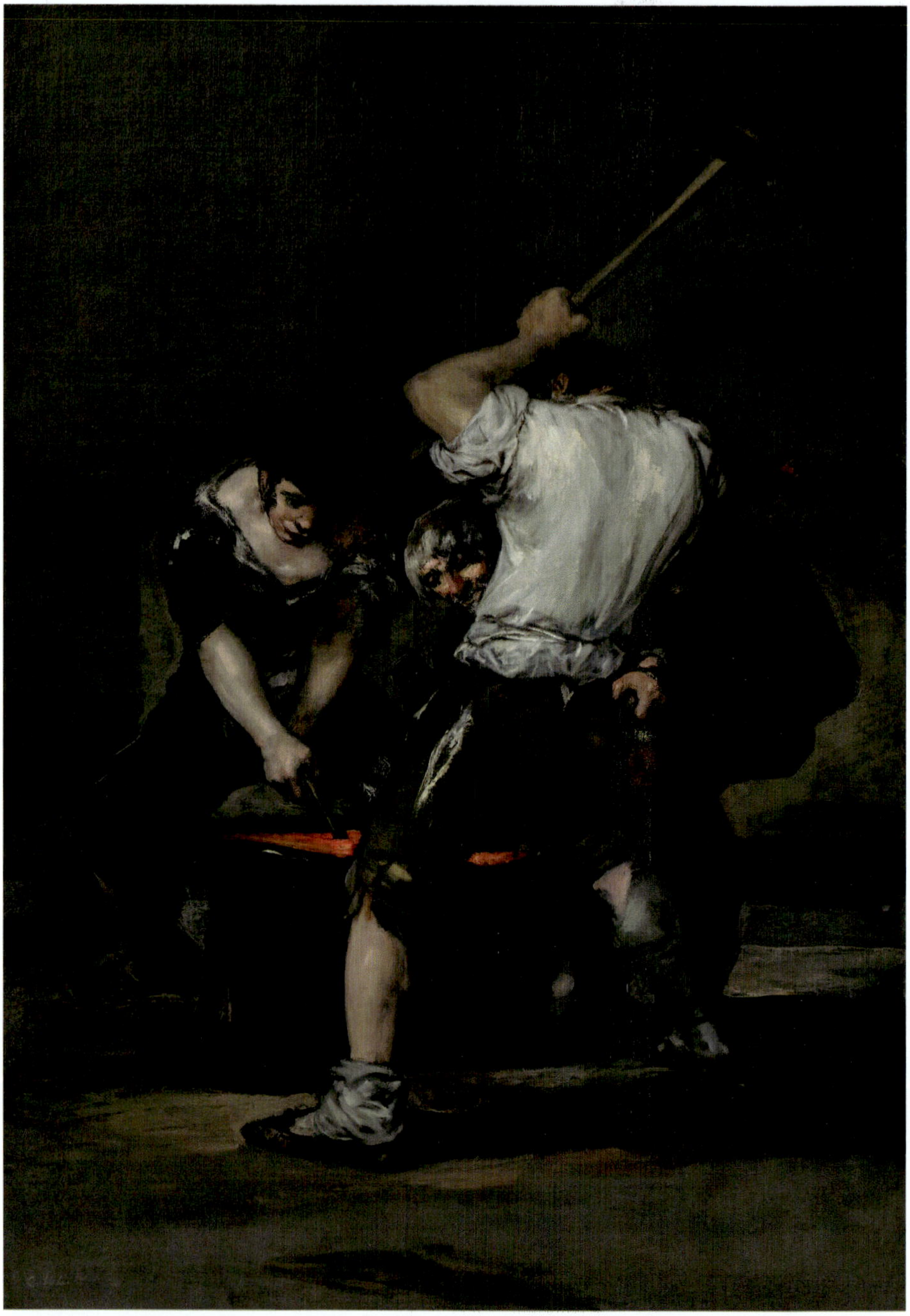

Figure 191: Francisco de Goya y Lucientes, Spanish, 1746-1828, *The Forge (La Fragua)*, oil on canvas, 181.6 × 125.1 cm. Copyright The Frick Collection, New York, Henry Clay Frick Bequest.

Conclusion

The result of this research, which may be surprising to some, is that from a technological standpoint, little changed in the building world from the earliest images of Roman masons and construction workers of the second and fourth centuries, seen in Figures 3, 86, 120, 121 and 122, to the scene of 1876 on a French building site, seen in Figure 173; a span of some fourteen hundred years. This is particularly so with tools, but also equipment, including access scaffolding, which changed minimally throughout the period. For instance wooden scaffolding was still being used during the construction of Freemasons' Hall in London, built between 1927 and 1933. Although most of the scaffold was steel, some photographs from the excellent record of construction show wooden scaffold poles and putlogs.[168] Wooden scaffold continues in use today in many parts of the world. (Figs. 192 and 193) Cranes and hoists did not change to any great degree, and we have seen in Chapter 2 that the treadwheel at Canterbury cathedral continued in use until the 1970s. The right-hand hoist on top of the building under construction in Figure 173 is little different to those shown in Figures 20, 26, 36, 89, 93, 94 and 132, ranging from 1330 to 1563. The simple wheel and rope shown in Figures 13, 16 and 114 is still widely in use today. Virtually every small to medium building site in Great Britain has one.

A modern masons' or bricklayers' trowel would be readily recognizable to a Roman, Medieval or Renaissance craftsman. The masons' axe, so often seen in Medieval images, can be seen in use in modern Palermo, Sicily for carving stone. (Fig. 194) The image from the building site in France of 1876 (Fig. 173) shows us also that carpenters tools were little changed over the period; chisels, mallets, hammers and planes have changed very little. A modern carpenter will have a kit of various power tools, but that kit will almost certainly include the former list. Taking planes as an example, their basic form and function have remained the same since Roman times. The only difference being that now most carpenters use a metal bodied plane, although in carpentry and joinery workshops wooden bodied tools are occasionally still in use. The only carpenter's tools unlikely to be found on a modern building site, certainly in the United Kingdom, are the axe and adze as most timber arrives on site machined to exact requirements. Exceptions would be special building repair and conservation projects where matching original work with traditional tools is required.[169]

Apart from highlighting these technological changes, or lack of them, this book brings a wider analysis to the study of architecture in painting by giving an insight into the process of creating these buildings, some of which are imaginary and some real, but all of which are based on the artist's experience and observation.

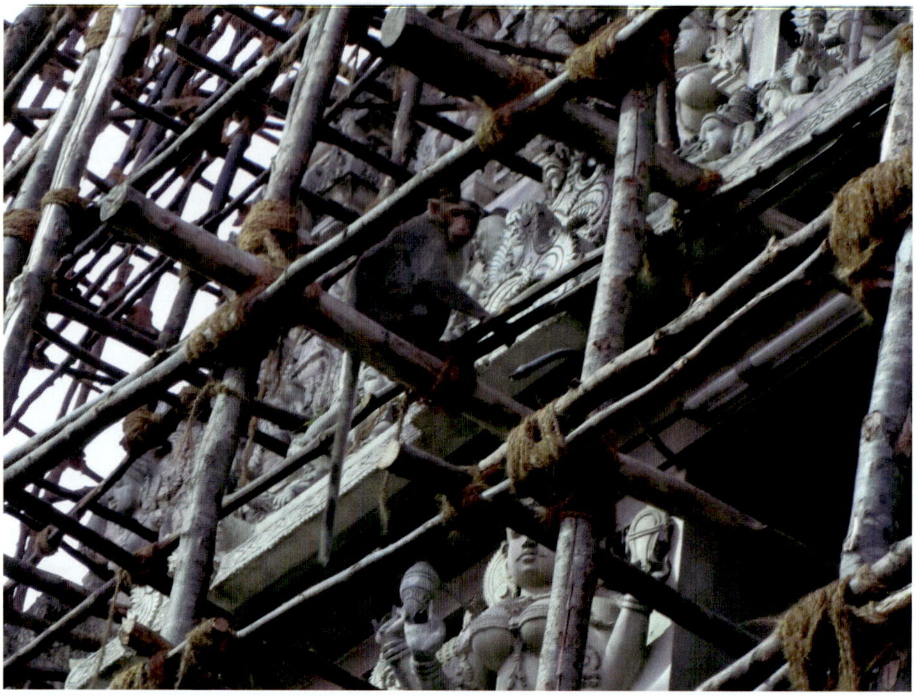

Figure 192: Wooden scaffolding in India in 2015, complete with guardian. (Author)

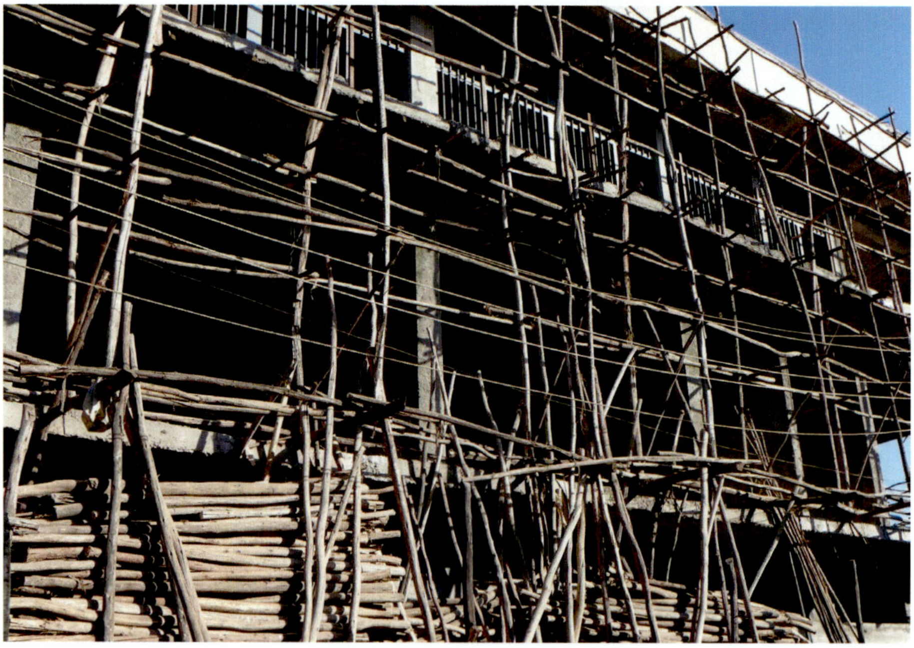

Figure 193: Wooden scaffolding in Gondar, Ethiopia in 2017. (Richard Wiseman)

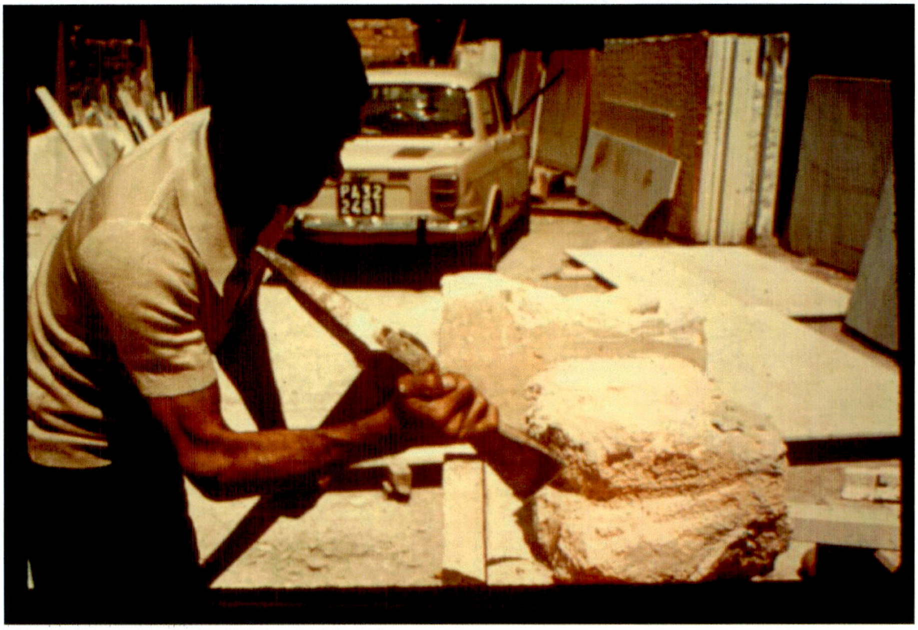

Figure 194: Carver working with an axe in modern Palermo. http://www.artofmaking.ac.uk/content/essays/2-stoneworking-tools-and-toolmarks-w-wootton-b-russell-p-rockwell/ (accessed 14/02/2019)

Notes

1. Exhibition catalogue: Paolo Galluzzi, *The Art of Invention: Leonardo and Renaissance Engineers*, 1996, 1997, 1999, Istituto e Museo di Storia della Scienza, Florence.
2. *c.*1470, Oil on canvas, 155 × 155 cm, Bartolini Salimbeni Collection, Florence. https://brunelleschi.imss.fi.it/genscheda.asp?appl=LIR&xsl=oggetto&lingua=ENG&chiave=100525. Last accessed 16/12/18.
3. See below figures 52 to 58, 63 to 68, 111 and 114.
4. Andrews, Francis B., *The Mediaeval Builder and his Methods*, 1992, New York, Dorset Press, p85.
5. See note 1 above.
6. Tempus Publishing Limited, translated by Alex Cameron, originally published in Germany as *Der Mittelalterliche Baubetrieb in Zeitgenössichen Abbildungen*, 2001, Konrad Theiss; and preceded by, with Norbert Nussbaum, *Der mittelalterliche Baubetrieb nörlich der Alpen in zeitgenössischen Darstellungen*, 1978, Wissenschaftliche Buchgesellschatf, Darmstadt.
7. State University of New York at Binghamton.
8. van Tyghem, Frieda, *Op en Om de Middeleeywse Bouwwerf de gereedschappen en toestellen gebruikt bij het bouwen van de vroege Middeleeuwen tot omstreeks 1600 studie gesteund op beeldende, geschreven en archeologische bronnen*, 1966, Paleis der Academiën, Hertogsstraat 1, Brussel.
9. Salzman, L. F., *Building in England down to 1540*, 1952, Oxford, The Clarendon Press.
10. Harvey, John, *Mediaeval Craftsmen*, 1975, London, B T Batsford.
11. http://www.bl.uk/manuscripts/ http://www.bl.uk/catalogues/illuminatedmanuscripts/welcome.htm https://imagesonline.bl.uk/?service=page&action=show_home_page&language=en
http://explore.bl.uk/primo_library/libweb/action/search.do?vid=BLVU1.
All last accessed 09/19.
12. http://www.enluminures.culture.fr/documentation/enlumine/fr/index3.html. Last accessed 27/12/18.
http://www.enluminures.culture.fr/documentation/enlumine/fr/rechguidee_00.htm. Last accessed 8/01/20
13. For a description of these various scripts see: Clemens, Raymond and Graham, Timothy, *Introduction to Manuscript Studies*, Chapter 10, 'Selected Scripts', 2007, Ithaca, New York and London, Cornell University Press, pp135-178.
14. Salzman, 1952, see bibliography, includes a chapter on 'Masonry: Technical Terms', VI, pp103-118
15. For a definition of the two terms, wood and timber see chapter 1 p. 37.
16. For the UK perspective see: Parsons, David, ed., *Stone Quarrying and Building in England AD 43-1525*, 1990, Chichester, Phillimore & Co; Clifton-Taylor, Alec and Ireson, A S, *English Stone Building*, 1983, London, Victor Gollancz; Stanier, Peter, *Stone Quarry Landscapes*, 2000, Stroud, Tempus Publishing; Peacock, D. P. S., *The Archaeology of Stone*, 1998, London, English Heritage.

17. Again for the UK perspective see: Lloyd, Nathaniel, *A History of English Brickwork*, 1925, London, H Greville Montgomery, (1983 edition, The Antique Collectors Club Ltd.); Wight, Jane A, *Brick Building in England: From the Middle Ages to 1550*, 1972, London, John Baker; Smith, Terence Paul, *The Brickmaking Industry in England 1400-1450*, 1985, Oxford, B.A.R.; Brunskill, R. W., *Brick Building in Britain*, 1990, London, Victor Gollancz.
18. Williams, Richard, *Limekilns and Limeburning*, 1989, Princes Risborough, Shire Publications; Holmes, Stafford and Wingate, Michael, *Building with Lime: A Practical Introduction*, 2002 (2nd revised ed), Bourton-on-Dunsmore, ITGD Publishing.
19. A useful chronology and Part 1 are invaluable in: Ashurst, John, *Mortars, Plasters and Renders in Conservation*, 2002 (2nd), The Ecclesiastical Architects' and Surveyors' Association (no place of publication available).
20. Lillie, Amanda, *Building the Picture: Architecture in Italian Renaissance Painting*, 2014, Online publication, London, the National Gallery; Rodríguez, D. and Borobia, M., *Arquitecturas Pintadas del Renacimiento al siglo XVIII*, 2010, exhibition catalogue, Madrid, Museo Thyssen-Bornemisza.; Curcic, Slobodan and Hadjitryphonos, *Architecture as Icon*, 2010, exhibition catalogue, Princeton, Princeton University Art Museum.; Pimental, A. F., ed. *A Arquitetura Imaginária: Pintura, escultura, artes decorativas*, 2012, exhibition catalogue, Lisbon, Museu Nacional de Arts Antiga and Marchi, A. and Vallazzi, M. R., eds. *La Città Ideale. L'Utopia del Rinascimento a Urbino tra Piero della Francesca e Raffaello*, 2012, exhibition catalogue, Milan, Galleria Nazionale della Marche, Urbino.
21. Millon, Henry A. and Magnago Lampugnani, Vittorio., eds., *The Renaissance from Brunellischi to Michelangelo: The Representation of Architecture*, 1994, London, Thames and Hudson.
22. John Harvey, *Mediaeval Craftsmen*, 1975, London, B T Batsford, *passim.*, but particularly Chapter 11 'Craftsmen of Construction (3) Timber, pp.147-158
23. For an invaluable source of drawings, watercolours etc, of carpentry, particularly roof structures, see David Yeomans, *The Architect & The Carpenter*, 1992, London, RIBA Heinz Gallery.
24. Oliver Rackham is essential for the understanding of woodland management. virtually all his published works are relevant, but see: *Trees and Woodland in the British Landscape*, 1976, London, J M Dent, *passim.*, but particularly Chapter 4, 'Classical Woodland Management: The Middle Ages and After', pp.66-95; *The History of the Countryside*, 1986, London, J M Dent, pp.64-90 and *passim.*, and *Woodlands*, 2006, Volume 100 in the New Naturalist Library, London, HarperCollins, *passim.*
25. Rackham, 1986, p.67.
26. Birch, W de G, *Cartularium Saxonicum*, London, 1885-93, quoted in Rackham, 1986, p.81.
27. The difference between coppice and pollard is associated with animals, both wild and domestic livestock. Animals in unprotected woodland will eat the young shoots of coppiced trees. Therefore pollarding would be utilized to the point when the tree was above the grazing line. Many woodlands were protected by a bank, often also with a hedge on top, to keep such animals out. See Rackham, 2006.
28. H M Colvin, ed., *The History of the Kings Works*, Vol I 'The Middle Ages', 1963, London, HMSO, p.328, note 6.

29. Colvin, 1963, p.349.
30. L.F. Salzman, *Building in England Down to 1540 A Documentary History*, 1952, Oxford, OUP, p.318.
31. *Ibid.,* p.319.
32. Thomas Corkhill, *A Glossary of Wood*, 1979, London, Stobart & Sons, pp.30 & 31.
33. http://romapedia.blogspot.co.uk/2014/02/hypogeum-of-trebius-justus.html, accessed 22/11/16.
34. Lavinium was the port city of Latium, south of Rome.
35. See for instance: https://commons.wikimedia.org/wiki/File:Palazzo_Pubblico.jpg, accessed 25/11/16.
36. Charles B McClendon, *The Origins of Medieval Architecture: Building in Europe, A.D. 600-900*, 2005, New Haven and London, Yale University Press, pp.108-120.
37. Paris, Bibliotheque National, ms. Fr. 6465, f. 96r. See also Erik Inglis, *Jean Fouquet and the Inventions of France*, 2011, New Haven and London, Yale University Press, p.143.
38. Salzman, 1952, p.237.
39. E M Escher, 1898-1972. Dutch graphic artist famous for his depictions of impossible buildings.
40. See https://en.wikipedia.org/wiki/Golden_Legend Accessed 7/01/20.
41. Christiane L Joost-Gaugier, *Raphael's Stanza della Segnatura: Meaning and Invention*, 2002, Cambridge, Cambridge University Press, p.69.
42. http://www.wga.hu/html_m/r/raphael/4stanze/1segnatu/2/00disput.html accessed 19/02/17.
43. Raymond Gillespie, *The Proctor's Accounts of Peter Lewis 1564-1565*, 1996, Dublin, Four Courts Press, *passim.* There are various spellings for his name. I am indepted to Dr. Stuart Kinsella, Research Advisor, Christ Church Cathedral Dublin, for bringing this publication to my notice.
44. https://www.clevelandart.org/art/1940.737.a accessed 8/01/20.
45. https://www.pubhist.com/w8077, accessed 6/12/16. In this engraving there is no scaffold.
46. Ralph Toledano, *Michele Marieschi L'opera completa*, 1988, Milan, Arnoldo Mondadori Editore, colour plate p.37. The painting is in a private collection.
47. For the Tate's catalogue entry for this painting, which makes no attempt at explaining this apparent discrepancy in scale, nor in the display caption, see: http://www.tate.org.uk/art/artworks/scott-an-arch-of-westminster-bridge-t01193, accessed 12/03/17.
48. Oliver Rackham, 1986, pp.64-65.
49. For the others see Francis Greenacre and Sheena Stoddard, *The Bristol Landscape. The Watercolours of Samuel Jackson 1794-1869*, 1986, City of Bristol Museum and Art Gallery.
50. He was commissioned by the London Bridge Committee to record the building of the new bridge, see: Peter Jackson, *George Scharf's London: Sketches and Watercolours of a Changing City, 1820-1850*, 1987, London, John Murray.
51. *Ibid,*, pp.7 & 15.
52. The cathedral was badly damaged by bombing in the Second World War and was rebuilt between 1948 and 1956.

53. Psaltery, an ancient stringed instrument of the zither family. David was a musician as well as being a warrior, statesman and monarch. This and similar instruments such as a harp, viol, lyre etc., are attributes of his and he is often depicted with one or the other instrument in Renaissance art.
54. For explanations of timber framing terminology see Alcock and others, *Recording Timber-Framed Buildings: An Illustrated Glossary*, 1996, York, Council for British Archaeology. For French terms see Jean-Marie Pérouse de Montclos, *Architecture Méthode et Vocabulaire*, 2009 (9th edition), Paris, Editions du Patrimoine, Centre des Monuments Nationaux, pp.138-160.
55. A 'trying' plane is the largest bench plane used by carpenters and joiners. Other terms used to describe such tools are 'long' and 'jointer', see R A Salaman, *Dictionary of Woodworking Tools*, revised edition, 1997, Lakeville, Minnesota, Astragal Press, pp.368-369.
56. Rodler, Hieronymus (d. 1539). *Eyn schön nützlich büchlin und underweisung der kunst des Messens mit dem Zirckel Richtscheidt oder Linial. Zu nutz allen kunstliebhabern furnemlich den Malern, Bildhawern, Goldschmiden, Seidenstickern, Steynmetzen, Schreinern, auch allen andern, so sich der kunst des Messens (Perspectiva zu latin gnant) zugebrauchen lust haben...* Simmern: Hieronymus Rodler, [24 July] 1531. See https://www.christies.com/lotfinder/books-manuscripts/rodier-hieronymus-eyn-schon-nutzlich-buchlin-5662831-details.aspx, accessed 27/8/18.
57. Dog: 'A strong iron fastening for heavy timbers, especially in temporary work. It consists of a length of iron with the ends bent at right angles and pointed.' Thomas Corkhill, *A Glossary of Wood*, 1979, London, Stobart & Son, p145.
58. Corkhill, *ibid.*, p92.
59. A Bolognese lawyer. Written *c.*1310.
60. An anonymous illuminator active in Bruges from 1470 to 1480, so named for *The Book of the Contemplative Soul*, a book for Margaret of York, now in the Royal Library of Belgium.
61. Bernard Bousmanne and Thierry Delcourt, eds., *Miniatures Flamandes 1404-1482*, 2011, Paris and Brussels, Bibliothèque National de France and Bibliothèque Royle de Belgique: Le Maître de Marguerite d'York, pp.295-296 and plate 210. See also: https://gallica.bnf.fr/ark:/12148/btv1b7100618w/f587.item, accessed 20/02/20.
62. *Ibid.*, 'Le Maître Brugeois de 1482', pp.345-347.
63. Ripsaw, a coarse toothed saw for ripping along or with the grain. Two-man pit and trestle saws were invariably for this purpose.
64. The cut or incision made in the wood by the saw.
65. Jaffe, Syson, Allen and Helvey, 'Ercole de' Roberti The Renaissance in Ferrara' in *The Burlington Magazine*, Vol. 141, No. 1153, April 1999, pp. i-xl and Cat. XI, p. xxxix, for a scholarly assessment of the artist and the painting.
66. Deuteronomy 8:2-3.
67. F. Streich and Dr K de Gerstenberg, *Méthode d'enseignement par l'aspect – les métiers et leurs outils*, 'Teaching methods by appearance – trades and their tools', 1876, Paris, Bonhoure, plate 10a.
68. Other writers attest to this custom, for instance Jean-Pierre Adam, *Roman Building: Materials and Techniques*, 1999, London and New York, Routledge., and Landels, J G,

Engineering in the Ancient World, 1978, London, Constable & Co., but little is written on the subject and much is down to folklore. There are two articles both relating to the US construction industry: John V Robinson, 'The "Topping Out" Tradition of the High-Steel Ironworkers' in *Western Folklore*, Vol. 60, No.4, 2001, pp. 243-262 and Scott L Melnick, 'Why a Christmas Tree?' in *Modern Steel Construction*, December 2000 (first published in the October 1995 issue), no page numbers.

69. This must be assumed to be formwork and the beams below him scaffold.
70. Gervase of Canterbury, *Tractatus de combustione et reparatione Cantuariensis ecclesie*, London, British Library, MS Cotton Vespasian B.XIX. The translation is from Salzman, 1952, appendix A, pp.369-376; see also: http://www.lancaster.ac.uk/staff/haywardp/hist424/seminars/Gervase.htm, accessed 10/12/16.
71. He is not to be confused with William the Englishman who succeeded him.
72. http://www.catholic.org/saints/saint.php?saint_id=184 Accessed 11/12/16.
73. See The Metropolitan Museum of Art, Gallery Label and Catalogue Entry: https://metmuseum.org/art/collection/search/437909?searchField=All&sortBy=relevance&ft=Antonio+Vivarini&offset=0&rpp=20&pos=1, accessed 24/09/18.
74. Quoted in A.G. Drachmann, 'A Note on Ancient Cranes', in Singer, Holmyard, Hall and Williams, eds. *A History of Technology*, Vol II, 1956, Oxford, Oxford University Press, pp.658-662.
75. *Ibid.*
76. Singer, Holmyard, Hall and Williams, eds. *A History of Technology*, Vol II, 1956, Oxford, Oxford University Press, pp.714-15
77. Plutarch, *Vitae parallelae: Marcellus*, xiv-xix. (Loeb ed., Vol.5, 1942, pp.468-486.) Quoted in *ibid.*, p.714.
78. Ingrid D Rowland and Thomas Noble Howe, eds. *Vitruvius Ten Books on Architecture*, 1999, Cambridge, Cambridge University Press, pp.120-122 and 294-298.
79. Vatican Musuems: Greforiano Profano Museum, Cat. 9997 and 9998. Collections online http://mv.vatican.va/3_EN/pages/xSchede/MGPs/MGPs_Sala01_06.html#top, accessed 01/09/2015.
80. http://commons.wikimedia.org/wiki/File:KranIBN.JPG, accessed 09/2015 and 24/09/2018.
81. J.G. Landels, *Engineering in the Ancient World*, 1997, London, Constable & Co., p.93 and Andrea Louise Matthies, 1992, 'Medieval Treadwheels: 'Artists' Views of Building Construction', in *Technology and Culture*, Vol. 33, No. 3, p.514.
82. Landels, 1997 and Jean-Pierre Adam, *Roman Building: Materials and Techniques*, 1999, London and New York, Routledge, p.47 & n.64.
83. Rowland, and Howe, 1999, and Coulton, J. J., *Ancient Greek Architects at Work*, 1977, Cornell University Press, p.141.
84. Peter Purton, *The Medieval Military Engineer: From the Roman Empire to the Sixteenth Century*, 2018, Woodbridge, The Boydell Press.
85. http://www.provincia.caserta.it/museocampano/articoli_dettaglio.asp?artID=156, accessed October 2017.
86. Andrea Louise Matthies, 1992. Although she does give clues for some twelth and thirteenth century documentary sources which would repay revisiting, pp.514-515.

87. Purton, 2018, particularly Chapter 2: 'Late Antiquity and the "Early Middle Ages": Were the Dark Ages Really Dark'. But also *passim.*
88. Roger Stalley, *Early Medieval Architecture*, 1999, Oxford, Oxford University Press, p.64, fig.33.
89. Charles B. McClendon, *The Origins of Medieval Architecture: Building in Europe, A.D. 600-900*, 2005. New Haven and London, Yale University Press. McClendon looks at buildings throughout Europe where many buildings are illustrated with large elements, many reused Roman features, high up on elevations, not to mention roof trusses, where mechanical lifting would have been necessary.
90. Blackinsell, William, G. C, *Medieval Windlasses at Salisbury, Peterbrough and Tewkesbury*, 1980, South Wiltshire Industrial Archaeology Society.
91. Tewkesbury: Eric Fernie, *The Architecture of Norman England*, 2000, Oxford University Press, p.160. Peterborough: N Pevsner, *The Buildings of England: Bedfordshire, Huntingdon and Peterborough*, 1968, London, Penguin, pp.303-317.
92. From the collection formed by J Pierpoint Morgan, now in the Morgan Library and Museum, New York: Pierpont Morgan Library & Museum. MS M.638, fol. 3r. The manuscript was possibly commissioned by Louis IX before embarking on his first crusade in 1248.
93. British Library. Detailed record for Royal mo DI: http://www.bl.uk/catalogues/illuminatedmanuscripts/record.asp?MSID=8327&CollID=16&NStart=200401, accessed 29/09/2018.
94. William R Purchase, *Practical Masonry: A Guide to the Art of Stone Cutting*, 1904 (5th edition), London, Crosby Lockwood & Son, p.8, fig.31 and plate II.
95. https://www.handshouse.org/czech-crane/, accessed 01/13/15 and 4/02/20.
96. At the time of writing, here a fortress is being built using exclusively Medieval building techniques, which started in 1977, although it is taking much longer than it would have done in the Medieval period. www.guedelon.fr, accessed 6/10/18.
97. Matthies, 1992, deals with treadwheels extensively and includes a list of extant 'Medieval' treadwheels, capstans and windlasses. They are: Belgium, Halle, O.-L.-Vrouwekerk; Lier, St. Gommaruskerk; Mechelen, St Rouboutskathedraal. England: Beverly Minster; Canterbury Cathedral; Durham Cathedral (fragments); Peterborough Cathedral; Salisbury Cathedral; Tewkesbury Cathedral. France: Auxerre, St Etienne; Beauvais, St Pierre; Colmar, St Martin; Locdure Abbey; Mont-Saint-Michel; Strasbourg, St Thomas; Thann, St Thiebaut. Germany: Hagenow, St Georges. Sweden: Stokholm, Storkyrkan. It is unlikely that any are all entirely Medieval. Many are restored, repaired and otherwise adapted over time. See also Blackinsell, 1980, and the main bibliography.
98. Diderot, Denis and le Rond d'Alembert, Jean-Baptiste, eds., *Encyclopédie, ou Dictionnaire raisonné des sciences, des arts et des métiers, etc.*, 1751-72, Paris, fig.7, plate XLVII.
99. A model of the crane is in the Science Museum, London, made in 1751. https://collection.sciencemuseum.org.uk/objects/co52620/model-scale-1-24-perronets-self-contained-crane-cranes-models, accessed 6/10/18.
100. The crane in the painting by Gerrit Berkeyde of *c.*1669, on the river Spaarne in Haarlem was still there 144 years later as depicted in a similar painting, *c.*1813, by

Wybrand Hendriks. Musee de la Chartreuse, Douai and Teylers Museum, Haarlem respectively.

101. Matthies, 1992. The treadwheel continued in use until the 1970s when health and safety requirements overtook it and it was taken out of service. Matthies interviewed a labourer who operated it in the early 1970s when it was used for the releading of the tower roof. The labourer claimed that he and sometimes a second person in the treadwheel could lift loads of 410kg or 900lbs or just under half a ton, pp.525, fig.7 and 540.

102. See the commentary by Klaus Demus, which may go some way in explaining these inconsistences: 'Tower of Babel 1563' in Wilfried Seipel, ed., *Pieter Bruegel the Elder at the Kunsthistorisches Museum in Vienna*, 1998, Vienna, Kunsthistorisches Museum and Milan, Skira editore, pp.56-7.

103. The chest, a narrow cabinet, has a pair of doors or wings. On the inside of these wings are portraits of masters from the seventeenth to the nineteenth centuries with the painting of the tower of Babel comprising the central internal panel. Such chests hung in guild houses in a closed state above the meeting table and were opened for official business.

104. https://www.koelner-dom.de/interessantes/domkran/?L=1 and https://commons.wikimedia.org/wiki/File:Cologne_Cathedral,_demolition_of_historic_crane,_March_1868.jpg, both accessed 14/10/18.

105. https://commons.wikimedia.org/wiki/Category:Cologne_Cathedral_(construction), accessed 14/10/18.

106. https://www.koelner-dom.de/interessantes/domkran/?L=1, accessed 14/10/18.

107. It is clearly seen in the restored and conserved fresco: https://www.wga.hu/html_m/d/domenico/bartolo/pellegri/1east.html, accessed 6/10/18.

108. Calman, 1974, Cat. no. 6, 'Scenes from the Aeneid', pp.54-55.

109. https://digital.bodleian.ox.ac.uk/inquire/Discover/Search/#/?p=c+1,t-+,rsrs+0,rsps+10,fa+,so+ox%3Asort%5Easc,scids+,pid+e38211bd-5854-4514-818d-ee9fba5ea15a,vi+a464586f-1015-4642-be5e-0899b530da01, accessed 28/10/18, and *op. cit.,* Millon, 1994, cat., nos. 97 and 98, pp.484 and 485.

110. Ulrich W Hiesinger, *Childe Hassam American Impressionist*, 1994, New York, Jordan-Volpe Gallery and Munich, Prestel, pp.128-131.

111. William R Purchase, *Practical Masonry: A Guide to the Art of Stone Cutting*, 1904 (5th edition), London, Crosby Lockwood & Son, pp.1-10; and Wootton W, Russell B, and Rockwell P, 2013, 'Stoneworking tools and toolmarks (version 1.0)', *The Art of Making in Antiquity: Stoneworking in the Roman World*: http://www.artofmaking.ac.uk/content/essays/2-stoneworking-tools-and-toolmarks-w-wootton-b-russell-p-rockwell/ (accessed October 2017).

112. Seton Lloyd and R J Forbes, Chapters 17 and 21 respectively, in Singer, Holmyard, Hall and Williams, eds., *A History of Technology*, Vol. I, 1954, Oxford, Oxford University Press.

113. *Ibid.*, p.478, and Dieter Arnold, *The Encyclopaedia of Ancient Egyptian Architecture*, 2003 (English edition), London, I B Tauris. p .211.

114. Mark Wilson Jones, *Principles of Roman Architecture*, Chapter VIII 'Trajan's Column', 2000, New Haven and London, Yale University Press, pp.161-174; and Lynne Lancas-

ter, 'Building Trajan's Column', 1999, *American Journal of Archaeology*, 103, No. 3, pp.419-439.

115. For the full photographic record see: <http://www.trajans-column.org/> and <https://commons.wikimedia.org/wiki/Category:Trajan%27s_Column_-_Cichorius_Plates> Both accessed September 2017.
116. Felix Marcu, 'The Construction of the Roman Forts in Dacia', 2011, *Dacia: Revue d'Archeologie et d'Historie Ancienne*, Nouvelle Serie, LV, Académie Roumaine, Bucarest, pp.123-135.
117. <https://www.flickr.com/photos/111072739@N03/sets/72157638580817145/with/11325542774/> Plate 93 right-hand, accessed October 2017
118. Wilson Jones, *op cit.*, p174. See also Martin Beckmann, *The Column of Marcus Aurelius: The Genesis & Meaning of a Roman Imperial Monument*, 2011, Chapel Hill, The University of North Carolina Press.
119. Wilson Jones, 2000, deals mainly with the stairs. Beckmann, *op cit.*, covers the carving more extensively in Chapters 5 and 6.
120. Museo Provinciale Campano di Capua, Sarcofagi (Sala III-IV), <http://www.provincia.caserta.it/museocampano/articoli_dettaglio.asp?artID=156>, accessed October 2017.
121. Bibliotheque Municipale de Valenciennes, France, ms. 0502.
122. Such a lack of scale is fairly common in early manuscripts. The inconsistency in the gable is possibly a repair to the parchment which has been glued in to cover a hole or other defect.
123. See the photograph of a mason carving with an axe in modern Palermo in: Wootton, Russell and Rockwell, 2013, p.21, Fig. 15.
124. Ezra 1:1.
125. For early wheelbarrows see Andrea L Matthies, 'The Medieval Wheelbarrow' in *Technology and Culture*, Vol. 32, No. 2, Part 1, April, 1991, The John Hopkins University Press, pp.356-364.
126. <http://www.bl.uk/manuscripts/FullDisplay.aspx?index=142&ref=Add_MS_15268> Accessed February 2018.
127. Ellen Callmann, *Apollonio di Giovanni*, 1974, Oxford, The Clarendon Press, Oxford University Press.
128. Erik Inglis, *Jean Fouquet and the Invention of France: Art and Nation after the Hundred Years War*, 2011, New Haven and London, Yale University Press.
129. *ibid.*, illustrated on p.194.
130. Genesis 4:17.
131. British Library, Royal ms 19 D III: http://www.bl.uk/manuscripts/FullDisplay.aspx?index=2315&ref=Royal_MS_19_D_III, accessed 12/3/18.
132. British Library, Add. MS 18850: <http://www.bl.uk/manuscripts/FullDisplay.aspx?ref=Add_MS_18850>, accessed October 2017, and Eberhard König, *The Bedford Hours: The Making of a Medieval Masterpiece*, 2007, London, The British Library.
133. Genesis, 11:4.
134. For further information see Stafford Holmes and Michael Wingate, *Building with Lime: A practical introduction*, 2002 (2nd revised ed.), ITDG Publishing.

135. For the Tower of Babel see Helmut Minkowski, *Turm zu Babel*, 1991, Luca Verlag Freren. Minkowski catalogues over 600 images from antiquity to the second half of the twentieth century.
136. http://www.enluminures.culture.fr/public/mistral/enlumine_fr?ACTION=CHERCHER&FIELD_98=REF&VALUE_98=D-013607, accessed 13/03/18.
137. http://www.enluminures.culture.fr/public/mistral/enlumine_fr?ACTION=CHERCHER&FIELD_98=REFD&VALUE_98=%27Amiens%20-%20BM%20-%20ms.%200107%27&DOM=All, accessed 14/03/18.
138. This central part of the page may be the result of a later collage or assemblage forming a patchwork. Catherine Yvard, *pers com.*
139. known as Filarete, *c.*1400-*c.*1469 a Florentine architect, craftsman and theorist.
140. Henry A Millon and Vittorio Magnago Lampugnani, eds., *The Renaissance from Brunellischi to Michelangelo: The Representation of Architecture,* 1994, London, Thames and Hudson, Cat. No. 87, pp.99 and 480.
141. Machtelt Israëls, *Sassetta's Madonna della Neve: an Image of Patronage,* 2003, Leiden, Primavera Pers, p.9 and *passim.*
142. *Ibid.*, p127.
143. Visible in the archive photograph taken *c.*1950, but see colour plates i and iv to x in *ibid.* for the altarpiece as it appears now.
144. Brian Sewell, 'Picture This' in *Evening Standard,* 12 February 1998, Hot Tickets magazine, p.44.
145. See Annick Born and Maximiliaan P J Martens, *Van Eyck in Detail*, 2012, Antwerp, Ludion, pp.40-41 & 182-184, and Till-Holger Borchert, *Masterpieces in Detail: Early Netherlandish Art from Van Eyck to Bosch*, 2014, Munich, London and New York, Prestel, pp.60-67, for magnified details.
146. See: Richard Williams, *Limekilns and Limeburning*, 1989, Princes Risborough, Shire Publications; John Ashurst, *Mortars, Plasters and Renders in Conservation*, 2002, EASA, for a useful chronology and Part 1, 'Materials', pp.5-41; and Stafford Holmes and Wingate, 2002.
147. Lawrence Gowing, 'Mantegna', in Jane Martineau, ed., *Andrea Mantegna*, 1992, Royal Academy of Arts, London and The Metropolitan Museum of Art, New York, exhibition catalogue, p3.
148. *Ibid.* Cat. 60, p244.
149. Collections du musée des beaux-arts de Dijon: Légende de saint Bertin: http://mba-collections.dijon.fr/ow4/mba/voir.xsp?id=00101-28267&qid=sdx_q0&n=2&e=, accessed March 2018.
150. https://www.wikiart.org/en/richard-parkes-bonington/abbey-of-st-berlin-near-st-omer-1823, accessed 22/02/20.
151. For a full discussion see: Dennis Geronimus, *Piero di Cosimo: Visions Beautiful and Strange*, 2006, New Haven and London, Yale University Press, pp.141-145; and Kathleen Weil-Garris Brandt, in 'The Relation of Sculpture and Architecture in the Renaissance', in Henry A Millon, 1994, where the painting is discussed in relation to the theories and treatises of Alberti, pp.86-98. However, I do not agree with various observations regarding the actual painting, particularly note 20, p.88, where it is asserted that there cannot be floors inserted yet because a guy rope for the hoist is

rigged around the band between the third and fourth level windows. Since this is the *piano nobile* this is surely a double height room with clearstory windows; otherwise judging from the figures on the balcony, if there were a floor or intended floor between the windows, anyone of slightly above average height would damage their heads on the exceptionally low ceiling every time they got up from a chair. Also, contrary to what is stated, there are no planks, bricks or reeds to be seen. There are certainly heavy baulks of large timbers and a possible stack of rubble stones, but there are no bricks, and no reeds anywhere.

152. Max J Friedlander, *From Van Eyck to Bruegel*, edited and annotated by F Grossmann, 1969 (3rd edition), New York, Phaidon, p.139.
153. Calculated using the average of two common limestones, sandstone and Italian marble. The calculation excludes granite which is heavier.
154. http://manchesterartgallery.org/collections/search/collection/?id=1924.30, accessed 9/04/18.
155. http://manchesterartgallery.org/fmb/docs/fmb_catalogue.pdf, accessed 13/04/18.
156. See http://www.tate.org.uk/art/research-publications/jmw-turner/joseph-mallord-william-turner-kirkstall-lock-on-the-river-aire-r1146209, accessed 8/04/18.
157. Streich, F and Dr K de Gerstenberg, *Méthode d'enseignement par l'aspect – les métiers et leurs outils* 'Teaching methods by appearance – trades and their tools', plate 10a, published by Bonhoure, Paris, 1876. See also the cover illustration of: Kit Wedd, *Victorian Housebuilding*, 2012, Shire Publications.
158. Bartholomew the Englishman (before1203-1272) a member of the Franciscan order: https://en.wikipedia.org/wiki/Bartholomeus_Anglicus#De_proprietatibus_rerum and: http://www.bl.uk/catalogues/illuminatedmanuscripts/record.asp?MSID=7691&CollID=16&NStart=150503. Both accessed 22/04/18.
159. For a full explanation of the chemistry and processes involved see Stafford Holmes and Michael Wingate, 2002; Richard Williams, 1989, and John Ashurst, 2002.
160. See https://commons.wikimedia.org/wiki/File:Whim_gin_beamish_2013.JPG, accessed 28/04/18, for a modern reproduction at Beamish Open Air Museum in the north of England.
161. See for instance, Arturs Lapins, 'Door and Gate Related Artefacts in Cesis Castle, Latvia', in W P Campbell et al., eds., *Water, Doors and Buildings: Studies in the History of Construction*, The Proceedings of the Sixth Conference of the Construction History Society, Queens' College Cambridge, 5-7 April 2019, pp.227-239.
162. Scripture quotation taken from *The Holy Bible*, New International Version® NIV® © 2011, Biblica Inc.™ Used by permission. All rights reserved worldwide. Published by Zondervan, Grand Rapids, Michigan. Book of Job, pp.822-23.
163. Various references for further information on early ironwork and the smith are: M T Richardson, ed, *Practical Blacksmithing*, 1899, reprinted 2017, New York, Chartwell Books; D W Crossley, 'Medieval Iron Smelting'; R F Tylecote, 'The Medieval Smith and his Methods'; Ian H Goodall, 'The Medieval Blacksmith and his Products', all in D W Crossley, ed. *Medieval Industry,* 1981, Council for British Archaeology Research Report, No. 40, pp.29-41, 42-50, 51-62, respectively and all with full and copious bibliographies. Also, Jane Geddes, 'Iron' in John Blair and Nigel Ramsay, eds. *English Medieval Industries*, 1991, London and Rio Grande, The Hambledon Press.

164. *The Saga of the Volsungs,* thirteenth century Icelandic author from earlier Norse poetry, trans. Jesse L Byock, 1990 (USA) and 1999, London, Penguin Books.
165. British Library, Additional Ms 47682: http://www.bl.uk/manuscripts/FullDisplay.aspx?ref=Add_MS_47682, accessed 7/03/20.
166. British Library, Sloane Ms 3938: http://www.bl.uk/manuscripts/FullDisplay.aspx?ref=Sloane_MS_3983, accessed 7/03/20.
167. http://www.nuernberger-hausbuecher.de, accessed 08/03/20.
168. Library and Museum of Freemasonry, Freemason's Hall, London WC2B 5AZ: libmus@freemasonry.london.museum
169. Of particular note is the Weald and Downland Living Museum, Singleton, West Sussex, UK, where a completely modern building, The Gridshell was built to house the museum's workshops where traditional building techniques are used and taught.

Glossary of Technical Terms

with references to further reading and information from the bibliography and other sources where appropriate

Note: within the entries where there are words in bold type, these refer to further entries within the glossary.

Adze: Although developed in antiquity for stone working, the Medieval adze was a carpenters tool used for squaring and trimming timber with the blade set at right angles to the haft or handle which would have been long or short depending on the work. Salaman lists some 69 different types. (Salaman 1997)

Aggregate: In construction a material, usually sand or gravel, formed from a mass of particles loosely compacted together. For making and mixing **mortar**, see also **sieve**.

Angle-try: A 'square' or mitre square; a carpenter's or mason's tool where a flat blade is set into a straight wooden stock at a given angle, say 45°. Some forms were made from a single sheet of metal by the blacksmith, particularly for the mason. Later types were adjustable such as the sliding bevel. (Purchase 1904 and Salaman 1997)

Architect: Broadly the person in charge of the design and scope of a building, although the person in charge of the actual construction may be different. Up until the Renaissance the term is a difficult one, and even then it is not straightforward. In the Medieval period the term and profession of architect did not exist and the closest equivalent is the **Clerk of Works**, **Master Mason** or **Master Carpenter**, all of whom (severally or together) could have been in charge of construction on site. But to complicate matters, the employer or owner could equally be involved. (Salzman 1952 and Kaye 1960)

Ashlar: Smooth faced stone-work precisely cut, dressed, squared with sharp arrises and laid in straight courses.

Auger: A boring tool for large holes in wood or timber, comprising a metal bit, the cutting end, fixed to a cross handle, the whole forming a 'T' shape. Salaman lists over 40 different types. (Salaman 1977)

Axe: One of the basic and fundamental tools of the carpenter (see also **adze**) used for both felling and shaping timber. Many types were developed for different trades utilising both straight and shaped handles. Salaman lists over 100 different types and patterns. (Salaman 1997)

Axe, mason's: The basic tool of the Medieval mason used for dressing and carving stone. They resembled the carpenter's axe and the modern pick-axe with one end pointed and the other chisel shaped. They appear unwieldy to modern eyes but were used with great skill and to great effect. (Purchase 1904)

Babewyn: see also **gryllus:** A hybrid zoomorphic figure, neither entirely human or animal. Also known as grotesques and drolleries. (Brown 2018) They often inhabit the margins and folios of Medieval manuscript illuminations. (Camille 1992 and Nishimura 2009)
Banksman: A worker on a construction site who supervises the operation of cranes and other equipment in order to avoid accidents and protect other workers.
Banker mason: see also **mason** and **fixer mason.** A banker is a bench either of timber or sometimes a block of stone on which the stone is worked or carved. Banker masons worked in stone-yards attached to, or remote from site and sometimes even at the quarry. Stone was therefore often delivered to site ready dressed to the dimensions given by the master mason. (Purchase 1904 and Ashurst and Dimes, 1977)
Barrow, see also **wheelbarrow:** A flat, rectangular frame used for carrying a load, especially with projecting shafts at each end for handles, carried by two men (Matthies 1991).
Bas-relief: Relief sculpture in which the figures or other detail project slightly from the background without undercutting.
Bast, Bass or ***bastropps***: Tough and fibrous inner bark of trees, particularly lime, which can be woven into twine and used as ties for scaffolding, see also '**withies**'. (Rackham 1986 and 2006)
Batter / battered: Receding upward slope of the outer face of a wall, embankment, earthwork or other structure.
Bible historiale / historiée: The biblical narrative in prose form, written by Guyart des Moulins and based on his translation into French (1291-94) of the *Historia scholastica* of Peter Comestor, interspersed with a French translation of the Bible produced in Paris around 1250. The illustrations accompanying the *Bible historiale* (usually in the form of column pictures) depict many scenes not normally found in the standard repertory of biblical images and also include representations of the compilation and translation of the text. (https://www.bl.uk/catalogues/illuminatedmanuscripts/GlossB.asp)
Blacksmith: In the pre-industrial world the blacksmith was responsible for the production of all iron and steel artefacts across the various markets; domestic, agriculture, weaponry and armour, fixtures and fittings in construction and not least the tools of the carpenter and mason. Most villages would have had a local blacksmith and towns often more than one depending on size. (Richardson 1897 and Crossley 1981)
Block and tackle: See **pulley**.
Boaster: Mason's wide tool or **chisel**, from 40 to 80mm wide, used for dressing stone down to a smooth face (Purchase 1904).
Boning-rod: A series of wooden T shaped devices used for establishing a level surface. Used from antiquity to the present day, particularly now in ground-works and paving. (https://www.pavingexpert.com/setout03)

Book of hours: A book, also called a *primer* or *horae*, for use in private devotions. Its central text, the Little Office of the Blessed Virgin (or Hours of the Virgin), is modelled on the divine office and represents a shorter version of the devotions performed at the eight canonical hours. The text, known from the tenth century, was originally read only by ecclesiastics; it entered into more popular use by the end of the twelfth century, often being attached to the Psalter, the book more commonly used for private devotions before the emergence of the 'book of hours'. The private recitation of the Little Office of the Virgin is an expression of the lay-person's desire to imitate the prayer-life of the religious, see **Psalter.**
(https://www.bl.uk/catalogues/illuminatedmanuscripts/GlossB.asp, and Brown 2018)
Box frame: In timber-framed construction a rectangular or square building where the **roof trusses** are supported on a frame of **posts**, tie-beams and **wall-plates.** (Alcock, 1996 and Pérouse de Montclos 2009)
Brace: (scaffold) A member placed diagonally with respect to the vertical or horizontal members of a scaffold and fixed to them to afford stability.
Brace: (tool) A tool for boring and drilling, comprising a chuck for holding the bit at the foot or drilling end, a head at the top for hand holding and between the two a crank for rotating. Salaman lists many different types (Salaman 1977).
Brick: A solid manufactured building block made primarily from clay and fired or burnt in a **kiln**, see below **brick kiln**. There are many different types and 'specials'. Additionally there are various patterns of laying bricks in a bond to form overlapping joints for a robust structure. (See Curl, James Stevens, *A Dictionary Architecture and Landscape Architectue*, 2006 (2nd edition), Oxford, Oxford University Press, for full descriptions)
Brick Kiln: A furnace or oven in which bricks are baked or burned to achieve permanent hardness. In some ancient civilisations without kiln technology, bricks were baked in direct sunlight. See also **Kiln, Lime Kiln.**
Buttress/ed: (scaffold) Similar to bracing but erected against the scaffold structure rather than within it to aid stability and support where the scaffold cannot be adequately attached or tied to the host building.

Calipers: also **Compasses, Dividers:** In their many forms these tools are interchangeable. They will be found in all carpenters' and masons' tool kits and are invariably referred to in the plural even for a single tool. They are used for transferring measurements from one piece of work to another and from or to drawings where these exist. They are particularly useful for measuring in cramped situations where it is not possible to use a rule, also for internal diameter and cross measurements of openings and holes. (Salaman 1977)
Chisel, Carpenters: There are many and varied types according to the type of work. A flat or curved tool with a steel blade for cutting wood, most have wooden handles

and are sharpened to a fine edge and can be used without percussive (a mallet) aides simply by paring and the strength of the carpenter's arm and wrist. Others used for heavier work such as cutting mortices will require striking by a mallet. Such tools have been used since antiquity for cutting and carving. (Salaman, 1977)

Chisel, Masons: See also **boaster, claw chisel, pitching tool** and **point.** Again there are many types and patterns, they are made of solid metal and are always struck with a mallet or hammer. Mallet-headed chisels have a bulbous head forged at the top of the tool and should only be struck by a wooden mallet. As opposed to carpenter's chisels the mason's tool does not so much cut the material but bursts it. Again, mason's chisels date from antiquity. (Purchase 1904)

Chisel, claw: A mason's chisel with teeth cut into the face for rapid dressing down of stone. (Purchase 1904)

Clays, or Clayes: See also **hurdles, wattlework.** Early **scaffold** platforms made of woven wattlework hurdles, usually of hazel rods or similar **underwood** also widely used for fencing. (Colvin 1963 and Tutton 2017)

Clerk of works: Unlike the architect, a profession traceable back to the Middle Ages. They were appointed by the Monarch and other Magnates to oversee construction works to fortresses and other important buildings. (Colvin 1963 and 1971, Salzman 1952 and Kaye 1960)

Collar: In roof construction a horizontal member tying together a pair of inclined components such as **truss** blades, principle or common **rafters.** Usually located about half-way up the truss, there may be up to three collars in a truss. (Cordingly, 1961, *British Historical Roof-Types and their Members: A Classification*, London, Ancient Monuments Society and Alcock 1996)

Colonnade: A series of columns in a straight line supporting an entablature.

Column: A detached vertical structure such as Trajan's Column in Rome or The Monument in London often with a statue at the top and essentially monumental. The Minaret and Victory Tower of Islam bear comparison, as do many early depictions of the construction of the Tower of Babel. (Tutton 2014 and Bloom, Jonathan M, *The Minaret*, 2013, Edinburgh, University Press)

Coppice: A managed woodland protected from livestock and wildlife, such as deer, where trees are harvested by being cut to near ground level every few years. The resulting harvest of young growth forming poles and rods was a valuable crop for many uses. Dating to the Roman period and throughout the Middle Ages to the nineteenth century when replaced by managed plantations. Protection was essential as animals, both wild and domestic, would eat the young shoots. Also a tree, having undergone this process. A coppice would also contain 'standard' or 'timber' trees, normally oak, which were not coppiced but felled at much greater intervals and used for structural elements in buildings. See also **pollard, timber** and **underwood.** (Rackham 1976, 1986 and 2006)

Cornice: The uppermost part of a wall such as a coping. In classical architecture the uppermost part of an entablature.
Crane: See also **hoist, treadwheel, gin wheel.** A lifting device comprising a **winch** and windlass connected to a jib. On many early building sites the winch and windlass comprised a tread wheel. (Glynn 1854 and Bachmann1997)
Crosstrees: A marine term associated with sailing ships; horizontal spars at the upper end of a mast which spread the rigging or guy-ropes
Cyclopean faced masonry: Rock or quarry faced masonry where the face projects with a rough surface as found undisturbed in the quarry or rock face. See also **Rusticated masonry.**

Dividers: See **Calipers.**
Dog (tool): Iron rods or flats with the ends turned down at 90° and sharpened to a tapered point or chisel shape. Carpenters use them for holding work, with one end being driven into the timber being worked on, and the other end into the bench or **Horse.** Usually used in pairs, the taper pulls the two together firmly. Also used to hold logs whilst sawing.

Fan (scaffold): Scaffold fans are structures attached to the perimeter of a building or scaffolding to prevent objects that may accidentally fall from the structure causing damage or injury to property or persons on the ground. The fan consists of a structure jutting out from the scaffolding, which is fitted with boards or decking to catch falling debris. They are usually inclined towards the scaffolding and can often be seen in towns and cities where the scaffolding is situated directly above the pavement or road.
Fixer mason: See also **mason** and **banker mason.** A stonemason, who fixes or lays the finished block into the building. (Ashurst and Dimes, 1977)
Forest: See **Plantation.**
Forge: A hearth used for heating metal by a **smith,** also the place of work of a smith, also called a smithy, see also **blacksmith.** (Richardson 2017)
Formwork: Temporary timber-work erected to support masonry during construction particularly of door and window openings, arches and vaults. (Heyman 1995)
Fresco: Form of wall painting where ground pigment mixed with water is painted directly into wet plaster. Dating to antiquity. (Cennino d'Andrea Cennini 15th c and Hale 1966)

Gantry: In construction a bridge-like overhead structure with a platform supporting equipment such as a **hoist** or **crane.** (Glynn 1854)
Gesso: (chalk) A mixture of whiting (ground chalk) and animal skin glue ('rabbit skin') applied hot to panels in several layers and widely used as the base or ground

for panel paintings and carved polychrome work. Essential as a ground for gilding where it will be burnished together with other finely ground layers of clays known as bole (also applied hot). (Mactaggart 1984, Hale 1996 and Cennino d'Andrea Cennini 15th c.)

Gin wheel: And rope, see also **pully.** A single pully rigged with a rope usually suspended from and overhanging a scaffold. Used as an aid for scaffold erection and striking and for lifting lightweight building materials. They are commonly seen on building sites and used frequently by builders, scaffolders and contractors.

Glazier: A craftsman who cuts and fits window glass, as opposed to the worker who constructs and fits the actual window, which is mason's or carpenter's work. Glaziers often worked in studios or workshops. Large building operations, such as cathedrals, because of the importance of glazing, had their own permanent glazing workshops under the supervision of a master craftsman.

Gryllus: See **Babewyn.**

Guy-rope: A rope or cable rigged to give tension and support to a post or mast. Also a loose rope attached to a load being lifted by a crane or hoist. This is handled by a **banksman,** who may be on the ground, part way up the scaffold, or at the level of the intended destination of the load. The rope is to steady the load and stop it swaying. When the load reaches the required level the rope is used to haul it onto the scaffold. Depending on the size and complexity of the load and the height to which it is being lifted, there may be several such ropes plus the necessary number of banksmen.

Hammer: A universal tool with a metal head and wooden handle for striking and driving metal objects such as nails. The basic carpenters hammer with the back of the head (the pein or pane) divided or bifurcated to form a claw for withdrawing nails; the joiner's hammer with a cross-pein; the engineers hammer with a ball-pein and a specific type for virtually all crafts and trades, from upholstery to **slate** roofing. (Salaman 1975)

Herm: A statue composed of a head, neck and shoulders joined to a downward tapering shaft or pedestal, generally proportioned to the height of the human body. A 'Term' additionally has a torso and waist.

Historiated Initial: In manuscripts a large initial combined into an image, often relating to the text and containing figures and other scenes. (Clemens 2007 and Brown 1994)

Hoarding: A robust fence dividing a building site from the public realm.

Hod: A container normally carried on the shoulder of a labourer, used for **mortar,** bricks and other building materials small enough for manual handling. Hod-carriers were essential for a constant supply of materials to fixer-masons and bricklayers working at the 'coal-face' of the building. They carried these materials normally

from the ground, where they were mixed and stored and formed almost continuous queues up and down the ladders on a construction site. See also **pannier**.
Hoe: Similar to a garden hoe, but more robustly made, for mixing mortar.
Hoist: A lifting device often without a **jib**. See also **crane**, **gin wheel** and **treadwheel**.
Horse: (Carpenter's) A carpenter's stool for resting work on. A large flat piece of timber about 1m long with splayed legs 60/70cm high, usually used in pairs for sawing.
Horse gin: An animal-powered machine where the animal, usually a horse or donkey, moves in a circle on a horizontal plane around a central pivot. Rotary motion is transmitted by gearing and can be applied to pumps, winding engines at mines and lime kilns, for hoisting on construction sites and to edge rollers for mixing mortar. Widely used in agriculture for threshing and for milling. (J. Kenneth Major, *Animal-Powered Machines*, 1985, Princes Risborough, Shire Publications)
Hurdle: see also **Clays** or Clayes, and **Wattlework**. Woven panels made of sufficiently supple rods, such as hazel. Used for **scaffold** platforms on early building sites. Also used in agriculture as fencing and animal enclosures.
Hypogeum: Literally 'underground', an underground tomb or temple.

Illumination: The miniature paintings and images that adorn illuminated manuscripts. From the latin *illuminare* meaning to enlighten or illuminate. Many are embellished with gilding and silver-leaf, which 'light up' the pages. (De Hamel 1986, Clemens 2007 and Brown 1994)
Incunabula / Incunable: A printed book produced before 1501. (Brown 1994)

Jamb: The side of a doorway or window opening.
Jib: The horizontal or inclined arm or boom of a **crane**, as opposed to the vertical member, which is the **mast**. (Glynn 1854 and Bachmann 1997)

Kerf: A cut or incision made by a saw in a piece of wood, or the width of such a cut left in the wood. Also a thin piece of wood inserted into such incisions in two pieces of wood across a join to reinforce the joint, particularly at a mitre.
Kiln: see also Limekiln and Brick Kiln. A kiln is a type of oven, that produces temperatures sufficient to complete various processes, such as hardening, drying, or chemical changes. Kilns have been used for millennia to turn objects made from clay into pottery, tiles and bricks.
Ledger: (Scaffold) A longitudinal member fixed parallel to the face of the building joined to the **standards**. It acts as a support for the **putlogs**, transoms and **braces**.
Level: (Tool) A carpenter's or mason's level consisting of a long timber **straight edge** made of durable hard-wood. At half way along its length an integral suspended plumb bob or pendulum is incorporated into a raised section. For testing the level

of work as it proceeds, such as wall tops, the precursor of the modern spirit level. The main body of the tool would be made by a carpenter, using a Jointing or **Trying plane**. (Salaman 1997)

Lift: (Scaffold) The assemblage of **scaffold** members forming each horizontal level, or working platform, of a scaffold.

Lime: (Mortar) Until the nineteenth century **lime**, calcium oxide (see below **Lime Kiln**), was the basis of virtually all construction mortars, plasters and renders, mixed together with aggregates such as sands of various grades. (Holmes 2002 and Ashurst 2002)

Lime Kiln: A **kiln** used for the calcination of limestone (calcium carbonate) to produce the form of lime called quicklime (calcium oxide). Slaked lime (calcium hydroxide), can be formed by mixing quicklime with water. The resulting reaction is highly vigorous and can be dangerous, the quicklime should always be added to the water, and appropriate health and safety measures should be adopted. (Holmes 2002 and Williams 1989)

Lodge, mason's: A mason's yard or workshop attached to a large construction site such as a fortress, palace or cathedral. Most cathedrals and large churches in Europe had permanent mason's workshops and many still exist today. Some of these also incorporated lodges where itinerant or journeyman masons, carpenters and other craftsmen found lodgings. The French organization *Compagnons du Devoir et du Tour de France* is similar and includes craftspeople of many different disciplines and not just those associated with construction. They maintain some 80 houses in France and operate a strict technical education based on apprenticeship. There are other similar organizations in Germany and other European countries. They still operate today (https://en.wikipedia.org/wiki/Compagnons_du_Devoir)

Mallet: A wooden **hammer**, they are used for driving and striking wooden objects such as tools with wooden handles which would fracture if struck with a metal hammer. Carpenter's mallets are normally made of hardwood such as beech and have flat faces and sides. A carvers mallet is shaped like a truncated cone and the heads are often made of exotic hardwoods such as box-wood and lignum-vitæ. Masons use carvers' mallets for striking all-metal mallet-headed chisels with bulbous heads, see **chisel, masons.** (Salaman 1975)

Manticore: see also **Babewyn.** An exotic monster, for instance, in medieval images of the prophet Jeremiah he may be accompanied by a manticore with human head, lions body and scorpion's tail. (Hall 1974)

Mason's lodge: See **Lodge, mason's.**

Masonry: In the context of this book, both stone and brick construction and the work produced by either masons or bricklayers. However, generally understood to be the craft of building with natural and artificial stone including quarrying, cutting, dressing and fixing.

Mast: (Crane) The main vertical member of a **crane**. See also '**Star**'. (Glynn 1854 and Bachmann1997)

Master Carpenter: With the master mason the closest we come to the profession of **architect** in the Medieval period. They were often in charge of all aspects of construction including design. See also **Architect.** (Salzman 1952, Kaye 1960 and Harvey 1987)

Master Mason: See **Master Carpenter** above. The Master Mason was possibly the more senior of the two but it largely depended on circumstances, the client and what was being built.

Mortar: The material used between elements of masonry construction, including bricks, see **masonry.** Essentially mortar is to keep these elements apart rather than bonding or sticking them together, although it does have a bonding effect, once cured. Until the twentieth century most mortar was based on limestone, see **lime**. Additives were often added known as pozzolans, such as brick dust and volcanic sands to produce more rapid setting. Pure lime mortar can take a considerable time to cure thus delaying construction, particularly in the winter months. (Holmes 2002 and Ashurst 2002)

Moulding/s: A continuous plain or decorated, raised and/or inset contour strip or profile, usually horizontal or vertical but also around arches, vaults and openings such as door-cases and windows. Forms, or is part of, architraves, entablatures, **string-courses**, around column bases and the base of capitals. They are manifest in virtually all periods and styles of architecture.

Navvy: A labourer associated with the canal and railway building periods from the seventeenth through to the end of the nineteenth centuries. Also a labourer employed in excavation work to the present day.

Overhand: A method of laying bricks or fixing masonry from inside the wall by leaning over it. It is more difficult to control the quality of workmanship and is only employed when access to the outer face of the work is not possible, e.g., where no external scaffold exists.

Pannier/s: A pair of containers slung over the shoulders, used as a type of **Hod**. They had the advantage of leaving both hands free for negotiating ladders. For larger loads they were fitted to beasts of burden such as donkeys for transporting building materials.

Peg: In timber-frame construction a fixing used to fix a joint of two members together. They are made of riven oak, square, round or irregular section and tapered.

They are often draw-bored, that is, driven through misaligned peg-holes to pull the timbers together in a tight joint. They are often left proud by a significant margin to allow later tightening as the frame timbers season; timber framing invariably used 'green' oak. (Alcock 1996)

Pit saw: see also **Saw** and **Two-man saw**. A saw for cutting logs and large baulks of timber by a team of two men. The log is placed lengthways over a pit, carefully dug with reinforced sides to avoid it collapsing, later pits were often brick lined. One operator works in the pit whilst his partner works above, standing astride the log and pit, or, more commonly, on the log. As the cut proceeds the log is moved along until the end is reached. (Salaman 1975)

Pitching tool: A mason's tool with a bevelled edge, used with the hammer. For pitching or knocking off irregularities and waste lumps of stone. (Purchase 1904)

Plane (tool) A carpenter's and joiner's tool first used in the Roman era consisting of a body or stock with a flat sole. Earlier plane stocks are made from wood, usually hardwood for maximum wear. The sole must be dead flat, except for some specialist planes. A steel cutting blade is held in the stock by a wedge, the blade protrudes slightly from the sole and is held firmly by tapping in the wedge. The blade can be set according to the finish required on the workpiece, the work can then be planed to give a true and flat surface as required. There are many different sizes and types of plane, upwards of 100 types. (Salaman 1975)

Plantation: A closely set stand of trees formed by planting them in straight lines, often mistakenly referred to as a forest. A legal **Forest** with a capital F is where the King or other great magnate had the right to keep deer. A physical Forest is common land within a Forest; forestry on the other hand is the practice of growing trees in plantations. Also confused with **Woods** and **Woodland** which are natural stands of trees, close enough together that their canopies meet, albeit that they were intensely managed in the past. See **Coppice** and **Pollard.** (Rackham 1976, 1986 and 2006 are essential for understanding these different terms)

Platform: (Scaffold) A **scaffold** lift or platform, composed of planks or boards, very expensive in Medieval and early Renaissance times, or Hurdles or Clays made of Wattlework. See also **Clays, Hurdle, Lift** and **Wattlework.**

Plumb line: A heavy weight suspended from a line, once it has stopped swinging it gives a true vertical. Used to check the verticality of work. Plumb lines were often specially made, by casting lead into a cone with a pointed end, at the top a ring eye was attached to take the line. They were an essential item in a mason's and carpenter's tool kit. Also produced in cast-iron.

Point: See also **Punch.** Mason's near pointed tool, with an edge barely 5mm across. Used with the **hammer** or **mallet** for evening up after using the punch. It leaves the surface in narrow ridges and furrows. (Purchase 1904)

Pointing: (brickwork) See also **Repointing.** The dressing or finishing of the mortar between joints in brickwork and stonework.
Pollard: In **woodland** where the ingress of livestock or deer cannot be controlled, trees are cut above the grazing line at 8-12ft above ground level They are then left to grow again to produce a harvest of wood products as in coppicing. The process is more difficult and labour intensive than coppicing. See **coppice.** (Rackham 1976, 1986 and 2006)
Post: In timber-frame construction a substantial vertical timber forming part of the main frame. (Alcock 1996) Also the main upright part of a crane, also called a **Mast.** (Glynn 1854 and Bachmann 1997)
Predella: A panel or series of panels below an altarpiece or triptych. The scenes depicted in the predella form part of the overall iconographic programme of the altarpiece.
Psalter: A biblical book of the Psalms. The psalter was the principal book for Christian private devotion before the emergence of the **Book of Hours** in the thirteenth century. (Brown 2018)
Psaltery: A stringed instrument consisting of a shallow sound box across which strings were stretched and plucked with the fingers. Known in the ancient world and popular later in Europe in the fourteenth and fifteenth centuries. It is the usual instrument of King David. (Hall 1974)
Pulley: One of the fundamental and most simple mechanical devices for pulling and lifting, known since antiquity. They consist of grooved wheels, around which a rope is passed, mounted on spindles within a block. A single fixed pulley is used for changing the direction of a rope or cable in a crane where it passes from the mast to the jib, and at the end of the jib where the rope then falls free. In a **hoist** composed of two blocks, known as a **'block and tackle'**: One block, the higher, is fixed whilst the lower is a moving block, both blocks normally have twin wheels. The mechanical advantage of such a block and tackle is 4:1. (Glyn 1854 and see https://en.wikipedia.org/wiki/Pulley accessed 30/04/20 for diagrams)
Punch: Mason's tool similar to the **point** but much heavier and used with the hammer only. For removing large amounts of waste stone before using the **point** or other dressing tools. (Purchase 1904)
Purlin: In roof construction the longitudinal horizontal beam supporting the **rafters.** They in turn are supported by the end walls, partition walls, props or roof-trusses. (Cordingley 1961)
Putlog: (Scaffold) The scaffold component at right-angles to the building attached to the **Standards** and built into the wall. They support the working platforms or **Lifts.** If for whatever reason they are not built into the wall, they are attached to inner Standards. (See below)

Putlog holes: The holes left by the removal of **Putlogs** (see above). They were often left in-situ and not filled to enable later **scaffold** to be erected for repair and maintenance.
Putti: (plural) (Italian: 'little boy') Representations of chubby, naked children, often winged little angels. They appear as subsidiary and anonymous figures in works of art. They have been a motif since classical antiquity and may have pagan origins and divine status. Cupids are the female Roman counter-part. (Erika Langmuir, *A Closer Look at Angels*, 2010, London, The National Gallery)

Quarry-faced masonry: See **Cyclopean faced masonry.**
Quicklime: See **Lime-Kiln** and **Lime.** Calcium oxide, limestone after being burnt or calcined in a **kiln.** (Holmes 2002 and Williams 1989)

Rafter: inclined timber, usually one of a pair, which support the battens or laths under the roof covering. There are variants, see Alcock 1996 and Cordingley 1961.
Ramp: A robust made-up long and wide timber board structure, made by the site carpenters, for walking up/down to attain different levels. In Medieval and early Renaissance construction, ramps appear to have been popular for gaining height rather than ladders, however they took up a lot of space.
Render: A plaster coat applied directly to brickwork or stonework. Traditionally the term is applied to the first coat of three, the second is the 'float' and the final the 'set'. In modern construction the term is generally applied to a sand and cement coating on the exterior of a building. (Ashurst 2002)
Repointing: (Brickwork) See also **Pointing.** In a masonry wall the prime material the brick or stone, is considered permanent, whilst the pointing is sacrificial and will need replacing at intervals, these can be as much as 100 years or more. The recent practice of repointing using hard sand and cement mortar has been responsible for a great deal of damage to brick and stonework, both visually and mechanically. **Mortar** should always be less mechanically strong than the prime material, if not then that prime material will tend to deteriorate more rapidly.
Ripsaw: A saw for cutting along the grain of timber rather than across it. A Ripsaw has larger teeth than a crosscut saw. See **Saw.** (Salaman 1975)
Rusticated masonry: Stone masonry of large rectilinear blocks with deeply sunk joints forming a channel, the raised face is decorated in some way to contrast with surrounding Ashlar masonry. Often comprising the lower courses on large buildings, also door and window surrounds and quoins. See **Cyclopean faced masonry.**

Sarking boards: In roofing closely butting boards laid over the rafters, upon which the battens were fixed. The precursor of roofing felt.

Saw A thin strip of hard steel with teeth cut into the bottom edge normally with a wooden handle, sometimes in a frame. Used for cutting wood and timber, metal and stone. They are normally associated with wood and timber, from felling trees to fine joinery work, so there are many types. They are also used in all stone-yards. (Salaman 1975 and Purchase 1904)
Scaffold The temporary access, dating to antiquity, required to carry out construction work. It is virtually impossible to build without it. Throughout the periods covered by this book scaffold comprised **wood** and sometimes **timber** components bound together with rope or twine.
Scribe: A person who writes or copies manuscripts and documents, often associated with monks in scriptoria (see below). Also in government administration and in great households. Some scribes were also artists producing illuminations. (Brown 2018)
Scribe: (tool) Metal tipped instrument with a sharp point used for marking work.
Scriptorium: A writing room or studio, where manuscripts were produced. Normally associated with a monastery.
Shear legs: Three legs or poles with the feet splayed apart and bound together at the top, from which a **block and tackle**, or a simple **pulley**, are suspended for hoisting. Some forms are just a pair of legs forming an A frame which is inclined, but these require a guy rope arrangement between them to stop them collapsing.
Shovel: a tool resembling a spade with a broad blade and upturned sides, with a long handle used on construction sites for moving mortar, aggregates, waste and spoil from excavations, or other materials and arisings
Sieve: A tool or piece of equipment consisting of a wire or fine wooden mesh held in a frame, used for separating coarser from finer particles of **aggregate** for mixing **Mortar**. On early construction sites, the frame, usually rectangular is held at an angle inclined away from the operator/s, who throw the raw aggregate at it using a **shovel**. The finer stuff passes through to fall behind the sieve and the coarse drops in front.
Sill beam: In timber-frame construction a horizontal timber at the bottom of a framed wall, into which **posts** and **studs** are tenoned, see also **wall plate**. (Alcock 1996)
Slake: See **Lime**.
Slate: a fine-grained grey, green, or bluish-purple metamorphic rock easily cleaved into smooth flat plates. Widely used as a roof covering.
Slate hammer: Specialist hammers used by the slater. There are three distinct types: The slater's hammer with an axe pien, the peck hammer with a long pointed pien used for holing slates for nails and cutting, finally the slater's chopper, a flat rectangular tool with an offset handle, similar to a trowel, also incorporates a curved point for holing. See **Hammer**. (Bennet and Pinion, *Roof Slating and Tiling*,

1948, London, The Caxton Publishing Co., reprinted 2001, Donhead Publishing, Shaftsbury)
Slater: A roffer who lays slates, see above two entries.
Smith: See **Blacksmith.**
Smoothing plane: The smallest of the carpenter's set of planes, see **Plane.** A joiner will have smaller planes for fine finishing. (Salaman 1975)
Square: See also **Try Square**, similar to the **angle-try** but with the blade set a 90° for marking and cutting timber to a square edge. (Salaman 1975)
Standard: (Scaffold) The principle upright members of a **scaffold**, often joined to achieve the required height.
Straight-edge: A dead flat length of iron, steel or hardwood, 2 to 4ft in length. Used by masons and carpenters for checking the true flatness of a work-piece.
Star: (*stella*) A crane consisting of a single mast held upright by guy-ropes, one at each corner, thus seen on plan, resembling a 'star'. Derived from ships rigging, as are many early lifting devices.
Strike: (Scaffold) To dismantle a **scaffold.**
Striker: See also **Blacksmith.** A blacksmith's assistant who strikes the hot work-piece with a heavy hammer, at the point where the blacksmith indicates with his own lighter hammer.
String-course: A horizontal course, band or **moulding** on the face of a building.
Stud: In timber-frame construction the secondary, smaller upright members in a wall or partition. They are set between the **posts**, and are normally vertical but can be curved or at an angle. (Alcock 1996)

Timber: The product of timber trees, or standards, capable of producing material for major structural elements in buildings such as beams or posts etc., or sawing into planks. See **Coppice** and **Pollard.** (Rackham 1976, 1986 and 2006).
Treadwheel: See also **Crane.** A windlass comprised of a large double wooden wheel joined by planks, resembling a water wheel, within which one or two humans walk in order to turn it. The axle extends out either side to the mounts, one side of which forms the winch around which the rope will wind or unwind. The rope usually extends upwards to a fixed **pulley** and then drops to the ground. They were widely used in Medieval construction particularly in large buildings such as cathedrals where many still exist. They could be used alone, or attached to a **jib.**
Trestle: A carpenter's trestle is a low stool or sawing bench to support work. Large trestles were used to support large timbers or logs whilst being rip sawn with a two handed saw. This involved the operator on top standing on the timber or log. See also **horse.** (Salaman 1975)
Tripod: The arrangement of legs or poles forming **shear legs.**

Trowel: The basic mason's or bricklayer's trowel are very similar. A diamond shaped steel tool with the leading part elongated, the back part has a raised offset with a wooden handle. The trowel is used for picking up and spreading mortar for laying bricks or stone. The handle is used for tamping the brick or stone into place. The bricklayer will also use the trowel to cut bricks, a sharp strike with the edge will produce a precise cut or fracture to give a brick of the required length. Trowels used for pointing and repointing are smaller.
Truss: (Roof) A row of heavy timber triangular frames within a pitched roof, they span transversely across a building supporting the **purlins.** The intervals of the trusses within a roof divide it into bays. See also **Collar.** (Cordingly 1961)
Try Square: See **Square.**
Two-man saw: A saw with handles at either end for cutting large timbers and logs, operated by a team of two. See **Saw** and **Pit-saw.** (Salaman 1975)
Trying plane: The largest of the carpenter's set of planes, also known as a jointer plane. See **Plane.** (Salaman 1975)

Underwood: The product of coppicing and pollarding. (Rackham 1976, 1986 and 2006)

Veduta: A painting or drawing of a landscape or townscape sufficiently accurate to allow topographical identification.

Wall plate: See **Sill beam.**
Wattlework: see **Clay/e** and **Hurdle**
Wheelbarrow: As a **barrow**, but with handles at one end only. At the other end is a wheel fitted on an axle fitted into simple forks.
Whetstone: A circular sandstone wheel mounted within a trough of water for sharpening mason's chisels it can be operated by a treadle, or a cranked handle. A simple flat piece of stone can be used but water is essential as a lubricant.
Winch: Hoisting apparatus around which a rope is wound. See **Treadwheel.**
Withies: Rope or twine scaffold ties used to bind and join two scaffold members together, such as **standards** and **putlogs**, or standards and **ledgers.**
Wood/ Woodland/s: See **Plantation.**
Wright: An early term for a carpenter.

Bibliography

Note: All entries are secondary sources. The only primary/archive sources are the works depicted in the illustrated images.

Adam, Jean-Pierre, 1999, *Roman Building: Materials and Techniques*, London and New York, Routledge.

Ahl, Diane Cole, 1997, *Benozzo Gozzoli*, Milan, Silvana Editoriale.

Ainsworth, Maryan W and Christiansen, Keith, eds. 1998, *From Van Eyck to Bruegel: Early Netherlandish Painting in The Metropolitan Museum of Art*, New York, Metropolitan Museum of Art.

Alcock, N W, Barley, M W, Dixon, P W and Meeson, R A, 1996, *Recording Timber-Framed Buildings: An Illustrated Glossary*, York, Council for British Archaeology.

Alexander, Jonathan and Binski, Paul, eds. 1987, *Age of Chivalry: Art in Plantagenet England 1200-1400*, exhibition catalogue, London, Royal Academy of Arts.

Alexander, Jonathan J G, 2016, *The Painted Book in Renaissance Italy*, New Haven and London, Yale University Press.

Andrews, Francis B, 1992, *The Mediaeval Builder and his Methods*, Dorset Press, New York.

Arnold, Dieter, 2003, *The Encyclopaedia of Ancient Egyptian Architecture*, (English edition), London, I B Tauris.

Ashurst, John and Dimes, Francis G, 1977, *Stone in Building*, London, The Architectural Press.

Ashurst, John, 2001, *Mortars, Plasters and Renders in Conservation*, (place of publication not identified), Ecclesiastical Architects and Surveyors Association.

Bachmann, Cohrs, Whiteman and Wislicki, 1997, *The History of Cranes*, Wadhurst, East Sussex, KHL International.

Barker, Kenneth, L, General editor, 2011, *The New International Version Study Bible*, Grand Rapids, Michigan, Zondervan.

Barringer, Tim, 2005, *Men at Work: Art and Labour in Victorian London*, New Haven and London, Yale University Press.

Baskins, Cristelle, 2009, *The Triumph of Marriage: Painted Cassoni of the Renaissance*, exhibition catalogue, Boston, Isabella Stewart Gardener Museum.

Beckmann, Martin, 2011, *The Column of Marcus Aurelius: The Genesis & Meaning of a Roman Imperial Monument*, Chapel Hill, The University of North Carolina Press.

Beddington, Charles, 2011, *Venice: Canaletto and His Rivals*, exhibition catalogue, London, The National Gallery.

Beltrami, Constanza, 2016, *Building a Crossing Tower: A Design for Rouen Cathedral of 1516*, London, Sam Fogg and Paul Holberton Publishing.

Binding, Günther and Nussbaum, Norbert, 1978, *Der mittelalterliche Baubetrib nördlich der Alpen in zeitgenössischen Darstellungen*, Darmstadt, Wissenschaftliche Buchgesellschaft.

Binding, Günther, 2001, *Medieval Building Techniques*, Stroud, Gloucestershire, Tempus.

Birch, Walter de Gray, ed., 1885-93, *Cartularium Saxonicum: A Collection of Charters Relating to Anglo-Saxon History*, London, publisher unknown.

Blackinsell, William G. C, 1980, *Medieval Windlasses at Salisbury, Peterbrough and Tewkesbury*, South Wiltshire Industrial Archaeology Society.

Blair, John and Ramsey, Nigel, eds., 1991, *English Medieval Industries*, London and Rio Grand, The Hambledon Press.

Blair, John, 2018, *Building Anglo-Saxon England*, Princeton and Oxford, Princeton University Press.

Borchert, Till-Holger, 2014, *Masterpieces in Detail: Early Netherlandish Art from Van Eyck to Bosch*, Munich, London and New York, Prestel.

Born, Annick and Martens, Maximiliaan P J, 2012, *Van Eyck in Detail*, Antwerp, Ludion,

Bousmanne, Bernard and Delcourt, Thierry, eds, 2011, *Miniatures Flamandes: 1404-1482*, exhibition catalogue, Paris, Bibliothèque National de France and Brussels Bibliothèque Royle de Belgique.

Brown, Michelle P, 2018, *Understanding Illuminated Manuscripts: A Guide to Technical Terms*, Los Angeles, The J Paul Getty Museum.

Brunskill, R W, 1990, *Brick Building in Britain*, London, Victor Gollancz.

Byng, Gabriel, 2017, *Church Building and Society in the Later Middle Ages*, Cambridge, Cambridge University Press.

Callmann, Ellen, 1974, *Apollonio di Giovanni*, Oxford, Oxford University Press.

Camille, Michael, 1992, *Image on the Edge: The Margins of Medieval Art*, London, Reaktion Books.

Campbell, Caroline, Korbacher, Dagmar, Rowley, Neville and Vowles, Sarah, 2018, *Mantegna & Bellini*, exhibition catalogue, London, The National Gallery.

Cennino d'Andrea Cennini, before1427, *Il Libro dell' Arte*, Florence, translated by Daniel V Thompson, Jr, 1933, 1954 and 1960, New York, Dover Publications.

Chapelot, Jean and Fossier Robert, 1985, translated by Henry Cleere, *The Village and House in the Middle Ages*, London, B T Batsford.

Clark, Kenneth, 2008, *One Hundred Details from The National Gallery*, London, The National Gallery.

Clarke, Somers and Engelbach, R, 1990 (reprint), *Ancient Egyptian Construction and Architecture*, New York, Dover Publications.

Clemens, Raymond and Graham, Timothy, 2007, *Introduction to Manuscript Studies*, Ithaca and London, Cornell University Press.

Clifton-Taylor, Alec and Ireson, A S, 1983, *English Stone Building*, London, Victor Gollancz.

Colvin, H M, ed., 1963, *The History of the Kings Works*, Vols I & 2, 'The Middle Ages', London, HMSO.

Colvin, H M, ed., 1971, *Building Accounts of King Henry III*, Oxford, Oxford University Press.

Cooper, Lisa H, 2011, *Artisans and Narrative Craft*, Cambridge, Cambridge University Press.

Cordingley, R A, 1961, *British Historical Roof-Types and Their Members a Classification*, London, The Ancient Monuments Society.

Corkhill, Thomas, 1979, *A Glossary of Wood*, London, Stobart & Sons.

Coulton, J. J, 1977, *Ancient Greek Architects at Work Problems of Structure and Design*, Ithaca, Cornell University Press.

Courtenay, Lynn T, ed., 1997, *The Engineering of Medieval Cathedrals*, Volume 1, Studies in the History of Civil Engineering, Aldershot, Ashgate Publishing.

Crossley, D W, ed. 1981, *Medieval Industry*, CBA Research Report No. 40, London, Council for British Archaeology.

Curcic, Slobodan and Evangelia Hadjitryphonos, 2010, *Architecture as Icon*, exhibition catalogue, Princeton, Princeton University Art Museum.

Czarnowsky, Charles, 1949, 'Engins de Levage dans les Combles D'Eglises en Alsace', *Les Cahiers Techniques de L'Art*, Strasbourg, Editions F X Le Roux.

Dagenhart, Bernhard and Schmitt, Annegrit, 1984, *Jacopo Bellini: The Louvre Album of Drawings*, New York, George Braziller.

Davey, Norman, 1961, *A History of Building Materials*, London, Phoenix House.

De Hamel, Christopher, 1986, *A History of Illuminated Manuscripts*, London, Phaidon.

Diderot, Denis and le Rond d'Alembert, Jean-Baptiste, eds., 1751-72, *Encyclopédie, ou Dictionnaire raisonné des sciences, des arts et des métiers, etc.*, Paris, André François le Breton.

Du Colombier, Pierre, 1973, *Les Chantiers des Cathédrals*, Paris, Editions A & J Picard.

Erlande-Brandenburg, Alain, 1995, *The Cathedral Builders of the Middle Ages*, English edition, London, Thames and Hudson.

Fernie, Eric, 2000, *The Architecture of Norman England*, Oxford, Oxford University Press.

Fitchen, John, 1961, *The Construction of Gothic Cathedrals*, Oxford, The Clarendon Press.

Folda, Jaroslav, 1976, *Crusader Manuscript Illumination at Saint-Jean d'Acre, 1275-1291*, Princeton, Princeton University Press.

Friedlander, Max J, 1969 (3rd edition), *From Van Eyck to Bruegel*, edited and annotated by F Grossmann, New York, Phaidon.

Frugoni, Chiara, 1988, *Pietro and Ambrogio Lorenzetti*, Florence, Scala.

Frugoni, Chiara, 2005, *A Day in a Medieval City*, Chicago, Chicago University Press.

Evans, M W, 1969, *Medieval Drawings*, London and New York, Paul Hamlyn.

Galluzzi, Paolo, 1996, *The Art of Invention: Leonardo and Renaissance Engineers*, exhibition catalogue, Florence, Istituto e Museo di Storia della Scienza, and London, Science Museum.

Gentili, Romanelli, Rylands and Nepi Scirè, 2002, *Paintings in Venice*, Boston, Bullfinch Press.

Geronimus, Dennis, 2006, *Piero di Cosimo Visions Beautiful and Strange*, New Haven and London, Yale University Press.

Gibney, Arthur, 2017,with Hurley, Livia and McParland, Edward, eds., *The Building Site in Eighteenth-Century Ireland*, Dublin, Four Courts Press.

Gies, Francis and Joseph, 1994, *Cathedral, Forge, and Waterwheel: Technology and Invention in the Middle Ages*, New York, Harper Collins.

Gillespie, Raymond, 1996, *The Proctor's Accounts of Peter Lewis, 1564-1565*, Dublin, Four Courts Press.

Gimpel, Jean, 1976, *The Medieval Machine: The Industrial Revolution of the Middle Ages*, New York, Holt Rinehart and Winston.

Gimpel, Jean, 1983, *The Cathedral Builders*, English edition, Salisbury, Michael Russell.

Giordano, Luisa, 1998, 'On Filarete's *Libro Architettonico*' in Vaughan Hart and Peter Hicks, *Paper Palaces: The Rise of the Renaissance Architectural Treatise*, New Haven and London, Yale University Press.

Glynn, Joseph, 1854, *Rudimentary Treatise on the Construction of Cranes and Machinery*, London, John Weale, reprinted 1970, Bath, Kingsmead Reprints.

Goldthwaite, Richard A, 1980, *The Building of Renaissance Florence – An Economic and Social History*, Baltimore, The John Hopkins University Press.

Goy, Richard J, 2006, *Building Renaissance Venice: Patrons, Architects and Builders, 1430-1500*, New Haven and London, Yale University Press.

Greenacre, Francis and Stoddard, Sheena, 1986, *The Bristol Landscape: The Watercolours of Samuel Jackson, 1794-1869*, Bristol, City of Bristol Museum and Art Gallery.

Hale, Gardner, 1966, *The Technique of Fresco Painting*, New York, Dover Publications.

Hall, James, 1996, *Hall's Dictionary of Subjects & Symbols in Art*, London, John Murray.

Harvey, John, 1975, *Mediaeval Craftsmen*, London, B T Batsford.

Harvey, John, 1987 (first 1952), *English Medieval Architects: A Biographical Dictionary down to 1550*, Gloucester, Alan Sutton.

Helmus, Liesbeth M and Seelig, Gero, 2011, *The Bloemaert Effect: Colour and Composition in the Golden Age*, exhibition catalogue, Petersberg, Michael Imhof Verlag.

Heyman, Jacques, 1995, *The Stone Skeleton: Structural Enginnering of Masonry Architecture*, Cambridge, Cambridge University Press.

Hiesinger, Ulrich W, 1994, *Childe Hassam American Impressionist*, New York, Jordan-Volpe Gallery and Munich, Prestel.

Hill, Donald, 1996, *A History of Engineering in Classical and Medieval Times*, London, Routledge.

Holmes, Stafford and Wingate, Michael, 2002 (2nd revised ed), *Building with Lime: A Practical Introduction*, Bradford, ITDG Publishing.

Husa, Václav, Petráň, Joseph and Subrtová Alena, 1967, *Traditional Crafts and Skills: Life and Work in Medieval and Renaissance Times*, London, Paul Hamlyn.

Icelandic author (thirteenth century) from earlier Norse poetry, *The Saga of the Volsungs*, translated by Jesse L Byock, 1990 (USA) and 1999, London, Penguin Books.

Inglis, Erik, 2011, *Jean Fouquet and the Invention of France: Art and Nation after the Hundred Years War*, New Haven and London, Yale University Press.

Israëls, Machtelt, 2003, *Sassetta's Madonna della Neve: An Image of Patronage*, Leiden, Primavera Pers.

Jackson, Peter, 1987, *George Scharf's London: Sketches and Watercolours of a Changing City, 1820-1850*, London, John Murray.

Jaffe, David, Syson, Luke, Allen, Denise and Helvey, Jennifer, 1999, 'Ercole de' Roberti The Renaissance in Ferrara', *The Burlington Magazine*, Vol. 141, No. 1153, April 1999, pp.i-xi.

James, Duncan, 2012 'Saw Marks in Vernacular Buildings and Their Wider Significance', *Vernacular Architecture*, Vol. 43, The Journal of the Vernacular Architecture Group and Maney Publishing, pp.7-18.

James, John, 1982, *The Master Masons of Chartres*, Leura, Australia, West Grinstead Publishing.

Jeannin, Emmanuelle, 2008 (2nd edition), *Chantiers d'Abbayes*, Moisenay, France, Editions Gaud.

Joost-Gaugier, Christiane L, 2002, *Raphael's Stanza della Segnatura: Meaning and Invention*, Cambridge, Cambridge University Press.

Kaye, Barrington, 1960, *The Development of the Architectural Profession in Britain*, London, George Allen & Unwin.

Kauffmann, C M, 2003, *Biblical Imagery in Medieval England, 700-1550*, London, Harvey Miller Publishers.

Koerner, Joseph Leo, 2016, *Bosch & Bruegel: From Enemy Painting to Everyday Life*, Princeton, Princeton University Press.

König, Eberhard, 2007, *The Bedford Hours: The Making of a Medieval Masterpiece*, London, The British Library.

Korres, Manolis, 2000, *The Stones of the Parthenon*, Los Angeles, The J. Paul Getty Museum.

Kowalczyk, Bozena A, ed., 2001, *Bernardo Bellotto and the Capitals of Europe*, New Haven and London, Yale University Press.

Lancaster, Lynne, 2009, 'Building Trajan's Column', *American Journal of Archaeology*, 103, No. 3, pp.419-439.

Landels, J G, 1997, *Engineering in the Ancient World*, London, Constable & Co.

Lapins, Arturs, 2019, 'Door and Gate Related Artefacts in Cesis Castle, Latvia', in J W P Campbell et al., eds., *Water, Doors and Buildings: Studies in the History of Construction*, The Proceedings of the Sixth Conference of the Construction History Society, Queens' College Cambridge, 5-7 April 2019, pp.227-239.

Lillie Amanda, 2014, *Building the Picture: Architecture in Italian Renaissance Painting*, Online publication, London, the National Gallery.

Lloyd, Nathaniel, 1925, *A History of English Brickwork*, London, H Greville Montgomery

Longnon, Jean, Cazelles, Raymond and Meiss, Millard, 1969, *The Très Riches Heures of Jean, Duke of Berry*, New York, George Braziller.

Mactaggart, Peter and Ann, 1984, *Practical Gilding*, Welwyn, Herts. Mac & Me Ltd.

Mainstone, Rowland, 1999, *Structure in Architecture: History, Design and Innovation*, Aldershot, Ashgate Publishing.

Marchi, A and Vallazzi, M R, eds., 2012, *La Città Ideale. L'Utopia del Rinascimento a Urbino tra Piero della Francesca e Raffaello*, exhibition catalogue, Milan, Galleria Nazionale della Marche, Urbino.

Marcu, Felix, 2011, 'The Construction of the Roman Forts in Dacia', *Dacia: Revue d'Archeologie et d'Historie Ancienne*, Nouvelle Serie, LV, Académie Roumaine, Bucarest.

Martin, Maryline and Renucci, Florian, 2011, *Guédelon: A Castle in the Making*, Rennes, Editions Ouest-France.

Martineau, Jane ed., 1992, *Andrea Mantegna*, exhibition catalogue, London, Royal Academy of Arts, and New York, The Metropolitan Museum of Art.

Matthies, Andrea L, 1984, *Perceptions of Technological Change: Medieval Artists View Building Construction*, unpublished PhD dissertation, State University of New York at Binghamton.

Matthies, Andrea L, 1991, 'The Medieval Wheelbarrow', *Technology and Culture*, Vol. 32, No. 2, Part 1, Baltimore, The John Hopkins University Press, pp.356-364.

Matthies, Andrea L, 1992, 'Medieval Treadwheels: Artists' Views of Building Construction', in *Technology and Culture*, vol. 33, No. 3, pp.510-547.

McClendon, Charles B, 2005, *The Origins of Medieval Architecture: Building in Europe, A.D. 600-900*, New Haven and London, Yale University Press.

McKendrick, Scott, 2003, *Flemish Illuminated Manuscripts, 1400-1550*, London, The British Library.

Melnick, Scott L, 1995, 'Why a Christmas Tree?' in *Modern Steel Construction*, December 2000 (first published in the October 1995 issue), no page numbers.

Millon, Henry A and Lampugnani, Vittorio Magnago, eds., 1994, *The Renaissance from Brunellischi to Michelangelo: The Representation of Architecture*, exhibition catalogue, London, Thames and Hudson.

Millon, Henry A, ed., 1999, *The Triumph of the Baroque Architecture in Europe, 1600-1750*, exhibition catalogue, New York, Rizzoli International Publications.

Minkowski, Helmut, 1991, *Turm zu Babel*, Freren, Luca Verlag.

Muschg, Walter, ed., 1941, *Die Schweizer Bilderchroniken des 15./16. Jahrhunderts*, Zürich, Atlantis Verlag.

Nègre, Valérie, 2018, *L'art du Chantier: Construire et démolir du xvi au xxi siècle,* exhibition catalogue, Paris, Editions Snoeck.

Nishimura, Margot McIlwain, 2009, *Images in the Margins*, Los Angeles, J Paul Getty Museum and London, The British Library.

Van Os, Kok, Luijten and Scholten, 2000, *Netherlandish Art in the Rijksmuseum, 1400-1600*, Amsterdam, Waanders Publishers.

Parsons, David, ed., 1990, *Stone Quarrying and Building in England, AD 43-1525*, Chichester, Phillimore.

Peacock, D P S, 1998, *The Archaeology of Stone*, London, English Heritage.

Pérouse de Montclos, Jean-Marie, 2009 (9th edition), *Architecture Méthode et Vocabulaire*, Paris, Editions du Patrimoine, Centre des Monuments Nationaux.

Pevsner, Nikolaus, 1968, *The Buildings of England: Bedfordshire and the County of Huntingdon and Peterborough*, London, Penguin Books.

Pimental, A F, ed., 2012, *A Arquitetura Imaginária: Pintura, escultura, artes decorativas*, exhibition catalogue, Lisbon, Museu Nacional de Arts Antiga.

Purchase, William R, 1904 (5th edition), *Practical Masonry: A Guide to the Art of Stone Cutting*, London, Crosby Lockwood & Son.

Purton, Peter, 2018, *The Medieval Military Engineer: From the Roman Empire to the Sixteenth Century*, Woodbridge, The Boydell Press.

Rackham, Oliver, 1976, *Trees and Woodland in the British Landscape*, London, J M Dent.

Rackham, Oliver, 1986, *The History of the Countryside*, London, J M Dent.

Rackham, Oliver, 2006, *Woodlands*, London, Harper Collins.

Ratté, Felicity, 2006, *Picturing the City in Medieval Italian Painting*, Jefferson, North Carolina, McFarland & Company.

Richardson, M T, ed., 1899, *Practical Blacksmithing*, 2017 edition, New York, Chartwell Books.

Richardson, Margaret, 1995, *Buildings in Progress: Soane's Views of Construction*, exhibition catalogue, London, The Soane Gallery.

Robinson, John V, 2001, 'The "Topping Out" Tradition of the High-Steel Ironworkers' in *Western Folklore*, Vol. 60, No. 4, 2001, pp.243-262.

Rodríguez, D and Borobia, M, 2010, *Arquitecturas Pintadas del Renacimiento al siglo XVIII,* exhibition catalogue, Madrid, Museo Thyssen-Bornemisza.

Rowland, I. D and Howe, T. N, 1999, *Vitruvius, Ten Books on Architecture*, Cambridge, Cambridge University Press.

Salaman, R A, 1997, *Dictionary of Woodworking Tools*, revised edition, Lakeville, Minnesota, Astragal Press.

Salzman, L F, 1923, *English Industries of the Middle Ages,* Oxford, The Clarendon Press.

Salzman, L F, 1952, *Building in England Down to 1540 A Documentary History,* Oxford, Oxford University Press.

Sciacca, Christine, 2010, *Building the Medieval World*, Los Angeles, Getty Publications.

Scolari, Massimo, 2012, *Oblique Drawing: A History of Anti-Perspective*, Cambridge, Massachusetts, The MIT Press.

Seipel, Wilfried, ed., 2007, *Pieter Bruegel the Elder at the Kunsthistorisches Museum in Vienna*, Milan, Skira.

Sellink, Manfred, 2007, *Bruegel: The Complete Paintings, Drawings and Prints*, Antwerp, Ludion.

Singer, Charles, Holmyard, E J, Hall, A R, and Williams, Trevor, eds., 1954, *A History of Technology*, 2 vols., Oxford, Oxford University Press.

Smith, Terence Paul, 1985, *The Brickmaking Industry in England, 1400-1450*, Oxford, B.A.R.

Stanier, Peter, 2000, *Stone Quarry Landscapes*, Stroud, Tempus Publishing.

Stalley, Roger, 1999, *Early Medieval Architecture*, Oxford, Oxford University Press.

Streich, F and Dr K de Gerstenberg, 1876, *Méthode d'enseignement par l'aspect – les métiers et leurs outils*, 'Teaching methods by appearance – trades and their tools', Paris, Bonhoure.

Surtees, Virginia, ed., 1981, *The Diary of Ford Madox Brown*, New Haven and London, Yale University Press.

Tanis, James R, ed., 2001, *Leaves of Gold: Manuscript Illuminations from Philadelphia Collections*, exhibition catalogue, Philadelphia, Philadelphia Museum of Art.

Toledano, Ralph, 1988, *Michele Marieschi L'opera completa*, Arnoldo Mondadori Editore, Milan.

Trost, Edit, 1991, *Eduard Gaertner*, Berlin, HenschelVerlag.

Tutton, Michael, 2015, 'A Brief Introduction to Tread and Hand Wheel Cranes and Hoists in Construction as Depicted in Western Art', in Bowen, Brian, Friedman, Donald, Leslie, Thomas and Ochsendorf, John, eds., 2015, *Proceedings: Fifth International Congress on Construction History*, 2015, Vol. 3, pp.481-489. Chicago, The Construction History Society of America.

Tutton, Michael, 2017, 'Stout Trees and Putlogs: Tales of Early Scaffolding up to *c*.1600', in Campbell, JWP, Nina Baker, Michael Driver, Michael Heaton, Yiting Pan, Treve Rosoman, Michael Tutton and David Yeomans eds. *Building Histories: Proceedings of the Fourth Conference of the Construction History Society*, 2017, Cambridge, The Construction History Society, pp.69-80.

Tutton, Michael, 2014, 'An Outline of the History and Development of the Staircase' in Campbell J W P and Tutton Michael, *Staircases: History, Repair and Conservation*, Abingdon, Routledge.

Van Tyghem, Frieda, 1966, *Op en Om de Middeleeywse Bouwwerf de gereedschappen en toestellen gebruikt bij het bouwen van de vroege Middeleeuwen tot omstreeks 1600 studie gesteund op beeldende, geschreven en archeologische bronnen*, Paleis der Academiën, Brussels.

Vasari, Giorgio, 1550, *Vasari on Technique. Being the Introduction to the three arts of Design, Architecture, Sculpture and Painting, prefixed to the Lives of the Most Excellent Painters Sculptors and Architects*, translated by Louisa S Maclehose, edited by G Baldwin Brown, 1960, New York, Dover Publications.

Vaughan, Richard, 1958, *Matthew Paris*, Cambridge, Cambridge University Press.

Vaughan, Richard, trans. and ed., 1993, *The Illustrated Chronicles of Matthew Paris: Observations of Thirteenth-Century Life*, Stroud, Alan Sutton Publishing.

Voronova, Tamara and Sterligov, Andrei, 2003, *Western European Illuminated Manuscripts 8th to 16th Centuries*, London, Greenwich Editions.

Vroom, Wim, 2010, *Financing Cathedral Building in the Middle Ages: The Generosity of the Faithful*, Amsterdam, Amsterdam University Press.

White, John, 1967 (2nd edition), *The Birth and Rebirth of Pictorial Space*, Boston, Massachusetts, Boston Book and Art Shop.

White, Lynn Jr, 1962, *Medieval Technology and Social Change*, Oxford, Oxford University Press.

Wilkinson, J Gardner, 1854, *A Popular Account of the Ancient Egyptians*, London, John Murray.

Williams, Richard, 1989, *Limekilns and Limeburning*, Princes Risborough, Shire Publications.

Wilton, Andrew and Lyles, Anne, 1993, *The Great Age of British Watercolours, 1750-1880*, exhibition catalogue, London, Royal Academy of Arts.

Wilson Jones, Mark, 2000, *Principles of Roman Architecture*, New Haven and London, Yale University Press.

Wight, Jane A, 1972, *Brick Building in England: From the Middle Ages to 1550*, London, John Baker.

Wootton W, Russell B, and Rockwell P, 2013, 'Stoneworking Tools and Toolmarks (version 1.0)', *The Art of Making in Antiquity: Stoneworking in the Roman World*: http://www.artofmaking.ac.uk/content/essays/2-stoneworking-tools-and-toolmarks-w-wootton-b-russell-p-rockwell/ (accessed October 2017).

Yeomans, David, 1992, *The Architect & The Carpenter*, London, RIBA Heinz Gallery.

Zenner, Marie-Thérèse, ed. 2004, *Villard's Legacy: Studies in Medieval Technology, Science and Art in Memory of Jean Gimpel*, Aldershot, UK, Ashgate.

Index

General (in bold type see also glossary)

Manuscripts

People including artists

For Product Safety Concerns and Information please contact our EU representative GPSR@taylorandfrancis.com Taylor & Francis Verlag GmbH, Kaufingerstraße 24, 80331 München, Germany

Batch number: 10722098

Printed by Printforce, the Netherlands